# Resonance Ionization Spectroscopy 1986

# Resonance Ionization Spectroscopy 1986

Proceedings of the Third International Symposium on Resonance Ionization Spectroscopy and its Applications held at the University College of Swansea, Wales, on 7–12 September 1986

Edited by G S Hurst and C Grey Morgan

Institute of Physics Conference Series Number 84

Institute of Physics, Bristol

CODEN IPHSAC 84 1–369 (1987)

*British Library Cataloguing in Publication Data*

International Symposium on Resonance Ionization Spectroscopy and Its Applications
  (*3rd: 1986: Swansea*)
  Resonance Ionization Spectroscopy 1986: proceedings of the Third International
  Symposium on Resonance Ionization Spectroscopy and Its Applications held at
  the University College of Swansea, Wales on 7–12 September 1986—
  (Conference series, ISSN 0305–2346; no. 84)
  I. Resonance Ionization Spectroscopy
  II. Hurst, G. F.      III. Morgan, C. Grey
  IV. Institute of Physics      V. Series
  535.8′4      QC454.R/

ISBN 0–85498–175–6

Sponsors
        EG&G Corporation, Los Alamos National Laboratory, University College of
        Swansea, University of Tennessee, US Air Force European Office of Aerospace
        Research and Development, US Air Force Office of Scientific Research (AFOSR),
        US Department of Energy, Office of Health and Environmental Research

Programme Committee
        H M Borella, P G Huray, G S Hurst, R A Keller, T B Lucatorto, C Grey Morgan,
        M G Payne, H W Schmitt, T J Whitaker, N Winograd

Honorary Editors
        G S Hurst and C Grey Morgan

Published under the Institute of Physics imprint by IOP Publishing Ltd
Techno House, Redcliffe Way, Bristol BS1 6NX, England

Printed in Great Britain by J W Arrowsmith Ltd, Bristol

# Preface

The Third International Symposium on Resonance Ionization Spectroscopy (RIS 86), held at the Department of Physics, University College of Swansea, Wales, from 7–12 September 1986, was truly international with 18 participating countries. By any standards, such as number of delegates, number of papers submitted, depth of subject matter and range of applications, RIS 86 represented splendid growth over the first two symposia. We were just able to avoid simultaneous presentations by making efficient use of poster presentations of very high quality. Trends such as the type of institution involved in RIS, their geographical distribution, and the professionalism reflected in the research suggest that future symposia will continue to show a pattern of growth.

The main sessions were devoted to the following: (1) Photophysics and Spectroscopy, dealing primarily with the basic physics of multiphoton and multiple photon ionization, with special emphasis on the role and application of Rydberg states; (2) to Noble Gas Atom Counting—reporting significant advances in techniques and results relevant to oceanographic, geological and astrophysical studies; (3) to Mass Spectrometry coupled to Resonance Ionization, RIMS, embracing wide ranging studies of physical and chemical phenomena involving the need for elemental, isobaric and isotopic selectivity and separation at the ultra trace level; (4) Materials and Surface Analyses dealing with sustained advances in the application of RIS and associated sample preparation techniques for the quantitative analysis of a variety of substances and substrates at the parts per billion level; (6) Small molecules—the extension of resonance ionization techniques, hitherto confined to atomic systems, to investigate the properties of molecules such as hydrogen isotopes, the important topic of cluster formation which effectively bridges the gap between the gaseous and condensed matter phases; and finally (7) to Medical and Environmental Applications.

The success of RIS 86 is due to a number of factors. Our sponsors include EG&G Corporation, Los Alamos National Laboratory, University College of Swansea, University of Tennessee, US Air Force European Office of Aerospace Research and Development, US Air Force Office of Scientific Reseach (AFOSR), US Department of Energy, Office of Health and Environmental Research, and we gratefully acknowledge their generous support. Programme arrangements were made by the staff of the Institute of RIS (University of Tennessee) under the general guidance of an advisory committee as follows: Dr G S Hurst, University of Tennessee, Mr H M Borella, EG&G Energy Measurements Group, Dr Keith Boyer, Los Alamos National Laboratory, New Mexico, Dr G Goldstein, US Department of Energy, Professor P G Huray, University of Tennessee; Dr R A Keller, Los Alamos National Laboratory, Professor T Kirsten, Max-Planck-Institut für Kernphysik, Dr T B Lucatorto, National Bureau of Standard, Professor C Grey Morgan, University College of Swansea, Dr M G Payne, Oak Ridge National Laboratory, Dr H W Schmitt, Atom Sciences, Inc., Professor B P Stoicheff, University of Toronto, Professor H Walther, Max-Planck-Institut für Quantenoptik,

Mr T J Whitaker, Pacific Northwest Laboratories, and Professor N Winograd, Pennsylvania State University.

One of us (GSH) wishes to express deeply felt appreciation to Professor Grey Morgan for the invitation to meet on the campus of the University College of Swansea. Both of us express appreciation to the City of Swansea for superb hospitality, and to Professor Frank Llewellyn-Jones for his scintillating recall of some of the history of Ionization Phenomena, as developed by J S Townsend and others of the Cavendish and Clarendon Laboratories. The scientific tradition carried out in Swansea in the areas of ionization phenomena, lasers, and mass spectroscopy, provided an excellent setting for our RIS conference.

**G S Hurst**
**C Grey Morgan**

# Contents

## Section 5: Small Molecules

## Section 6: Medical and Environmental Applications

**Section 7: Resonance Ionization and Materials Separation**

**Section 8: Elementary Particles and Nuclear Physics**

**Section 10: Summary**

*Inst. Phys. Conf. Ser. No. 84*
*Paper presented at RIS 86, Swansea, Wales, 7–12 Sept. 1986*

# Trends in resonance ionization spectroscopy

G. S. Hurst,[‖] Institute of Resonance Ionization Spectroscopy
University of Tennessee, Knoxville, TN 37996, U.S.A.

## 1. Introduction

This conference on RIS, the third since 1981, is truly an international event with 13 countries represented on our program. All of us are grateful to Prof. C. Grey Morgan and to his staff and colleagues at the University College of Swansea for their invitation to meet here and for their preparations for this five-day event. The Welsh language is an old and respected one, free of the modern technical jargon such as "resonance ionization spectroscopy." Research by Prof. J. H. Beynon shows that "spectroscopeg ioneiddio drwy atsain" is a suitable translation.

I wish to mention the origin of RIS and then comment on the delineations of RIS with reference to many related laser processes. The substance of the talk will deal with the trends in RIS and especially how the needs for sensitive analytical methods have overshadowed the original plan to study excited species. We can only speculate on what we may be missing in the interesting field of radiation physics and chemistry.

## 2. Origin of RIS and Related Processes

The term "resonance ionization spectroscopy" (RIS) originated in connection with an experiment designed to measure the population of metastable states, $He(2^1S)$, created by the interaction of proton beams with helium gas (Hurst et al 1975, Payne et al 1975). The ORNL group in radiation physics had a background in the nonselective ionization associated with the passage of charged particles through matter. The importance of the RIS process was recognized when simple calculations showed that selective ionization could be done in single laser pulses with efficiencies approaching 100%.

Our background in radiation dosimetry had taught us that single electrons could be detected with proportional counters, as first demonstrated by Curran et al (1949). Thus, a stress was placed on the use of RIS with atoms in their ground states so that one-atom detection could be demonstrated (Hurst et al 1977) and applied to a variety of problems in

[*]Research sponsored by the Office of Health and Environmental Research, U.S. Department of Energy under contract ACO5-840R21400 with Martin Marietta Energy Systems, Inc.

[‖]Consultant to the Oak Ridge National Laboratory, Oak Ridge, TN 37831, U.S.A.

analytical chemistry.  Supporting research on the photophysics, dissociation of free atoms from molecules, chemistry at the one-atom level, diffusion of atoms, and other demonstrations received much attention from the ORNL group (Hurst et al 1979).

Of course, the idea of RIS could not have been carried out had pulsed lasers with suitable specifications not been available.  Furthermore, we were not alone in novel applications of pulsed lasers for photoionization. The Institute of Spectroscopy, USSR Academy of Sciences in Moscow pioneered the use of "multistep ionization" for isotope enrichment; see, for example, a paper by Ambartsumyan et al (1971) and Hurst et al (1979) for other early references.  And we realize today, of course, that RIS is just a subset of the field known as multiphoton ionization (MPI) which will be discussed by Professor Mainfray following this talk. A more complex process known as the optogalvanic effect involves the combination of resonance ionization with some other agency to create a plasma with excited species, as illustrated by Professor Hirose in Session IX.

3. Analytical Applications of RIS

Modern applications of elemental analyses require improved sensitivity and selectivity; thus RIS is of considerable interest.  A fundamental feature of RIS is the requirement that the atom to be ionized must be in a free state.  All RIS schemes depend on the known spectroscopy associated with nearly isolated atoms.  Thus, any approach to the elemental analysis of solids using RIS must provide a way to "atomize" the sample.  In this way the inherent features of RIS--Z-selectivity and high efficiency--are retained.

There are many well known methods for the atomization of a solid sample. Thermal evaporation by simply heating a sample is the best known. Electrical discharge is another method, but it has secondary problems produced by high density ionization that would be a background to the resonance ionization.  It is noted, however, that the important subject of optogalvanic phenomena has dealt with this problem in a clever way.  It is also possible to focus laser light on a solid and produce atoms by laser ablation.  This method, too, can have the problem of background ionization (even strong plasmas), which complicates the RIS analysis.  Sputtering of solids with energetic ion beams is a highly developed field and has now been used extensively as atomization for RIS.

Essentially all of the RIS methods now under development for analyses of solids utilize a combination of lasers and mass spectrometers, and the entire field could be called "resonance ionization mass spectrometry" (RIMS).  Although very different in its objective, an early proposal to combine laser ionization and mass spectroscopy was made by Ambartsumyan and Letokhov (1972) and further discussed by Letokhov (1976).  These concepts, however, dealt with improved characterization and detection of complex molecules.  A combination of Z-selection using RIS and A-selection with a mass spectrometer was recognized at the beginning of RIS. Therefore, RIMS always involves (1) atomization of the solid to liberate neutral atoms, (2) resonance ionization, (3) mass analysis of positive ions, and (4) sensitive detection of the ions following mass analysis. The use of a mass spectrometer is a key factor in improving selectivity since it reduces background due to multiphoton ionization of nonresonance types.  Viewed from the perspective of mass spectrometry, the use of resonance ionization sources in a mass spectrometer provides isobaric resolution and decreases interference due to molecular ions.

The most logical way to discuss the use of RIS for elemental analyses of solids is a classification according to the method of atomization. Thus, RIMS could be retained as a general acronym; and we could append LA, TH, and SI for laser ablation, thermal heating, and sputter initiated, respectively. However, tradition often runs against logic and acronyms proliferate. Currently, RIMS refers only to the thermal methods. Laser ablation goes by LARIS and sputter initiated by SIRIS. Sessions IV and V will describe exciting new developments on RIMS and SIRIS.

## 4. RIS Studies of Excited States

We now return to explore, briefly, the possible use of RIS for studying excited states. The studies involving $He(2^1S)$ gave absolute yield due to charged particle excitations and gave lifetimes in specified environments. From these it was possible to resolve questions of long standing on the phenomenon known as the "Jesse effect." Apparently, in spite of this, no additional applications have been made of RIS to study problems in radiation physics.

When radiation interacts with matter, a variety of effects are produced. These include (1) ionization of the atoms or molecules; (2) excitation of electronic states in atoms; (3) excitation of rotational, vibrational, or electronic states in molecules; (4) dissociation of molecules into neutral species, or free radicals in excited states or ground states. These direct excitation and ionization steps are followed with a host of reactions that determine new chemical and thermodynamic states of the system. And, of course, all biological effects are a consequence of these early steps. Little is known about many of these early stages, especially processes (2), (3), and (4) involving neutral species.

In principle, RIS can be used to obtain information on neutral species in specified quantum states which may be produced as a consequence of radiation interactions. Noble gases provide the elementary testing grounds for new theories and new experimental methods. Thus, the work on $He(2^1S)$ could be followed by work on $He(2^3S)$, using the simple apparatus of Fig. 1 which was developed at ORNL to measure the yield of $He(2^3S)$ due to $\alpha$-particles in helium gas, in support of a magnetic monopole detector (Hurst et al 1985). No doubt, such elementary methods could be used to measure the yield and kinetic behavior of other neutral species created by $\alpha$-particles in atomic and molecular gases.

## 5. Comments on RIS-86

The Program Committee for RIS-86 observed that RIS is developing into a truly international specialty and is used in most parts of the world. Further, the number of papers was estimated by R. A. Keller to be more than 300 per year. The Committee was especially pleased that more than 90 papers of high quality were submitted for RIS-86; many of these are presented in this meeting as posters. The Committee felt that all papers, including the poster presentations, were of a quality deserving publication in the proceedings. Fortunately, James Revill of Adam Hilger Ltd. agreed.

The Organizing Committee joins me in gratitude to Professor C Grey Morgan, the University College of Swansea, and the City of Swansea for hosting RIS-86.

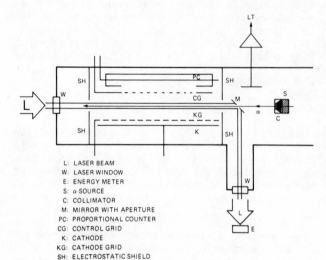

Fig. 1. Schematic of apparatus for exciting He($2^3$S) states in helium with alpha particles. Kinetics information for a monopole detector can be obtained by using the same RIS process that would be used in an actual monopole detector.

L: LASER BEAM
W: LASER WINDOW
E: ENERGY METER
S: α-SOURCE
C: COLLIMATOR
M: MIRROR WITH APERTURE
PC: PROPORTIONAL COUNTER
CG: CONTROL GRID
K: CATHODE
KG: CATHODE GRID
SH: ELECTROSTATIC SHIELD
LT: LASER TRIGGER

In addition, we thank Gerald Goldstein (Office of Health and Environmental Research, U.S. Department of Energy) for his active role in all of the RIS symposia to date.   C. R. Richmond (Oak Ridge National Laboratory) has provided continuing support and interest in the success of these symposia. Henry M. Borella (EG&G Energy Measurements Group) deserves special acknowledgment for his superb management of RIS-81 and RIS-84.   Finally, staff members Robert R. Rickard and Norma Brashier of the Institute of Resonance Ionization Spectroscopy deserve much credit for their work in arranging the program and in coordinating their efforts with those of Professor C. Grey Morgan and Elizabeth Hughes of the University College of Swansea.

## References

Ambartsumyan R V, Kalinin V N, and Letokhov V S 1971 Zh. Eksp. Teor. Fiz. Pis'ma Red. 13(6) 305
Ambartsumyan R V and Letokhov V S 1972 Appl. Opt. 11(2) 354
Curran S C, Cockroft A L, and Angus J 1949 Philos. Mag. 40 929
Hurst G S, Jones H W, Thomson J O, and Wunderlich Rainer 1985 Phys. Rev. A 32 1875
Hurst G S, Nayfeh M H, and Young J P 1977 Appl. Phys. Lett. 30 229
Hurst G S, Payne M P, Kramer S D, and Young J P 1979 Rev. Mod. Phys. 51 767
Hurst G S, Payne M G, Nayfeh M H, Judish J P, and Wagner E B 1975 Phys. Rev. Lett. 35 82
Letokhov V S 1976 in Tunable Lasers and Applications eds A Mooradian, T Jarger, and P Stokseth (Berlin: Springer) p 122
Payne M G, Hurst G S, Nayfeh M H, Judish J P, Chen C H, Wagner E B, and Young J P 1975 Phys. Rev. Lett. 35 1154

*Inst. Phys. Conf. Ser. No. 84: Section 1*
*Paper presented at RIS 86, Swansea, Wales, 7–12 Sept. 1986*

# Multiphoton ionization of atoms

G. Mainfray

Centre d'Etudes Nucléaires de Saclay, Service de Physique des Atomes et des Surfaces, F-91191 Gif sur Yvette Cedex, France.

## 1. Introduction

An atom with an ionization energy $E_i$ can be ionized by photons which have an energy much less than $E_i$, if the photon flux is strong enough, which, from a practical point of view, can only be achieved with laser radiation. The atom has to absorb several photons from the laser radiation in order to be ionized. This can be done using two different methods with two very different laser intensity ranges.

The first method named multistep ionization is the basis of Resonance Ionization
Spectroscopy. Figure 1 shows schematically
this process, using as an example the ionization of an atom through the absorption of
three photons of different energies $E_1$, $E_2$,
$E_3$. In the first two steps, each absorbed
photon matches the energy difference between
two atomic states to excite an atom initially from its ground state $E_a$ to an excited
state $E_c$. For the different jumps, each
absorbed photon has a platform to step on,
which enormously facilitates the transition.
The lifetime of the intermediate atomic
states is typically $10^{-8}$ s. A bound continuum step to ionize the atom completes the
multistep ionization process. This method
is very well known for its selective and
efficient ionization (Hurst et al. 1975,
1979). It utilizes tunable-wavelength
lasers delivering different wavelengths with
an intensity of about 1 kW cm$^{-2}$.

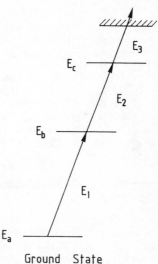

Fig.1 – Resonance ionization spectroscopy scheme.

In contrast, the second method designated multiphoton ionization requires a much higher laser intensity, and can be performed with a single
laser. Figure 2 shows schematically the four-photon ionization of an
atom. One of the most essential features of a multiphoton ionization
process is that it occurs through laser-induced virtual states which are
not eigen states of the atom. In principle, such a multiphoton ionization
process does not require any intermediate atomic state. Laser-induced
virtual states act as atomic states except for the corresponding lifetimes
which are much shorter. We may roughly regard the atom as spending a time
$\tau$ in a laser-induced virtual state. This time $\tau$ is of the order of one
optical cycle, typically $10^{-15}$ s. As a result, the absorption of photons

through laser-induced virtual states must occur in a time $< 10^{-15}$ s. Therefore, the photon flux has to be strong enough for there to be a large number of photons within $10^{-15}$ s. We thus understand why multiphoton ionization can only be achieved with intense laser radiation.

The N-photon ionization rate is given by $W = \sigma_N I^N$, where $\sigma_N$ is the generalized N-photon ionization cross section. W is in $s^{-1}$, and $\sigma_N$ is expressed in $cm^{2N} s^{N-1}$ units when the laser intensity I is expressed in numbers of photons $cm^{-2} s^{-1}$. Typically the two-photon ionization cross section $\sigma_2 \simeq 10^{-49} cm^4$ s, and the three-photon ionization cross section is of the order of $10^{-81} cm^6 s^2$ (Morellec et al. 1982, Mainfray and Manus 1984). Since $\sigma_N$ rapidly decreases as N increases, N-photon ionization with large N values can be observed by simply increasing the laser intensity. As an example, figure 3 shows that, by using a coherent 30 ps laser pulse at 1064 nm, the 4-photon ionization of Cs occurs at $10^{10}$ W $cm^{-2}$, the 11-photon ionization of Xe at $10^{13}$ W $cm^{-2}$ and the 22-photon ionization of He at nearly $10^{15}$ W $cm^{-2}$.

Ground state

Fig. 2 - Four-photon ionization scheme

The investigation of multiphoton ionization of atoms in a high laser intensity, ranging from $10^{10}$ to $10^{15}$ W $cm^{-2}$, gives rise to the observation of high intensity effects which are not observed in RIS experiments. The present paper is devoted to the analysis of these high intensity effects.

## 2. Resonant Multiphoton Ionization of Atoms

Resonant multiphoton ionization of atoms is an interesting example which lies between RIS and pure (non resonant multiphoton ionization. In a multiphoton ionization scheme, as shown

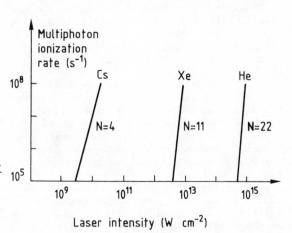

Fig.3 - Laser intensity required in order to observe N-photon ionization

in Fig. 2, the next to last photon is absorbed in the dense part of the atomic energy spectrum. As a result, an atomic state is generally located not too far from a laser-induced virtual state, as shown in Fig. 4a. The ionization time, which was as short as $10^{-15}$ s in the non-resonant case, is now given by $\tau = \dfrac{1}{\delta E}$ , where $\delta E$ is the energy difference between the energy of an atomic state and the energy of an integral number of photons. $\tau = 10^{-11}$ s for $\delta E = 3$ $cm^{-1}$. By tuning the laser frequency close to $\delta E = 0$, the multiphoton ionization rate exhibits a typical resonant character. For $\delta E = 0$, the ionization time $\tau$ is given by the lifetime of the resonant state induced by the laser field, $\tau = \frac{1}{\sigma I}$ , where $\sigma$ is the photoio- nization cross section of the resonant state. In a weak laser intensity $(I < 10^7$ W cm$^{-2})$, the characteristic time of a resonant multiphoton ionization is mainly determined by the natural life- time of the resonant state, typi- cally $10^{-8}$ s, while in a high laser intensity the characteris- tic ionization time is dramatical- ly shortened, down to $10^{-13}$ s at $10^{13}$ W cm$^{-2}$. As a result, reso- nance effects in the ionization rate are expected to play a very important role in a weak laser field, and expected to become un- important in a high laser field.

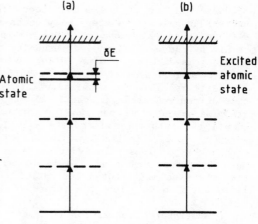

Fig.4 - Four-photon ionization scheme.
a) quasi-resonant process,
b) resonant process.

This picture is confirmed by experimental results. First, an experiment has been performed at about $10^7$ W cm$^{-2}$ to investigate the four-photon ionization of Cs through the three photon excitation of 6F levels (Petite et al. 1979). The ionization rate exhibits a resonance profile which has an amplitude of over three orders of magnitude. Second, resonance enhancement of less than one order of magnitude of the 11-photon ionization rate has been observed for Xe at about $10^{12}$ W cm$^{-2}$ (Alimov and Delone 1976). Finally, no resonance enhancement has been observed for Kr when the laser intensity was increased up to $10^{13}$ W cm$^{-2}$ (Lompré et al. 1980).

High laser intensity effects in resonant multiphoton ionization mainly consist of large broadening of the resonant atomic state induced by the strong coupling between the resonant state and the continuum. The dis- appearance of resonance effects at high laser intensity is due to this damping term. As a conclusion, for high laser intensity, intermediate atomic states appear to be of little significance in the multiphoton ionization rate. The selective excitation of an atomic state, which is the basis of RIS in weak laser fields, becomes ineffective at high laser intensities.

## 3. The Production of Multiply Charged Ions

A number of further high intensity effects have been observed in multipho- ton ionization of atoms. Two illustrative examples are now considered: the production of multiply charged ions in multiphoton ionization of many- electron atoms, and the energy spectrum of electrons generated in multi- photon ionization of atoms.

The interaction between an intense laser pulse and many-electron atoms leads to the removal of several electrons and the production of multiply charged ions (L'Huillier et al. 1982). Up to Xe$^{5+}$ ions have been observed at 1064 nm and $10^{14}$ W cm$^{-2}$ (L'Huillier et al. 1983), and up to Xe$^{8+}$ ions at 193 nm and $10^{16}$ W cm$^{-2}$ (Luk et al. 1985). The production of doubly charged ions has been investigated in detail. Doubly charged ions can be

produced, either by the simultaneous removal of two electrons from the
neutral atom, or by a stepwise process via singly charged ions. The conditions under which one of these two processes dominates are determined
mainly by the wavelength and intensity of the laser field. At long laser
wavelengths such as 1064 nm, doubly charged ions are produced through the
simultaneous removal of two electrons from the neutral atom when the laser
intensity is below the saturation intensity $I_S$ corresponding to the depletion of neutral atoms in the interaction volume. On the other hand, the
stepwise process becomes dominant when the laser intensity is increased
beyond the saturation intensity $I_S$. In contrast, at short laser wavelengths, double ionization always proceeds in a stepwise process via
singly charged ions. In addition, the stepwise process strongly depends
on the atomic number Z (Lompré et al. 1985a). Figure 5 is a typical
result of the multiphoton ionization of Xe at 532 nm (L'Huillier et al.
1983b). Up to $Xe^{5+}$ ions are formed.

In multiphoton ionization of a many-electron atom, one can no longer
consider the interaction of the laser field with a single electron as one
could for a one-electron atom such as atomic hydrogen. Two electrons can
be simultaneously excited and removed. More generally speaking, electron
correlation effects can lead to a collective response of the outer shell
irradiated by an intense laser pulse. Several electrons of the closed
shell could be excited while the first, or first and second electrons are
removed from the shell. Multiple ionization could proceed via highly
excited intermediate states. This would explain why an additional electron can be removed by increasing further the laser intensity by only 50%
as shown in Fig. 5. Finally, an additional effect should be considered.

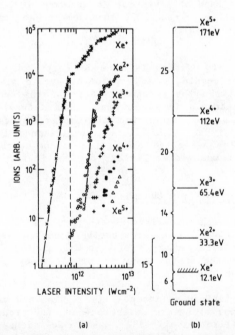

(a)                    (b)

Fig.5 (a) A log-log plot of the variation in the number of Xenon ions
formed at 532 nm as a function of the laser intensity. (b) Schematic
representation of the number of photons involved in the production of
multiply charged ions.

A 5p or 4d electron of Xe does not experience directly the external laser field, but instead an effective field screened by the other 5p electrons (Wendin et al. 1986). The stripping of the 5p-shell will successively reduce the effects of screening. Therefore, the effective field will increase at each step of the stripping of the outer shell. This might partly explain the relative ease with which highly charged ions are produced in rare gases.

4. Electron Energy Spectra

An N-photon ionization of a neutral atom produces an electron-ion pair within a time of some optical cycles of the laser field, i.e. of the order of $10^{-14}$ s. As the laser pulse duration is much longer, an interaction of the laser field with the electron as well as with the singly charged ion occurs. The latter process gives rise to stepwise multiply charged ions, as was described in the section above. Now, we will consider the interaction history of the electron. The electron is released with an initial energy $E_o = Nh\omega - E_i$, where $E_i$ is the ionization energy, and N is the minimum number of photons required to reach the ionization threshold. The outgoing electron can absorb addition photons, while it is sufficiently close to the parent ion, so that the electrostatic interaction can supply the extra momentum to the electron. The absorption of additional photons in the continuum can be described by the continuum-continuum transitions picture (Deng and Eberly 1984, Edwards et al. 1984, Bialynicka-Birula 1984), or better still by a semi inverse Bremsstrahlung picture in which the electron absorbs photons from the laser field in the field of its own ion (Lompré et al. 1986). Furthermore, when the electron goes far away from the residual ion, the effective absorption cross section becomes smaller. This decrease depends on the photon energy, the ion potential, and the initial electron energy $E_o$.

This physical picture is far from complete because the electron is not born in vacuum. It is born in the e.m. field, and as a result, it acquires a quiver energy $\Delta$ as it oscillates in the e.m. field. The value of $\Delta$ averaged over a period of the e.m. field is $\dfrac{e^2 \mathscr{E}^2}{4m\omega^2}$ , where $\mathscr{E}$ and $\omega$ are respectively the laser field and laser frequency. The absorption of additional photons corresponding to an energy less than $\Delta$ is thus energetically impossible. In other words, the impossibility of absorbing photons corresponding to energy less than $\Delta$ can be viewed as a new selection rule which arises at high laser intensity and long laser wavelength. For example, at 1064 nm, $\Delta \simeq 0.1$ eV at $10^{12}$ W cm$^{-2}$ and 10 eV at $10^{14}$ W cm$^{-2}$.

The above physical picture is confirmed by experimental results upon measuring energy spectra of electrons generated in multiphoton ionization of atoms. Electron energy spectra generally consist of a series of peaks evenly spaced by an amount equal to the photon energy. The number of peaks strongly depends on laser wavelength and laser intensity. The electron energy distribution is very much changed when the laser intensity is gradually increased (Kruit et al. 1983). This change in the relative amplitude of the first electron peaks can be explained in terms of continuum-continuum transitions which induce a coupling between the laser-induced states in the continuum. Furthermore, the further selection rule which arises at high laser intensity and prevents the absorption of photons in a range of energy less than $\Delta$ has been confirmed in multiphoton ionization of He at 1064 nm and $10^{15}$ W cm$^{-2}$ (Boreham and Luther-Davies 1979, Lompré et al. 1985). The electron energy distribution extends to 100 eV, with no electrons observed in the first 30 eV range.

In conclusion, electron energy spectra give information on various aspects of the history of the electron, while the electron is interacting with the laser field as long as it remains in the close vicinity of its parent ion. Electron energy spectra do not give so much information on the interaction of the laser field with ions or neutral atoms.

## 5. Conclusion

The usual picture of an atom in a weak laser field becomes quite the reverse in a very high laser field. In a weak field, intermediate atomic states play a very important role in RIS and enhance dramatically the ionization rate, while the continuum of ionization is structureless. On the contrary, at high laser field, the selective excitation of atomic states becomes ineffective, so that intermediate atomic states appear to be of little significance in the multiphoton ionization rate. On the other hand, a periodic structure evenly spaced by an amount equal to the laser photon energy is induced at high laser field and superimposed onto the continuum. In addition, the beginning of this laser-induced structure is shifted towards higher energy in the continuum when using a long laser wavelength at very high intensity.

## References

Alimov D and Delone N 1976 Zh. Eksp. Theor. Fiz. 70 29 (1976 Sov. Phys. JETP 43 15)

Bialinicka-Birula Z 1984 J. Phys. B 17 3091

Boreham B and Luther-Davies B 1979 J. Appl. Phys. 50 2533

Deng Z and Eberly J 1984 Phys. Rev. Lett. 53 1810

Edwards M, Pan L and Armstrong Jr L 1984 J. Phys. B 17 L515

Hurst G, Payne M, Nayfeh M, Judish J and Wagner E 1975 Phys. Rev. Lett. 35 82.

Hurst G, Payne M, Kramer S and Young J 1979 Rev. Mod. Phys. 51 767

Kruit P, Kimman J, Muller H and Van der Wiel M 1983 Phys. Rev. A 28 248

L'Huillier A, Lompré L-A, Mainfray G and Manus C 1982 Phys. Rev. Lett. 48 1814

L'Huillier A, Lompré L-A, Mainfray G and Manus C 1983a J. Phys. B 16 1363

L'Huillier A, Lompré L-A, Mainfray G and Manus C 1983b Phys. Rev. A 27 2503

Lompré L-A, Mainfray G and Manus C 1980 J. Phys. B 13 85

Lompré L-A, L'Huillier A, Mainfray G and Manus C 1985a Phys. Lett. 112a 319

Lompré L-A, L'Huillier A, Mainfray G and Manus C 1985b J. Opt. Soc. Am. B 2 1906

Lompré L-A, Mainfray G, Manus C and Kuperstuych J 1986 J. Phys. B to be published

Luk T, Johann U, Egger H, Pummer H and Rhodes C 1985 Phys. Rev. A 32 214

Mainfray G and Manus C 1984 in Multiphoton ionization of atoms ed Chin S.L and Lambropoulos P. Academic Press 7

Morellec J, Normand D and Petite G 1982 Adv. At. Mol. Phys. 18 97

Petite G, Morellec J and Normand D 1979 J. Physique 40 115

Wendin G, Jönsson L and L'Huillier A 1986 Phys. Rev. Lett. 56 1241.

*Inst. Phys. Conf. Ser. No. 84: Section 1*
*Paper presented at RIS 86, Swansea, Wales, 7–12 Sept. 1986*

# Experiments with the single-atom maser

Gerhard Rempe and Herbert Walther

Sektion Physik der Universität München and Max-Panck-Institut für
Quantenoptik, D-8046 Garching, Fed. Rep. of Germany

## 1. Introduction - Rydberg Atoms and Radiation Interaction

The development of tunable lasers opened up the spectroscopy of highly
excited atoms, the so-called Rydberg atoms (Haroche 1981, Kleppner et al
1983, Haroche et al 1984, Gallas et al 1984). These are atoms with a valence electron excited into an orbit of very high principal quantum number,
i.e. far from the ionic core. Because of their hydrogenic nature, the
energies of these highly excited levels are well described by the Rydberg
formula. Due to their high principal quantum numbers, and since the relative energy difference between neighbouring levels is small, Rydberg
atoms are expected to exhibit a number of classical properties. In particular, according to Bohr's correspondence principle, the transition
frequency between neighbouring levels approaches the orbital frequency of
the electron. On the other hand, these systems also represent an almost
ideal testing ground for some of the most fundamental models and predictions of low energy quantum electrodynamics. Some of the reasons for this
are as follows.

The dipole interaction between electromagnetic radiation and Rydberg atoms
is very large. The matrix element between neighbouring levels scales as $n^2$,
where n is the principal quantum number. For high enough n, stimulated
effects can overcome spontaneous emission for very small photon numbers. As
a consequence, Rydberg atoms are very sensitive, e.g. to black-body radiation (Haroche 1981, Haroche et al 1984, Gallas et al 1984).

Because of the large wavelength of the radiation emitted in Rydberg transitions, it is possible to modify physically the nature of the environment
into which they decay, using, for example, conducting walls, or a resonator
with a high quality factor to impose on the electromagnetic field boundary
conditions different from the usual, free space ones. One is then in a
position to observe the consequences of the discrete mode structure of the
electromagnetic field (Purcell 1946). These include diminishing (Drexhage
1974, Kleppner 1981, Gabrielse et al 1984, Gabrielse et al 1985, Hulet et
al 1985) or enhancing (Goy et al 1983) the rate of spontaneous emission,
depending upon the cavity being tuned off or on resonance with a transition
frequency, as well as modifying the Lamb shift of Rydberg levels (Dobiasch
et al 1985, Lütken et al 1985, Barton 1987) and the anomalous magnetic moment of the electron (Fischbach et al 1984, Brown et al 1985, Svozil 1985).

For cavities with high quality factors, the photon emitted during a Rydberg
transition remains stored inside the resonator long enough for the atom to
reabsorb it with a finite probability. In this way, it is possible in
principle to build a single-atom maser (Meschede et al 1985) which is the
subject of the following sections.

## 2.  The Jaynes-Cummings Model and the Single-atom Maser

The situation realized in the one-atom maser approaches the idealized case of a two-level atom interacting with a single mode of a radiation field (Jaynes et al 1963, Cummings 1965, Stenholm 1973, von Foerster 1975, Meystre et al 1975, Knight et al 1980). It is therefore now possible to perform experiments on the dynamics of the atom-field interaction predicted by this theory. Some of the features are explicitly a consequence of the quantum nature of the electromagnetic field: the statistical and discrete nature of the photon field leads to new dynamic characteristics such as collapses and revivals in the Rabi nutation. The first experimental observations of the predicted effects are reviewed in this paper.

First we should like to summarize the main results of the Jaynes-Cummings model. We consider a two-level atom which enters a resonant cavity with a field of n photons. The time development of the probability $P_{e,n}$ of the atom in the excited state is then given by

$$P_{e,n}(t) = 1/2 \ [1 + \cos \ (2\Omega\sqrt{n + 1} \ t)],$$

where $\Omega$ is the single-photon Rabi frequency.

In the realistic case there will always be a fluctuating number of photons initially present in the cavity and the quantum Rabi solution has to be averaged over the probability distribution $p(n)$ of having n photons in the mode at t = 0:

$$P_e(t) = 1/2 \sum_{n=0}^{\infty} p(n) \ [1 + \cos \ (2\Omega \sqrt{n + 1} \ t)]$$

At a low atomic beam flux the cavity contains essentially thermal photons and their number is a random quantity conforming to Bose-Einstein statistics. In this case $p(n)$ is given by

$$p_{th}(n) = n_{th}^n \ / \ (n_{th} + 1)^{n+1},$$

with the average number of thermal photons $n_{th} = [\exp \ (h\upsilon/kT) - 1]^{-1}$. In this case the distribution of Rabi frequencies results in an apparent random oscillation $P_{e,th}$.

At higher atomic beam fluxes the number of photons produced by stimulated emission in the cavity will increase and the statistics changes. If there were no initial thermal field present, $p(n)$ would correspond to a Poissonian distribution representing a coherent field:

$$p_c(n) = \exp(-\langle n \rangle) \cdot \langle n \rangle^n/n!$$

As first shown by Cummings (1965), the Poisson spread in n gives dephasing of the Rabi oscillations, therefore $P_{e,c}(t)$ first exhibits a collapse. This is described in the resonant case by the approximate envelope $\exp(-\Omega^2 t^2/2)$ and is independent of the average photon number (Eberly et al 1980). This independence does not hold for nonresonant excitation. After the collapse there is a range of interaction times for which $P_{e,c}(t)$ is independent of time. Later $P_{e,c}(t)$ exhibits recorrelations (revivals) and starts oscillating again in a very complex way. As has been shown by Eberly and co-workers (Eberly et al 1980, Narozhny et al 1981, Yoo et al 1981, Yoo et al 1985) the recurrences occur at times given by

$$t = k \ T_R \ (k = 1,2,3 \ \ldots), \ \text{with} \ T_R = 2\pi \ \sqrt{\langle n \rangle} \ /\Omega.$$

Both collapses and revivals in the coherent state are purely quantum features and have no classical counterpart. Eberly and co-workers have shown
that the physical reason for the collapse is the Poisson photon number distribution, leading to a "granularity" of the quantized radiation field
present even at high average photon numbers.

The inversion also collapses and revives in the case of a chaotic Bose-
Einstein field (Knight et al 1982). Here the photon number spread is far
larger than for the coherent state and the collapse time is much shorter.
The revivals completely overlap and interfere to produce a very irregular
time evolution. A classical thermal field represented by a Gaussian distribution of amplitudes also shows collapse, but the revivals are absent in
this classical version. Therefore, the revival can be considered as a clear
quantum feature, but the collapse is less clear-cut as a quantum effect.

We conclude this section by noting that in the case of Raman type two
photon processes the Rabi frequency turns out to be $2\Omega n$ rather than $2\Omega\sqrt{n+1}$,
enabling the sums over the photon numbers in $P_e(t)$ to be carried out in
simple closed form. In this case the inversion revives perfectly with a
completely periodic sequence (Knight 1986).

## 3. Experiments with the Single-atom Maser

For the single atom maser experiment (Meschede et al 1985) an atomic beam
of highly excited Rydberg atoms was used. The atoms pass through small
apertures into the liquid helium cooled part of the apparatus. There they
are excited to the upper maser level and enter the cavity. Behind the
cavity the atoms in Rydberg states were monitored by field ionization (see
Fig. 1).

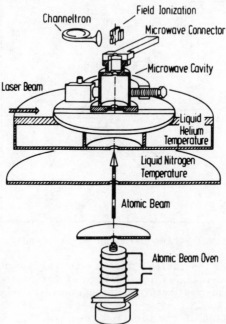

Fig. 1: Experimental setup for the single atom maser experiment. The micro-
        wave cavity is cooled to liquid-helium temperature

The Rydberg states were populated by using the frequency-doubled radiation of a continuous wave ring dye laser. The constant stream of Rydberg atoms is ionized in an inhomogeneous dc electric field of a plate capacitor. The atoms reach the point of maximum field strength in front of a hole in the anode through which the ejected electrons pass and reach a channeltron multiplier. If the field strength is adjusted properly, mainly the atoms in the upper maser level are monitored. Transition from the initially prepared state to the lower maser level are thus detected by a reduction of the electron count rate. The cylindrical cavity (diameter 24.8 mm, length 24 mm) was manufactured from pure niobium rods. It is enclosed in a cryoperm shield to reduce the influence of ambient magnetic fields. The temperature of the cavity could be varied from 4.3 to 2.0 K, corresponding to Q factors of $1.7 \times 10^7$ and $8 \times 10^8$, respectively. The atomic beam passes through the cylindrical cavity along its axis, where only the $TE_{1np}$ and $TM_{1np}$ modes possess a nonvanishing linear polarized electric field. For our experiment the $TE_{121}$ mode was used. The transversal electric field of this mode is linearly polarized and forms a half-wave along the axis of the cavity. It is doubly degenerate in an ideal cylindrical cavity. The degeneracy is removed by a slight deformation of the circular cross section into an oval shape, which then determines the direction of polarization of the field mode. The deformation is achieved by squeezing the cylinder with a screw and a piezoelectric transducer for fine tuning. (0.5 MHz/1500 V). The upper maser level was the $63p_{3/2}$ level of $^{85}$Rb. The fine structure splitting between $63p_{3/2}$ and $63p_{1/2}$ amounts to 396 MHz (see Fig. 2). It is therefore no problem to excite a single fine stucture level with the narrow-band ultraviolet radiation ($\Delta\upsilon \sim 2$ MHz).

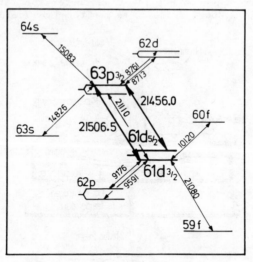

Fig. 2:    Rubidium level scheme with the maser transition

To demonstrate maser operation, the cavity was tuned over the $63p_{3/2}$ - $61d_{3/2}$ transition by changing the voltage of the piezoelectric transducer; the field ionization was recorded simultaneously. Transitions from the initially prepared $63p_{3/2}$ state to the $61d_{3/2}$ level (21.5065 GHz) are detected by a reduction of the electron count rate.

In the case of measurements (Meschede et al 1985) at a cavity temperature of 2 K, a reduction of the $63p_{3/2}$ signal could be clearly seen for atomic fluxes as small as 800 atoms/s. An increase in flux caused power broadening

and finally an asymmetry and a small shift (Fig. 3). This shift is attributed to the ac Stark effect, caused predominantly by virtual transitions to
the $61d_{5/2}$ level, which is only 50 MHz away from the maser transition. The
fact that the field ionization signal at resonance is independent of the
particle flux (between 800 and 22 x $10^3$ atoms/s) indicates that the transition is saturated. This, and the observed power broadening show that
there is a multiple exchange of photons between Rydberg atoms and the
cavity field.

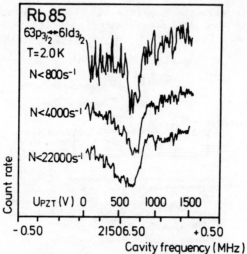

Fig. 3:   Maser resonances at a temperature of 2 K for three different
          atomic beam densities

With an average transit time of the Rydberg atoms through the cavity of 80
µs and a flux of 800 atoms/s one obtains that 0.06 Rydberg atoms are in the
cavity on the average. According to Poisson statistics this implies that
more than 99 % of the events are due to single atoms. This clearly demonstrates that single atoms are able to maintain continuous oscillations of
the cavity.

Since the transition is saturated, half of the atoms initially excited in
the $63p_{3/2}$ state leave the cavity in the lower $61d_{3/2}$ maser level. The decay to other levels can be neglected for the average transit time of 80 µs.
The energy radiated by those atoms is stored in the cavity for its decay
time, increasing the average field strength. The average number of photons
left in the cavity by the Rydberg atoms is given by $n_m = T_c N/2$ where $T_c$ is
the characteristic decay time of the cavity and N the number of Rydberg
atoms entering the cavity in the upper maser level per unit time. For the
highest particle flux used in our experiment N = 22 x $10^3$ atoms/s, one
finds $n_m$ = 55 photons at 2 K ($T_c$ = 5 ms). This value is larger than the
average number of black-body photons $n_{th}$ = 1.5 at 2 K. For N = 800 atoms/s
one obtains $n_m$ = 2, which means that the radiation generated by the Rydberg
atoms in the cavity has about the same energy as the black-body radiation.

The experimental setup described above is suitable to test the Jaynes-
Cummings model describing the dynamics of the interaction of a single atom
with a single cavity mode. An important requirement is, however, that the

atoms of the beam have homogeneous velocity so that it is possible to observe the Rabi nutation in the cavity directly. In a modified setup a
Fizeau type velocity selector is inserted (Fig. 4) between atomic beam oven
and cavity, so that a fixed atom-field interaction time is obtained. Changing the selected velocity leads to a different interaction time and leaves
the atom in another phase of the Rabi cycle when it reaches the detector
(Rempe et al 1987).

The Fizeau velocity selector consists of 9 discs rotating with the same
velocity. Each disc has 1486 radial slits close to the outer edge (width
and distance 0.2 mm). The width of the velocity distribution of the atoms
was 4 %.

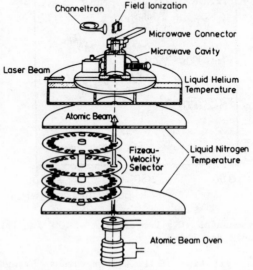

Fig. 4:    Single-atom maser setup including the Fizeau velocity selector

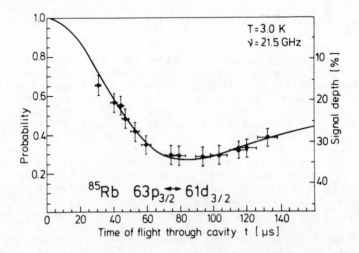

Fig. 5:    The measured probability of finding the atom in the upper maser
level $63p_{3/2}$ for the cavity tuned to the $63p_{3/2}$ - $61d_{3/2}$ transition of $^{85}Rb$, compared to the theoretical prediction (solid line)

The experimental results obtained for the $63p_{3/2}$ - $61d_{3/2}$ transition are shown in Fig. 5. In the figure the ratio between the field ionization signals on and off resonance is plotted versus the interaction time of the atoms in the cavity. The solid curve was calculated using the Jaynes-Cummings model. The total uncertainty in the velocity of the atoms is 10 % in this measurement. The error in the signal follows from the statistics of the ionization signal and amounts to 4 %. The measurement was made with the cavity at 3 K and $Q = 6 \times 10^7$ ($T_c = 500$ μs). There are on the average 2.5 thermal photons in the cavity. The number of maser photons is small compared with the number of black-body photons. In the present setup the thermal velocity distribution of the atomic beam does not allow measurements for interaction times larger than 130 μs. In order to observe the evolution of the excited state further on, it is necessary to switch to a transition having a larger Rabi frequency $\Omega$. Suitable for this purpose is the neighbouring transition $63p_{3/2}$ - $61d_{5/2}$.

The experimental results for this transition are shown in Figs. 6 to 8. The quality factor of the cavity was $Q = 2.7 \times 10^8$, this corresponding to $T_c = 2$ ms and a cavity temperature of $T = 2.5$ K, which results in $n_{th} = 2$. The experimental result shown in Fig. 6 was obtained with very low atomic beam flux ($N = 500$ s$^{-1}$ and $n_m = 0.5$). The solid curve represents the result of the Jaynes-Cummings model, which is in very good agreement with the experiment. When the atomic beam flux is increased, more photons are piled up in the cavity. Measurements with $N = 2000$ s$^{-1}$ ($n_m = 2$) and $N = 3000$ s$^{-1}$ ($n_m = 3$) are shown in Figs. 7 and 8. The maximum of $P_e$ (t) at 70 μs flattens with increasing photon number, thus demonstrating the collapse of the Rabi nutation induced by the resonant maser field. Figures 7 and 8 together show that for atom-field interaction times between 50 μs and about 130 μs $P_e$(t) does not change as a function of time. Nevertheless, at about 150 μs $P_e$(t) starts to oscillate again (Fig. 8), thus showing the revival predicted by the Jaynes-Cummings model.

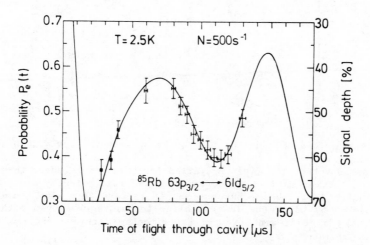

Fig. 6: The measured probability of finding the atom in the upper maser level $63p_{3/2}$ with the cavity tuned to $63p_{3/2}$ - $61d_{5/2}$ transition of $^{85}$Rb. The flux of Rydberg atoms is $N = 500$ s$^{-1}$

In order to measure the continuation of $P_e$(t) to times larger than 130 μs it was necessary to increase the number of slow Rydberg atoms. The efficiency of the second harmonic generation was therefore increased by a factor of three by using a LiJO$_3$ instead of an ADA crystal. In this way 600 µW of radiation at 297 nm was produced, leading to a more efficient population of the upper maser level, so that measurement of P$_e$(t) was possible up to 170 µs. This allowed one to observe the revival shown in Fig. 8.

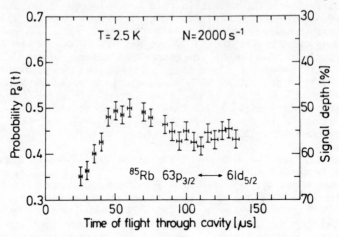

Fig. 7:   Same as Fig. 6, but the flux of Rydberg atoms is N = 2000 s$^{-1}$

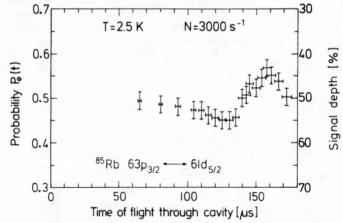

Fig. 8:   Same as Fig. 6, but the flux of Rydberg atoms is N = 3000 s$^{-1}$

In order to study the atom field evolution for times even longer than this, one cannot select atoms from the original Maxwellian velocity distribution of the effusive atomic beam, but one has to prepare very slow atoms. This can be achieved by using the laser cooling technique developed within the last few years by different groups. Experiments in this direction are in progress in our laboratory.

## 4. Conclusions

The experimental results presented in this paper show clearly the collapse and revival predicted by the Jaynes-Cummings model. The variation of the Rabi nutation dynamics with increasing atomic beam fluxes and thus with increasing photon numbers in the cavity generated by stimulated emission is obvious from Figs. 6 to 8. The results also represent the change of the photon statistics starting with the chaotic Bose-Einstein field (Fig. 6). It is clear that not under all conditions a coherent field is generated, also a sub-Poissonian photon number distribution is possible (Filipovicz et al 1986a). In addition, squeezing (Meystre et al 1982) and the generation of number states seems to be feasible (Filipovicz et al 1986b). In order to characterize the cavity field completely the determination of the inversion of the atom behind the cavity is not sufficient: the phase of the field has also to be measured. This can be performed by means of an additional coherent and stable electromagnetic field which is applied before the atom enters the cavity (Krause et al 1986).

References

Barton G 1987 to be published
Brown L S Gabrielse G Helmerson K and Tan J 1985 Phys. Rev. Lett. 55 44
Cummings F W 1965 Phys. Rev. 140 A1051
Dobiasch P and Walther H 1985 Annales No6 Alfred Kastler Symposium (Editions de Physique) p 825
Drexhage K H 1974 Progress in Optics 12 (Amsterdam North Holland) p 165
Eberly J H Narozhny N B and Sanchez-Mondragon J J 1980 Phys. Rev. Lett. 44 1323
Filipovicz P Javanainen J and Meystre P 1986a Opt. Comm. 58 327
Filipovicz P Javanainen J and Meystre P 1986b J. Opt. Soc. Am B3 906
Fischbach E and Nakagawa N 1984 Phys. Rev. D 30 2356
von Foerster T 1975 J. Phys. A8 95
Gabrielse G and Dehmelt H 1985 Phys. Rev. Lett. 55 67
Gabrielse G van Dyck Jr R Schwinberg J and Dehmelt H 1984 Bull. Am. Phys. Soc. 29 926
Gallas J A C Leuchs G Walther H and Figger H 1984 Advances in Atomic and Molecular Physics 20 ed D Bates and B Bederson (Academic Press: New York) pp 412-466
Goy P Raimond J D Gross M and Haroche S 1983 Phys. Rev. Lett. 50 1903
Haroche S 1981 Atomic Physics 7 ed D Kleppner and F C Pipkin (Plenum Press: New York and London) pp 141-165
Haroche S and Raimond J M 1984 Advances in Atomic and Molecular Physics 20 ed D Bates and B Bederson (Academic Press, New York) pp 350-411
Hulet R G Hilfer E S and Kleppner D 1985 Phys. Rev. Lett. 55 2137
Jaynes E T and Cummings F W 1963 Proc. IEEE 51 89
Kleppner D 1981 Phys. Rev. Lett. 47 233
Kleppner D Littman M G and Zimmermann M L Rydberg States of Atoms and Molecules ed R F Stebbings and F B Dunning (London: Cambridge Univ. Press) pp 73-116
Knight P L and Milonni P W 1980 Phys. Rep. 66C 21
Knight P L and Radmore P M 1982 Phys. Lett. 90A 342
Knight P L 1986 Physica Scripta T12 51
Krause J Scully M O and Walther H. 1986 Phys. Rev. A34 2032
Lütken C A and Ravndal F 1985 Phys. Rev. A31 2082
Meystre P Geneux E Quattropani A and Faist A 1975 Nuovo Cimento B25 521
Meystre P and Zubairy M S 1982 Phys. Lett. 89A 390
Meschede D Walther H and Müller G 1985 Phys. Rev. Lett. 54 551
Narozhny N B Sanchez-Mondragon J J and Eberly J H 1981 Phys. Rev. A23 236
Purcell E M 1946 Phys. Rev. 69 681
Rempe G Walther H and Klein N 1987 to be published
Stenholm S 1973 Phys. Rep. 6C 1
Svozil K 1985 Phys. Rev. Lett. 54 742
Yoo H I Sanchez-Mondragon J J and Eberly J H 1981 J. Phys. A14 1383
Yoo H I and Eberly J H 1985 Physics Reports 118 239

*Inst. Phys. Conf. Ser. No. 84: Section 1*
*Paper presented at RIS 86, Swansea, Wales, 7–12 Sept. 1986*

21

# Hydrogen in strong dc electric fields—atomic engineering

Munir H. Nayfeh

Department of Physics, University of Illinois at Urbana-Champaign, 1110 West Green Street, Urbana, IL   61801

   Abstract   We use strong external dc electric fields to manipulate and control the atomic structure of highly excited atomic hydrogen (atomic engineering). We can construct nearly one-dimensional atoms whose electronic distribution are highly extended along the field, and which may have enormous dipole moments ("giant dipole" atoms). The nuclear charge $Z_1$ that defines the energy and other properties of the "new" atom is a fraction of the proton charge. The fractions 0, 1/4, 1/2, 3/4 and 1 define four quarters that classify some properties of the atoms. The dipole moment is found to be opposite to the field in the first and third, and in its direction in the second and fourth quarters, and zero at the boundaries. Those atoms are unstable against ionization, however at $E < 0$ but $E > -2\sqrt{F}$ one can populate "giant dipole" atoms whose potential barriers are large enough (tunneling small enough) to render their lifetimes quite long. Hydrogen is a special case because these states are not mixed with less stable ones and so their "giant dipoles" survive long enough to be studied. We have used what we call "charge shape tuning" utilizing a three-photon process to excite them without the excitation of the overlapping continuum.

It is interesting to consider sufficiently high fields such that the electronic interactions with the external and the Coulomb fields are comparable, that is when none of them dominates. However, it is almost impossible to create fields in the laboratory which are strong enough to disrupt atoms in their normal states: for instance, the electric field on an electron in the ground state of the hydrogen atom has a strength of 5 x $10^9$ volt/cm. On the other hand, for the n=30 state in hydrogen, the Coulomb field of the nucleus can be overcome by an external electric field of only 5 kV/cm. Thus entry of atomic physics into the strong-field regime has been accomplished by dealing with highly excited states, rather than by generating enormous laboratory fields. The same statement also applies to the interaction of atoms with external magnetic fields.

These points are clearly illustrated by examining the potential energy of the electron of a hydrogen atom placed in an external electric field $\vec{F}$ (along the z axis) $V = -e^2/r - e\vec{F}\cdot\vec{r}$. The Coulomb term dominates at small r (normal atomic size), where as at very large r (highly excited atoms), the external field dominates. At some intermediate distances the two potentials become comparable and none of them dominates. It is in this intermediate regime that the concept of atomic engineering has been realized.

To explain the concept of atomic engineering we utilize Fig. la which shows the potential of a hydrogen atom for a cut along the z axis in the presence of an external electric field along the z axis. First, it is evident that the potential is not spherically symmetric; the electric field leads to a lowering of the ionization potential of the atom and creation of a potential barrier in the z > 0 half space.

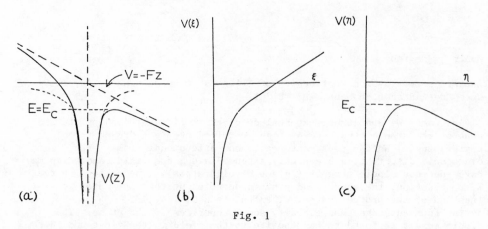

Fig. 1

Thus the state of energy higher than the top of the barrier can classically escape, that is ionize; in other words the motion for this state in this half space is unbounded. On the other hand in the z < 0 half space the electric field leads to a rise in the potential towards infinity, consequently the motion of the electron is bounded for all energies including positive energies in this half space, something that does not occur in the isolated atom case.[1]

Although the potentials in the z = 0 plane are very useful in bringing out some features of the interaction, they are not very useful for quantitative calculation. This is because the nonspherical symmetry of the potential makes the interaction non separable: that is, it cannot be separated into independent one dimensional motions in spherical coordinates. The interaction, however, is separable in parabolic coordinates $\xi = r + z$, $\eta = r-z$, and the azimuthal angle $\phi$, with quantum numbers $n_1$, $n_2$, and $m_\ell$ respectively. The effective potentials for the $\xi$ and $\eta$ motions shown in Fig. 1b in fact, have good resemblance to that of the z cut in the z < 0 and z > 0 regions respectively, and hence govern the energy of the system (location of the energy) and the ionization lifetimes of these levels respectively. The quantum number $m_\ell$ is common to both parabolic and spherical descriptions, and the principle quantum number $n = n_1 + n_2 + |m_\ell| + 1$. The spherical $\ell$ and parabolic $n_1$, $n_2$ quantum numbers do not have a one-to-one correspondence: a state with definite values of $n_1$ and $n_2$ is composed of many different values of $\ell$.

One important property of the atom that comes out of this procedure is the fact that only a fraction of the nuclear charge $Z_1 < 1$ drives the $\xi$ motion and hence dictates the energy of the system, while the rest of the charge $Z_2 = 1-Z_1$ drives the free $\eta$ motion and hence dictates its lifetime. Thus the presence of an external electric field provides us with a situation where the nuclear charge that drives the bounded motion can be varied, in a near continuous fashion. Considering the fact that the physical and mm x 10 mm slot cut into it and covered by a fine mesh to allow the passage of ions. This gives a limit of 1 μs on the detection time. Ions travel through a 100 cm long, field-free drift tube which provides mass analysis and are detected using an 18 stage venetian blind electron multiplier capable of single ion detection. The data are collected and analyzed using an LSI-11 computer system. A slightly different scheme where the n=2 state is excited using one photon at 1215 Å has been recently reported.[5]

The ability to create focussed dipoles as intermediates is the key to the success of the multi-stage shaping operation.[4] This is explained in Fig. 2 for a two-stage process using n=2 of hydrogen as an intermediate. These are labelled by their parabolic quantum numbers $(n_1, n_2, m_\ell)$ as 100, 001 and 010, where $n = n_1 + n_2 + |m_\ell| + 1$. Because the fine structure level splittings in n=2 of hydrogen are small enough ($.3$ cm$^{-1}$) such that an electric field imposed on the atom which is larger than 5 kV/cm will be able to mix all of these sublevels and hence their charge distributions (each has a zero dipole), which are shown to the left of the figure, to produce distinct dipole distributions, shown to the right of the figure, needed for the shaping process. Our calculations using a field of 16.8 kV/cm show that by utilizing the upfield extended dipole of n=2 as an intermediate (shown as a solid line in Fig. 2), the efficiency can be increased from 10 percent to 30 percent, where as by utilizing the down field extended dipole (shown as a dashed line in Fig. 2) the efficiency is reduced to 1 percent. These were confirmed in our hydrogen experiment as shown in Figs. 3a and 3b respectively. Our further calculations using the bluest components of higher n states ($n_1$ = n-1) whose charge can be focussed along the field more easily (as shown in Fig. 4) and hence can be matched or tuned more closely to the charge of the giant dipoles showed dramatic effects on the efficiencies.[3] Figure 5, which gives normalized lineshapes for excitation via the 000, 100, and 800 components of the n=1, 2, and 9 manifolds respectively, shows that excitation via n as high as 9 rejects almost completely the excitation of the continuum. This rejection is accompanied by an enhancement of the excitation cross section of the giant dipole; in specific we find the peak in the case of n=9 is about a factor of 300 larger than that of the case of excitation from the ground state. Moreover, the visibility of the giant dipole states, defined here as $V = (S_{max}-S_{min})/(S_{max}+S_{min})$, reach near 85 percent for n=9.

The enhanced efficiency is very nice, but it is found that it is not possible to increase the lifetimes in this positive energy region. Such inability is related to the fact that the bound motion of all of these giant dipoles in this region are driven by nearly the same charge, most of the nuclear charge $Z_1 \sim 1$, which also dictates very similar orbits where the nucleus is located at the lower tip of the cigar. However, it is found that such enhanced efficiency can be extended to the negative energy region where it is also possible to produce giant dipoles that live quite long. Therefore we will now discuss[3] such promising negative energy regions between E = 0 and E = -2 $\sqrt{F}$. In this region the giant dipole atoms take on different properties than the one in the positive energy region. Firstly, the fraction of the charge that drives the bound motion can be varied from 0 to 1 by varying the energy of the system, and consequently the position of the nucleus inside the cigar can also be controlled. In Fig. 6 we plotted an indicator of the location of the nucleus inside the cigar as a function of $Z_1$ along the sketches of some possible orbits (the nucleus is shown as a white dot). This indicator which is the ratio chemical identity of isolated atoms is defined by the nuclear charge, then it is clear that we have at our hand a means for creating new "types" of atoms.

We will now discuss the preparation and nature of the new types of atoms. The motion of the electron in these states is nearly one-dimensional with the electronic charge distribution highly extended and resembling a cigar whose axis has specific directions.[2,3] The atoms have enormous size and may have enormous electric dipole moments ("giant dipole" atoms), but they spontaneously ionize in about $10^{-12}$s in the positive energy region. Moreover, in general, one cannot exclusively prepare these types of states in the $E \geq 0$ region without preparing the highly excited normal state of the atom (continuum state) since, first of all, the excitation has to start from the ground state of the normal atom which is only weakly affected by electric fields and secondly both fields Coulomb and Stark will have to compete. For example, the efficiency of excitation or the "visibility" of the giant dipoles, which is a measure of how much they rise above the accumulated smooth continuum, tends to be very small (4% at 5 kV/cm). Moreover, for $E < 0$ but $E > -2 \sqrt{F}$ other effects arise due to the scattering of the active electron from the rest of the electrons. This causes mixing and smearing of the Coulomb-Stark interaction to the degree of preventing selective excitation of the giant dipoles, causing mixing among states of short and long lifetimes, resulting in shortening of the lifetimes of the long lived ones.[1]

Considering the shortness of their lifetimes, and the low efficiency of excitation it is clear that experimentation with these "new atoms" will not be easy unless these two properties are enhanced. Our recent theoretical and experimental work on atomic hydrogen has shown that it is possible by using multi-step excitation via resonant intermediate Stark states to selectively excite and enhance the giant dipole without the excitation of the continuum states.[3] Also we were able to prepare a variety of these giant dipoles with a variety of ionization lifetimes ranging from $10^{-7} - 10^{-12}$s and to find systematic classification of some of their properties thus making them easier to work with.

The scheme we devised for this purpose relies on a process we call multi-stage shaping or charge shape tuning. In one photon excitation from the ground state one starts from a spherically symmetric charge (zero dipole moment), and tries to mold it by a single operation into a giant dipole whose charge is highly focussed along the field. On the other hand in multi-stage shaping one uses one photon to create from a ground state a not too large dipole of charge distribution that is focussed along the field at an intermediate state followed by another photon absorption from this intermediate state that produces larger dipole whose charge is even more focussed along and so on till one excites the giant dipole in a highly focussed distribution along the field.

We will briefly describe the experimental set up used in these studies.[4] Simultaneous absorption of two photons from a single tunable pulsed laser beam at 243 nm results in excitation from 1s to n=2, and some photo-ionization of the resulting n=2 population. A second pulsed beam near 365 nm excites states near the continuum from the n=2 state. An atomic beam is formed by effusion from a Wood discharge tube through a multicollimator assembly composed of 25 small glass capillaries. The beam is loosely collimated, but produces a density of about $10^{11}$ $H°/cm^3$; the background gas density is on the order of $10^{12}/cm^3$. One of the Stark plates has a 3

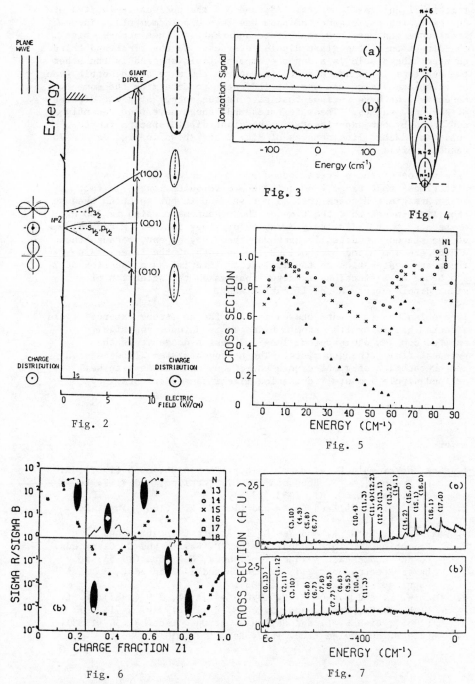

Fig. 2

Fig. 3

Fig. 4

Fig. 5

Fig. 6

Fig. 7

of the excitation cross section from 010, and 100 to the giant
dipoles which corresponds to the two processes labelled by dashed
and solid lines in Fig. 2 respectively. This ratio is related to
the ratio of the area of the orbit below the nucleus to that above
the nucleus. The figure shows a remarkable property: for the

fractional charges $Z_1$ = 1/4, 1/2, and 3/4 the nucleus is located at the center of the cigar (the atom has zero dipole moment). These fractional charges thus constitute demarkation lines across which the direction of the giant dipole reverses. In the first and third quarters the dipole is along the imposed field whereas in the other two quarters it is opposite to it. Given the size of the orbit (can be calculated), one can use the indicator to determine the moment. Moreoever, we have recipes that give us the energy of giant dipoles of given $Z_1$ values. These features and others have been recently confirmed by our experiments.[3] The giant dipole spectra for the solid and dashed line schemes of Fig. 2 are shown in Fig. 7a and 7b respectively.

Examination of the spectra indeed shows a variety of widths (lifetimes) that range from quite short to quite long. In fact there are giant dipoles that do not show up in our spectrum because they live longer than the time of the measurement which is 100 ns, or because they radiatively decay before they ionize. Moreover some of the states are actually narrower than they appear, because their widths are dominated by the 1 $cm^{-1}$ bandwidth of the laser. Also there are systematics to the ionization lifetime as a function of $Z_1$ and hence as a function of energy that makes the selection of a giant dipole of given specification easy.

In conclusion, this work showed that as far as strong external field effects, highly excited atomic hydrogen is unique, and highly excited complex atoms become less and less hydrogenic as the external field strength rises. The hydrogen system allows design and preparation of giant dipole one-dimensional atoms with well defined dipole moments, ionization lifetimes and orientation.

References

1. See for example D. A. Harmin, Phys. Rev. A 26, 2656 (1982); R. R. Freeman, N. P. Economou, G. C. Bjorklund, and K. T. Lu, Phys. Rev. Lett. 41, 1463 (1978).
2. Nayfeh, M. H. Hillard G. B. and Glab W. L. (1985) Phys. Rev. A 32, 3324.
3. Nayfeh, M. H. Ng K. and Yao D. in Atomic Excitation and Recombination in External Fields, M. Nayfeh and C. Clark eds., Gordon and Breach Science Publishers, New York (1985); in Laser Spectroscopy VII, T. Hänsch and Y. Shen eds., Springer-Verlag, Berlin (1985).
4. Glab W. L. and Nayfeh M. H. Phys. Rev. A 31, 530 (1985); W. L. Glab, K. Ng, Yao D. and Nayfeh M. H. Phys. Rev. A 31, 3677 (1985); Glab W. L. and Nayfeh M. H. Opt. Lett. 8, 30 (1983).
5. Rottke H. and Welge K. H. Phys. Rev. A 33, 301 (1986).

*Inst. Phys. Conf. Ser. No. 84: Section 1*
*Paper presented at RIS 86, Swansea, Wales, 7–12 Sept. 1986*

# Anomalous linewidths and peak height ratios in continuous-wave RIMS

T. J. Whitaker, B. D. Cannon, G. K. Gerke, and B. A. Bushaw

Pacific Northwest Laboratory, Box 999, Richland, WA 99352

Abstract. This paper describes effects of optical pumping in certain RIMS experiments. We show that anomalous peak height ratios and broadening effects can result from optically-induced redistribution of magnetic substate populations. We also show how lineshapes and peak heights in RIMS can be altered by branching to metastable electronic states.

## 1. Introduction

Continuous wave (cw) lasers have been shown to have several advantages in resonance ionization mass spectrometry (RIMS) (Whitaker 1986, Cannon 1985, Miller 1985, Keller 1986). These advantages include unity duty cycle, low background, and extremely high resolution. In some cw experiments, it is possible for an atom to undergo several excitation/spontaneous-emission cycles before the excited-state atom is ionized. This can lead to optical pumping processes which alter lineshapes and peak heights in resonance ionization spectra. We have investigated these phenomena in the $6s6p$ $^1P_1^\bullet$ $\leftrightarrow$ $6s^2$ $^1S_0$ transition of Ba and have discovered aspects of optical pumping which have not been previously reported. An understanding of these results are necessary to obtain quantitative information from resonance ionization spectroscopy (RIS) and RIMS measurements. We show that magnetic substate population redistribution in the ground state and weak branching of the $^1P_1^\bullet$ state to intermediate $^1D$ or $^3D$ metastable states explain the observed phenomena.

## 2. Experimental

The experimental system has been described previously (Bushaw 1986). Briefly, it consists of an effusive Ba atomic beam (14 mrad full-angle divergence) which perpendicularly intersects a 1.6 mm diameter beam from a single-frequency cw dye laser in the ionization region of a quadrupole mass spectrometer. The frequency of the dye laser beam can be scanned or tuned to an individual isotopic line of the $6s6p$ $^1P_1^\bullet$ $\leftrightarrow$ $6s^2$ $^1S_0$ Ba transition. Frequency drift is controlled to better than 1 MHz/h by referencing to a single-frequency He:Ne laser. A 1-watt ultraviolet (UV) beam (75 $\mu$m diameter) from an argon ion laser overlaps the center of the dye laser beam within the ionizer. The UV beam can photoionize the $^1P_1^\bullet$ state in

---

* This work was supported by the Office of Health and Environmental Research and the Office of Basic Energy Sciences of the U. S. Department of Energy under Contract DE-AC06-76RLO 1830.

barium but cannot photoionize any lower energy states. Ions formed in the ionization region pass through the mass spectrometer and are detected by conventional means.

## 3. Theory and Results

$^{137}$Ba has a nuclear spin of 3/2, producing three hyperfine levels in the $^1P_1^*$ state which have hyperfine quantum numbers $F$ = 1/2, 3/2, and 5/2. The $^1S_0$ ground state has only a single hyperfine level, $F'$=3/2. Thus the $^{137}$Ba $^1P_1^* \leftrightarrow {}^1S_0$ transition has three hyperfine components which are identified in this paper by the upper state hyperfine quantum number. Figure 1 shows RIMS spectra of the $^1P_1^* \leftrightarrow {}^1S_0$ $^{137}$Ba hyperfine structure taken using different polarizations of the dye laser beam. Note that randomly polarized light produces the expected peak height ratios for the 5/2:3/2:1/2 peaks of 3:2:1. With linear polarization, the F=1/2 peak vanishes; with circular polarization, the F=1/2 and F=3/2 peaks vanish. These phenomena are easily explained by considering the magnetic substates of the hyperfine levels. Each hyperfine level will have 2F+1 degenerate magnetic sublevels identified by magnetic quantum numbers, $M_f$ = F,F-1,...,-F. Plane polarized ($\pi$) light produces only vertical transi-

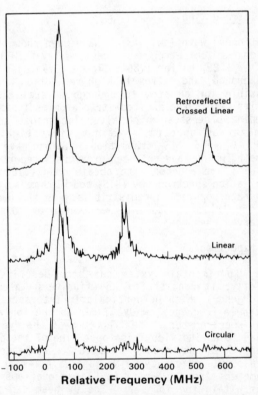

tions, i.e. those for which $\Delta M_f$=0. Circularly polarized ($\sigma$) light produces diagonal transitions, i.e. those for which $\Delta M_f$=±1 (the +1 and -1 values correspond respectively to right-hand [$\sigma_+$] and left-hand [$\sigma_-$] circular polariza- tion). Spontaneous emission can occur for $\Delta M_f$=0,±1 and therefore repeated excita- tion to the F=1/2 state with $\pi$ polarized light pumps all atoms into the $M_f$=±3/2 sub- levels of the lower state from which further vertical excitation cannot occur. Thus the F=1/2 peak height goes to zero. Likewise, excitation to the F=3/2 state with $\sigma_+$ polarized light will cause the substate population to reside in the $M_f$=3/2 state from which further excitation is not allowed (the same would be true for $\sigma_-$ polarized light except the $M_f$=-3/2 state would be populated) and the F=3/2 peak will "disappear". The F=1/2 peak also vanishes with $\sigma$ excitation because the magnetic substate population is pumped equally to the two positive $M_f$ values ($\sigma_+$) or the two negative $M_f$ values ($\sigma_-$). Note in Figure 1 that with $\pi$ polarized

**Figure 1.** CW RIMS spectra of $^{137}$Ba $^1P_1^* \leftrightarrow {}^1S_0$ hyperfine structure using different light polarization

Table 1. Steady-state population of lower level magnetic substates

| Light Polarization | Normalized Population of Lower Level Substates | | | | Relative Peak Intensity |
|---|---|---|---|---|---|
| | $M_f=-3/2$ | $M_f=-1/2$ | $M_f=1/2$ | $M_f=3/2$ | |
| **F=5/2 ↔ F=3/2 Transition** | | | | | |
| Random | 1/4 | 1/4 | 1/4 | 1/4 | 1/2 |
| $\pi$ | 1/10 | 4/10 | 4/10 | 1/10 | 14/25 |
| $\sigma_+$ | 0 | 0 | 0 | 0 | 1 |
| **F=3/2 ↔ F=3/2 Transition** | | | | | |
| Random | 1/4 | 1/4 | 1/4 | 1/4 | 1/3 |
| $\pi$ | 1/20 | 9/20 | 9/20 | 1/20 | 3/25 |
| $\sigma_+$ | 0 | 0 | 0 | 1 | 0 |
| **F=1/2 ↔ F=3/2 Transition** | | | | | |
| Random | 1/4 | 1/4 | 1/4 | 1/4 | 1/6 |
| $\pi$ | 1/2 | 0 | 0 | 1/2 | 0 |
| $\sigma_+$ | 0 | 0 | 1/2 | 1/2 | 0 |

excitation the ratio of the 5/2 to 3/2 peak heights is modified from the expected 3:2. This is due to redistribution of the magnetic substate population, but in this case no substate population is completely depleted.

Table 1 shows the steady-state populations of the magnetic sublevels in the lower state assuming low-intensity excitation (upper state population is negligible). The relative peak heights are found by summing the product of the appropriate transition probability (depending on polarization) and corresponding substate population over all substates. The transition probabilities are given by the square of the Clebsch-Gordon (CG) coefficients which can be found in any of several standard texts, e.g. Tinkham (1965). Note in Table 1 that under steady state conditions with $\pi$ polarized excitation, the 5/2 to 3/2 peak height ratio should be 14/3.

Figure 2 shows the change in peak height ratios at different laser intensities ($\pi$ polarization). At very low intensities, no optical pumping effects are observed and the expected 3:2:1 ratios are observed. At moderate powers, steady-state conditions are

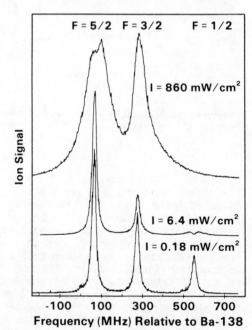

F = 5/2  F = 3/2  F = 1/2

I = 860 mW/cm$^2$

Ion Signal

I = 6.4 mW/cm$^2$

I = 0.18 mW/cm$^2$

-100  100  300  500  700
**Frequency (MHz) Relative to Ba-138**

**Figure 2. CW RIMS spectra of $^{137}$Ba $^1P_1^o$ ↔ $^1S_0$ hyperfine structure using different laser intensities**

achieved and the 1/2 peak goes nearly to zero, and the 5/2 to 3/2 peak height ratio is about 4. At higher powers, saturation effects dominate and this ratio approaches 1.0. The small dip in the 5/2 peak is due to weak branching to the metastable D states and is discussed later. We have used rate equations to model the magnetic substate redistribution in $^{137}$Ba as a function of laser intensity and interaction time. A detailed description of this model is given in a publication submitted to Phys. Rev. Lett. and only the results will be given here. The equations of the model can be solved at steady state for linearly polarized light but must be numerically integrated to include the approach toward steady state. To determine peak height ratios and linewidths, several factors besides the redistribution and saturation were considered. These include the velocity distribution for an effusive atomic beam, Voigt lineshape of the transition, intensity distribution of the laser, and weak branching to the metastable D states. These are known or measurable quantities; no fitted parameters are used to determine the calculated ratios or linewidths. Figure 3 shows the good agreement between the calculated ratio of 5/2 to 3/2 peak heights as a function of laser intensity and the experimental values.

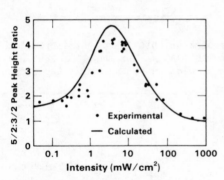

Figure 3. 5/2:3/2 peak height ratio as a function of laser intensity.

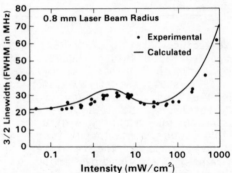

Figure 4. Line broadening due to unequal substate redistribution on and off line center

From Table 1, it is obvious that redistribution significantly reduces the line strength of the 3/2 transition. An interesting broadening effect occurs at laser intensities high enough to produce significant redistribution on line center but not high enough to produce significant redistribution a half-width away from line center. Under these condition, the 3/2 peak is suppressed more on line center than at frequencies off line center, leading to an apparent broadening effect. Figure 4 shows the measured linewidth of the 3/2 transition as a function of laser power and the linewidth predicted by our model. The broadening around 3 mW cm$^{-2}$ is due to this reduction of line strength on line center. This broadening decreases at higher intensities where the degree of redistribution in the tails is comparable to redistribution on line center. We are not aware of any previous disclosure of this effect in the literature. It should be noted the same effect can cause line narrowing if the line strength significantly increases with redistribution. The small increase in the 5/2 line is not enough to show this effect. However, if the ground state is prepared in the $M_f$=-3/2 state by optical pumping with $\sigma_-$ polarized light, redistribution with $\sigma_+$ polarized light increases the line strength of the 5/2 transition by a factor of 10. Our numerical model predicts a 25% narrowing of this line, thus giving a width less than the natural width.

The broadening in Figure 4 at higher intensities is primarily due to saturation but branching to the metastable D states contributes by flattening the top of the peaks. This flattening is due to the fact that more atoms are in the upper $^1P_1^{\bullet}$ state when the laser wavelength is exactly tuned to line center and therefore more atoms decay to the D states. This removes atoms from the two-level system and reduces the ionization signal at line center (the asymmetry will be discussed below). At very high intensities, this reduction appears as a dip in the lineshape even when the branching ratio is small as in the case of barium.

We have developed an equation that describes the lineshape distortions caused by D state trapping. A full description of the equation is not possible here and will be the subject of a future publication. The equation considers spatial distribution and velocity distribution of the atoms. Saturation effects are included and Poisson statistics are assumed. Figure 5 shows a comparison between the calculated lineshape of the $^1P_1^{\bullet} \leftrightarrow {}^1S_0$ $^{138}$Ba transition and the experimental values. The asymmetry in Figure 5 is caused by the fact that untrapped atoms, on the average, absorb many more photons than trapped atoms. For each absorbed photon that is followed by spontaneous emission, $h/\lambda$ momentum is transferred to the atom in the direction of laser beam propagation. When the photon is spontaneously emitted, $h/\lambda$ recoil momentum is transferred back to the atom, but this is a symmetric process and the recoil momentum averages to zero over many

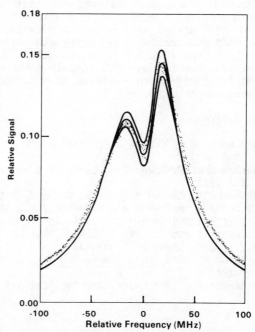

Figure 5. Lineshape distortion due to branching of the $^1P_1^{\bullet}$ state to $^1D$ and $^3D$ states

photons. The net momentum shows up as a Doppler shift (higher laser frequency is required to excite the deflected atoms). This effect is included in the calculated lineshape. The three solid lines in Figure 5 are calculated lineshapes for our experimental conditions and branching ratios of 440, 460, and 480 respectively for the bottom, middle, and top line. The dots are experimental data taken over three separate scans. For the experiments, part of the dye laser beam was split off and expanded to a "pump" beam which intersected the barium atomic beam just before the photoionization area. Atoms traversing the pump region were cycled hundreds of times through the excitation/spontaneous emission cycle to assure a significant fraction of the atoms wound up in the D state traps. The best fit is for R=450 ± 40 where R is defined as the number of atoms radiatively decaying to the $^1S_0$ state divided by the summation over all atoms decaying to the D states. This value is significantly lower (indicating more branching to the D states) than the lower limit of 700 found by Bernhardt (1976). Our value is in reasonable agreement with a theoretical value of 340 found by McCavert (1974). Lewis, Kumar, Finn, and Greenlees (private communication) have recently measured R using the intensity dependence of resonance fluorescence from the Ba $^1P_1^\bullet$ state and found R to be about 280 ± 30.

## 4. Conclusions

Continuous wave RIMS has several significant advantages over pulsed laser RIMS. It is important to understand the benefits and pitfalls of such a powerful process. We have shown that optical pumping can lead to anomalous linewidths and peak heights but that these effects can be described adequately by a mathematical model. The model has no fitted parameters and still agrees very well with experimental data. We have also demonstrated a new method of measuring weak branching ratios. These measurements are important to cw RIMS and other techniques such as photon-burst spectroscopy (Bushaw 1985) and laser-enhanced electron-impact ionization (Bushaw 1986) which cycle an atom through many excitation/spontaneous-emission cycles.

## 5. References

Bernhardt A F, Duerre D E, Simpson J R and Wood L L 1976 J. Opt. Soc. Am. 66 416
Bushaw B A, Whitaker T J, Cannon B D and Warner R A 1985 J. Opt. Soc. Am. B 2 1547
Bushaw B A, Cannon B D, Gerke G K and Whitaker T J 1986 Opt. Lett. 11 422
Cannon B D, Bushaw B A and Whitaker T J 1985 J. Opt. Soc. Am. B 2 1542
Keller R A and Snyder J J 1986 Laser Focus 22:3 86
McCavert P and Trefftz E 1974 J. Phys. B 7 1270
Miller C M, Engelman R and Keller R A 1985 J. Opt. Soc. Am. B 2 1503
Tinkham M, 1965 Group Theory and Quantum Mechanics (New York:McGraw-Hill) p 124
Whitaker T J 1986 Lasers and Applications 5:8 66

*Inst. Phys. Conf. Ser. No. 84: Section 1*
*Paper presented at RIS 86, Swansea, Wales, 7–12 Sept. 1986*

# Coherent excitation and photoionization of Rydberg states using picosecond laser pulses

R. Beigang
Institut für Quantenoptik, Universität Hannover,
3000 Hannover, FRG

and

D. Krökel and D. Grischkowski
IBM T. J. Watson Research Center, Yorktown Heights/USA

Abstract. Coherent Excitation of high lying Rydberg states
with picosecond laser pulses leads to the formation of a
spatially localized wave packet, which shows close simi-
larities with Rydberg electrons in a classical Kepler
orbit. The time dependence of this wave packet can be
probed with time delayed photoionization.

## 1. Introduction

The investigations of Rydberg states of atoms and molecules
have gained considerable interest over the past few years
(see e. g. Stebbings and Dunning, 1983). The rather exotic
properties of highly excited atoms gave rise to numerous
interesting experiments like, e. g. , the one atom maser
(Walther et al 1986), inhibited spontaneous emission
(Kleppner et al 1981) or the investigation of a single atom
in a high Q cavity (Goy et al 1983, Meschede et al 1985).
In particular properties of atoms which scale with high
powers of the principal quantum number n can be studied
conveniently using Rydberg atoms. Table I summarizes typical
scaling laws of of some properties of Rydberg atoms together
with data for principal quantum numbers n = 1, 10.

| Property | n-dependence | n = 1 | n = 100 |
|---|---|---|---|
| Mean radius $\langle r \rangle$ | $n^2 a_o$ | $5.3 \cdot 10^{-9}$ cm | $5.3 \cdot 10^{-5}$ cm |
| Binding Energy | $R/n^2$ | 13.6 eV | 1.36 meV |
| Level spacing $\Delta E_n$ | $2R/n^3$ | 10 eV | $1 \cdot 10^{-5}$ eV |
| Period of electronic motion $t_n$ | $n^3 \cdot t_1$ | $1.5 \cdot 10^{-16}$ s | $1.5 \cdot 10^{-10}$ s |
| RMS velocity of Rydberg electrons | $v_o/n$ | $2.2 \cdot 10^8$ cms$^{-1}$ | $2.2 \cdot 10^6$ cms$^{-1}$ |

Table I Typical scaling laws of properties of Rydberg atoms

In this contribution we will focus on the large orbit time of highly excited states and discuss how the classical orbit time can be "measured" experimentally.

## 2. Coherent Excitation of Rydberg States

According to the uncertainty principle a single Rydberg state cannot be excited with a laser pulse of a duration $t_L$ shorter than the classical round trip time $t_n$ of the Rydberg electron ($t_n \gg t_L$). The finite bandwidth $\hbar/t_L$ leads to the excitation of several neighbouring Rydberg states as their level spacing $\Delta E_n$ is small compared to the spectral width of the laser pulse:

$$\hbar/t_L \gg \Delta E_n$$

if

$$t_n \gg t_L .$$

This coherent excitation of neighbouring Rydberg states can be treated as quantum beats between states with different principal quantum number n. The superposition of the excited wave functions leads to the formation of a spatially localized wave packet, which oscillates with a frequency given by the mean energy difference $\Delta E_n$ of the excited Rydberg states. This has been first shown quantitatively by Parker and Stroud (1986) and recently by Alber et al (1986). The wave function of the excited Rydberg electron can be written as (Parker and Stroud 1986):

$$\psi(r,t) = \sum_n a_n(t) \exp(-iw_n t) R_{nl}(r)$$

Here, $a_n$ and $\omega_n$ are amplitudes and transition frequencies of the excited Rydberg states, respectively. $R_{nl}(r)$ are hydrogenic radial wave functions and the sum extents over all excited states. For the following calculations five Rydberg states with principal quantum numbers n=29 to 33 are coherently excited with a laser pulse of 0.4 ps duration. The electron density distribution $r^2|\psi(r,t)|^2$ after the laser pulse is shown in Fig.1. A well defined wave packet is formed which moves towards the outer classical turning point $r_{max} = 2n^2 a_0$. After the maximum is reached the wave packet moves back to

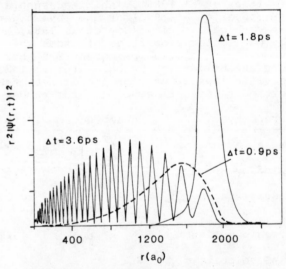

Fig.1 $r^2|\psi(r,t)|^2$ as a function of radius r for three selected times

the atomic core where it becomes broad and diffuse. This
oscillation takes place with a frequency determined by the
mean energy separation between the excited Rydberg states.
This frequency is identical with the oscillation frequency of
an electron in a classical Kepler orbit with a mean radius
$\langle r \rangle = n^2 a_0$. For n=31 the classical period of electronic motion
is 3.6 ps in agreement with the calculated oscillation of the
wave packet. After a few oscillations the wave packet will
spread out as expected from quantum theory. However, the
wave packet will reform after a time $t_R$ whenever the
individual orbit times $t_n$ of the excited states n are
simultaneously integer multiples of $t_R$ .

## 3. Time delayed photoionization

The oscillation of the wave packet may be detected by use of
a second time delayed picosecond pulse which interacts with
the coherently excited Rydberg atoms. This method relies on
the fact that the interaction of photons with the excited
atoms is largest near the inner turning point where the
acceleration of the electron has a maximum. Instead of using
a Raman type de-excitation as discussed quantitatively by
Alber et al (1986), a time resolved photoionization from the
excited states can also be used to detect the time development of the wave packet. Fig. 2 shows the calculation of
relative ionization
cross sections as a
function of time de-
lay between exciting
and ionizing laser
pulse. For the cal-
culations the over-
lap of the wave
packet with the core
was taken as a mea-
sure of the ioniza-
ion signal. The
pulse width is again
0.4 ps and five Ryd-
berg levels (n=29..
..33) are coherently
excited. The ioni-
zation signal shows
maxima after each
round trip in quali-
tative agreement with
the oscillation of
the wave packet. The
relative ionization
signal decreases with
increasing numbers of
oscillations as the
wave packet decays.
The same behaviour

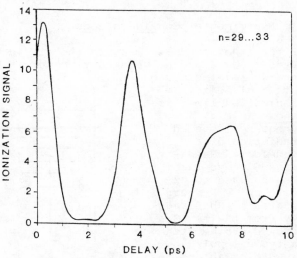

Fig. 2 Relative ionization signal
as a function of time delay between exciting and ionizing laser
pulse

was observed in the case of the Raman type de-excitation
where the fluorescence after de-excitation was used for the
detection of the wave packet (Alber 1986).

For long time delays the ionization signal recovers and shows the expected "revival" as the wave packet forms again.   This is illustrated quali- tatively in Fig 3. where the relative ionization signal is displayed up to delay times of 100 ps corresponding to approximately 35 orbits. For this calculation the photoionization was assumed to be the only process to de- populate the exci- ted Rydberg states. No other relaxation processes, like sti- mulated or sponta- neous emission to lower lying states have been taken into account.

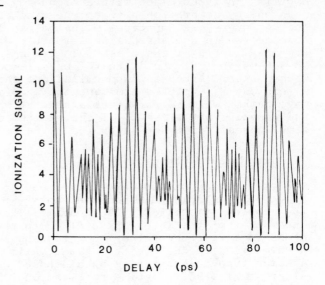

Fig. 3 Relative ionization signal as a function of time for long time delays

## 4. Experimental Considerations

First experiments to measure the time development of a Rydberg wave packet were carried out in Rb. A simplified energy level diagram of Rb close to the ionization limit is shown in Fig. 4. Frequency doubled radiation of a modelocked Rhoda- mine 6 G dye laser can be used to ex- cite $np$ $^2P$ Rydberg states of Rb direct- ly from the 5s $^2S_{1/2}$ ground state. With a typical pulse length of about 7 ps in the visible the pulse shortens to nearly 5 ps after frequency doubling. This corresponds to a spectral width of $\Delta\nu = 3$ cm$^{-1}$ in the uv. A well suited range of Rydberg states for these parameters of

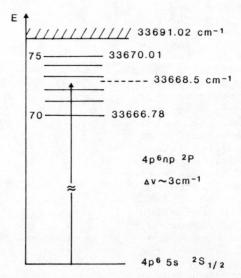

Fig. 4 Relevant Rydberg levels of Rb for coherent excitation

the exciting laser is around n = 72. The mean orbit time is approximately a factor of ten larger than the pulse length and the band width of the laser is large enough to excite more than five states coherently. On the other hand, the spectral width of the laser is small compared to the energy separation from the ionization limit so that there is no overlap with continuum states. Our experimental set up is shown schematically in Fig. 5.

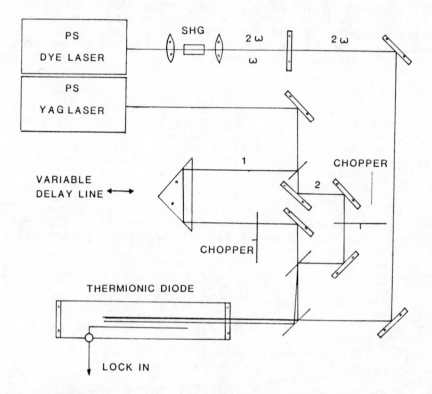

Fig. 5 Schematic diagram of the experimental set up

The Rb atoms are excited in a Pyrex glass cell which contains a thermionic diode for the detection of ions produced via photoionization. Due to the high duty cycle of the modelocked laser lock in techniques can be applied for the measurement of the ion current. In a first experiment the fundamental wave length of the dye laser was used for ionization. A variable delay line provided the required time delay between exciting (second harmonic) and ionizing (fundamental) radiation. It turned out, however, that no photoions could be measured above principal quantum numbers n > 10. In order to increase the photoionization process a modelocked Nd:YAG laser with higher peak power (and shorter wavelength) was used to ionize the excited atoms. With a double beam (see Fig. 5) technique the background due to non-resonant four-photon ionization could be reduced considerably and photo-ionization of excited Rydberg atoms was detected. Part of the Nd:Yag laser beam (2 in Fig. 5) was sent through the cell

before the exciting uv pulse producing only four-photon
ionization with no contribution due to ionization from
Rydberg states. Thus with the uv pulse turned off the
background produced by beams 1 and 2 could be leveled off.
The signal to noise ratio was still to poor to detect
ionization from states above n > 30, so that no time delayed
ionization experiment was performed.  The main source of
noise was still the four-photon ionization caused by the
powerful Nd:YAG laser. With an improved balance between the
two Nd:YAG laser beams it should be possible to suppress the
background even better and to extent the ionization from
Rydberg states up to the required range of quantum numbers n
for the detection of Rydberg wave packets.

## 5. Conclusions

The coherent excitation of Rydberg states with picosecond
laser pulses leads to the formation of a spatially localized
wave packet, which oscillates, decays and reforms. The
oscillation, decay and revival of this wave packet can, in
principle, be measured using time delayed photoionization.
First experiments with Rb Rydberg states indicate that
photoionization from Rydberg states with principal quantum
numbers n > 35 can be performed using state of the art
modelocked lasers. By use of a more powerful modelocked
Nd:YAG laser for the photoionization process and an improved
detection scheme the observation of Rydberg wave packets can
be performed experimentally.

## 6. References

Alber G Ritsch H and Zoller P 1986 Phys. Rev. A to be
    published
Goy P, Raimond J M, Gross M and Haroche S 1983 Phys.   Rev.
    Lett. 50 1903
Kleppner D 1981 Phys. Rev. Lett. 47 233
Meschede D, Walther H and  Müller G 1985 Phys. Rev. Lett. 54
    551
Parker J and Stroud Jr. C R 1986 Phys. Rev. Lett. 56 716
Stebbings R F and Dunning F B 1983 Rydberg states of atoms
    and molecules (Cambridge: Cambridge University Press)

*Inst. Phys. Conf. Ser. No. 84: Section 1*
*Paper presented at RIS 86, Swansea, Wales, 7–12 Sept. 1986*

# Multiphoton spectroscopy of autoionising states and AC Stark shifts in strontium atoms

P. Agostini, A. L'Huillier, G. Petite
Service de Physique des Atomes et des Surfaces
CEN Saclay 91191 Gif sur Yvette

X. Tang, P. Lambropoulos[+]
University of Southern California, Los Angeles Ca 90089-0484
[+]and Research Center of Crete, Iraklion, Crete 71110, Greece

## 1. Introduction

Evidence accumulating over the last 10 years or so points to a complex set of processes taking place when lasers with intensities above $10^{10}$ W cm$^{-2}$ interact with atoms. The most recent activity has focused upon the multiple ionisation of atoms for intensities between $10^{12}$ and $10^{16}$ W cm$^{-2}$ and has provoked widespread debate as to the underlying mechanisms (L'Huillier 1986).

One of the important features inherent in these processes is the participation of doubly and multiply excited electronic states. Doubly excited states have been known to play a role in traditional weak field photo-absorption as well as electron scattering. Under a strong laser field and the resulting multiphoton excitation a completely new set of phenomenon appear. First, not only the participating states will usually differ from those known in traditional spectroscopy in total angular momentum, but also entirely new configurations will come in. Second, these states imbedded as they are in the strong field undergo substantial distortion which leads to behavior that cannot be anticipated on the basis of the bare, (field-free) atoms.

Alkaline-earth atoms have been chosen as experimental tests for several reasons (Agostini and Petite 1984, 1985). First, their ionisation potentials are rather low and they can be doubly ionised with less than 10 photons of visible light. Therefore it is possible to use tunable lasers which is a great help in disentangling the different mechanisms. Second, the spectroscopy has been well studied and, at least, some information about doubly excited state is available. Third, they are easily handled in thermal atomic beams. The purpose of this paper is to report and discuss recent data on multiphoton ionisation (MPI) of Strontium and especially on resonant MPI. Two important points have been highlighted in this work: a) the importance of singlet-triplet mixing which was found to be a necessary element to understand the observed Stark shifts ; b) the influence of autoionising states from the 5p5d configuration which play a significant role in the shifts and which are, for the first time directly appearing in an experiment. Detection of these states had eluded, so far, a number of UV experiments although much indirect and theoretical evidence suggested their presence (Connerade et al., 1980).

The paper is organised as follows: in section 2 the theoretical framework will be outlined. In section 3 the experiment is briefly described while section 4 will be devoted to presenting and discussing the results.

## 2. Theoretical Framework

The experimental results discussed and interpreted in this paper require the calculation of 3- and 4-photon ionisation for wavelengths such that 3 photons are energetically sufficient for ionisation. The fourth photon will therefore be involved in transitions between states above the first threshold with or without the participation of doubly excited autoionising states depending on the photon energy. Since two-photon resonances with intermediate bound states are also encountered in this wavelength region, the apparatus of resonant MPI theory becomes necessary including the laser induced AC stark-shifts of the resonant states (Tang 1986). Furthermore, the singlet-triplet mixing of the atomic state is indispensable to the interpretation of the observations. Its significance is obviously reflected upon the fact that two-photon resonant transitions to triplet states, such as $5s5d(^3D_2)$ and $5p^2(^3P_{0,2})$ exhibit prominent peaks. Given that the ground state $5s^2(^1S_0)$ is singlet, only intercombination transitions could account for the presence of these resonances which can, in turn, be used as information input in evaluating the strength of singlet-triplet mixing. All states must here be understood as properly antisymmetrised two-electron states. For example, consider a two-photon transition from $5s^2(^1S_0)$ to a state of the form $5s\ ns(^1S_0)$ (singlet-singlet transitions). The usual general expression for the transition amplitude:

$$r^{(2)}_{ba} \equiv \sum_n \frac{<b|r|n><n|r|a>}{E_n - E_a - \hbar\omega} = \int d^3\vec{r}_2 \int d^3\vec{r}_1 \psi_b(\vec{r}_2) r_2 G(\vec{r}_1,\vec{r}_2;\Omega) r_1 \psi_a(\vec{r}_1) \quad (1)$$

will involve matrix elements of the type $<5s\ mp|r_\alpha + r_\beta|5s^2>$ where $|5s\ mp> = \mathscr{Y}_{5s\ mp}(\vec{r}_\alpha,\vec{r}_\beta)$ are the two electrons (labels $\alpha$, $\beta$) wavefunctions. Here $G(\vec{r}_2,\vec{r}_1;\Omega)$ is the Green's function

$$G(\vec{r}_2,\vec{r}_1;\Omega) = \sum_n \frac{\psi_n^*(\vec{r}_2)\psi_n(\vec{r}_1)}{E_n - \hbar\Omega} \quad (2)$$

related to the radial Green's function $g(r_1,r_2;\Omega)$ through:

$$G(r_1,r_2;\Omega) = \sum_{\ell m} \frac{g_\ell(r_1,r_2;\Omega)}{r_1,r_2} Y^*_{\ell m}(\Theta_1, \varphi_1) Y_{\ell m}(\Theta_2, \varphi_2) \quad (3)$$

Finally the two-photon matrix element between $|5s^2>$ and $|5s\ ns>$ is written as

$$<5s^2|r^{(2)}|5s\ ns> = \sqrt{2} \int dr_1 \int dr_2\ \psi_{ms}(r_2) r_2\ g(r_2,r_1;\Omega) r_1\ \psi^*_{5s^2}(r_1)$$

$$\int dr\ \psi^*_{5s^2}(r)\ \psi_{5s}(r) \quad (4)$$

where the Green's function is evaluated at the appropriate energy $\hbar\Omega = E_a + \hbar\omega$. When the energies are not known from previous spectroscopic measurements they have to be calculated. To this end, and also to calculate final state energies, a multiconfiguration Hartree-Fock calculation, which can require up to 100 wavefunctions and 10 configurations as inputs, has been performed. As it will be shown below, the resulting energies are then tested by different observations like the stark shifts or, when possible, the positions of resonance peaks.

Because of the relatively high value of Z (38), the distinction between singlet and triplet is not strict in Sr. Neither LS or jj coupling is rigourously applicable in the description of all states. For instance, state 5snp which would be separated in ($^1P_1$) and ($^3P_0, ^3P_1, ^3P_2$) in pure LS coupling are coupled with the consequence that neither $^1P_1$ is pure singlet nor $^3P_1$ is pure triplet (Sobelman 1972). In fact states can be written as a linear superposition of purely singlet and triplet states. In principle the coefficients $\alpha, \beta$ of the superposition can be obtained through ab initio calculations. In practice, if the energies of the pure states are known, the value of $\rho \equiv \beta/\alpha$ is obtained through approximate equations like the Pauli-Houston equation. When the necessary spectroscopic information is not available we may use $\rho$ as a parameter for fitting some of the observations. The formalism which has been outlined above must then be extended to include the singlet-triplet mixing in the calculation of multiphoton transitions. Using the linear superposition for the mixed states, the two-photon matrix element can be written as:

$$\langle 5p^2(^3P_2)|r^{(2)}|5s^2(^1S_0)\rangle =$$

$$\sum_n \frac{1}{\alpha_n^2+\beta_n^2}\left[\alpha_n\beta_n\left(\frac{1}{E_n^{(1)}-E_{5s^2}-\hbar\omega} - \frac{1}{E_n^{(3)}-E_{5s^2}-\hbar\omega}\right)\right.$$

$$\left. \langle 5p^2(^3P_2)|\vec{r}|5s\ np(^3P_1)\rangle\langle 5s\ np(^1P_1)|\vec{r}|5s^2(^1S_0)\rangle \right] \tag{5}$$

($E_n^{(1)}$ and $E_n^{(3)}$ denote the intermediate singlet and triplet energies respectively).

In (5) only singlet-singlet and triplet-triplet matix elements appear and it is clear that if either $\alpha_n$ or $\beta_n$ (coefficients of the singlet-triplet superposition) is zero the two-photon transitions between a singlet and a triplet state is totally forbidden. Note that physical triplet states will also contribute, in principle to a tow-photon transition between two singlet states. Eq.(5) also shows that the importance of a triplet intermediate state is determined not only by the amount of singlet it contains but also by the proximity of its energy to a single photon resonance.

## 3. Experiment

Fourier limited pulses (20 ps) from a Nd:Yag laser or from a synchronously pumped dye oscillator-amplifiers system are focused into a vacuum chamber on a thermal effusive beam of Strontium (Petite and Agostini 1986). Intensities up to $10^{12}$ W cm$^{-2}$ can be produced at wavelengths 532 nm and 558-574 nm. Singly and doubly charged ions (Sr$^+$ and Sr$^{++}$) signals can be separated in a time-of-flight analyser and monitored. The photoelectrons are energy-analysed in a time-of-flight spectrometer carefully avoiding space charge effect (Yergeau et al., 1986). The resulting spectra are composed of lines with definite energies which are associated with 3- and

4-photon ionisation of Strontium . Three lines have been well identified: at 0.1 eV characterising a 4-photon ionisation leaving the ion in the $5p^2P$ state ; at 0.9 eV characterising the 3-photon ionisation to the $Sr^+(5s^2S)$ state ; at 1.3 eV characterising the 4-photon ionisation to the $Sr^+(4d^2D)$ states (Petite and Agostini 1986). A few more lines can be identified at some laser wavelengths like the ones corresponding to the 4-photon ionisation to the $Sr^+(5s^2S)$ states or the ones corresponding to sequential double ionisation. In the tuning range available (558-574 nm) three types of resonances could be observed either on the ion $Sr^+$ signal or on electron lines. When monitored as a function of the laser wavelength at constant intensities. (i) intermediate two-photon resonances; (ii) intermediate 3-photon resonances in 4-photon ionisation; (iii) final state 3-photon resonances in 3-photon ionisation. Due to the limited available space we will restrict the discussion to two aspects of the observation: the first one is the AC Stark shifts of the two-photon resonant three-photon ionisations. The resonant states are the $5p^2 \, ^3P_{0,2}$ and the $5s5d^3D_2$. This study will enlight both the importance of the singlet-triplet mixing and the influence of autoionising states of the configurations 5p5d and 5p6s. The second one will be the identification of states from the 5p5d configuration with three-photons resonances displayed on the $Sr^+$ signal and the four-photon ionisation electron signals.

## 4. Results and Discussion

### 4.1 AC Stark shifts of two-photon resonances

When the laser is tuned through the two-photon resonances, sharp peaks appear on the $Sr^+$ signal. When the laser intensity is increased the resonance peaks are shifted and broadened. The measured shifts are linearly dependent on the laser intensity and the coefficients ($\alpha$) can be expressed in $cm^{-1}/GW \, cm^{-2}$ (see Table 1). The shift of one resonance is the sum of the shifts of the ground state and of the resonant state.

The shift of a specific level $|n>$ can be written as:

$$\Delta E_n = \alpha I = A \left[ \sum_i |\mu_{ni}|^2 \frac{2 \, \Omega_{in}^3}{\Omega_{in}^2 - (\hbar\omega)^2} \right] I \qquad (6)$$

where the sum runs over a complete set of states $|i>$. $\mu_{ni}$ are the dipole matrix elements, $\Omega_{in} = E_i - E_n$, $\hbar\omega$ is the photon energy and I is the laser intensity. It is therefore clear that the shifts depend critically on the atomic wavefunctions and level energies. A simple inspection of the energy denominators of (6) allows to select out the dominant terms of the sum, but, in order to account for the observed shifts, a complete calculation including both configuration mixing and singlet-triplet mixing is necessary. Configuration mixing of the 5p6s, 5p5d, 4d4f autoionising states is discussed in some details in the next subsection. The importance of singlet-triplet mixing is illustrated in table 1 showing calculated $\alpha$'s with and without the mixing together with measured values. The coefficients of this mixing could not be calculated ab-initio but were fitted from available angular distribution data (Feldman and Welge 1982) and used in the shifts calculations. As seen from table 1, the agreement between calculated and observed shifts is excellent when the mixing is taken into account.

| State | Experiment | Calculation 1 | Calculation 2 | Calculation 3 |
|---|---|---|---|---|
| $5p^2\ {}^3P_0$ | $- 2.5 \pm 0.5$ | $- 4.26$ | $- 2.2$ | $- 2.41$ |
| $5p^2\ {}^3P_2$ | $- 0.03 \pm 0.1$ | $- 0.91$ | $- 0.021$ | $- 0.0225$ |
| $5s5d\,{}^3D_2$ | $0.7 \pm 0.3$ | | $1$ | |

Table 1. Observed and calculated coefficients of the shifts ($cm^{-1}/GW\ cm^{-2}$): in calculation 1 only the 5p ns configuration is taken into account for the $5p^2$ states. Calculation 2 includes singlet-triplet mixing and calculation 3 includes also the configurations 5p5d and 4d4f. For the $5s5d\,{}^3D_2$ state configurations 5s np and 5s nf have been included.

## 4.2 Configuration mixing for autoionising states

Autoionising states eigher playing an important role in the shifts of two-photon resonances or showing in three and four-photon ionisation are states of the 5p6s, 5p5d and 4d4f configurations. In recent experiments (Feldmann and Welge 1982, Petite and Agostini 1986), three-photon resonances were observed in the $Sr^+$ signal but could not be identified. Supplementary information from resonances observed on energy-analysed electron signals together with the multiconfiguration Hartree Fock calculation allow a much more complete identification. We first tried a single configuration HF calculation for the 5p5d P,D,F states which yielded energies different from observed values by about $6000\ cm^{-1}$. We then proceeded to MCHF including a number of configurations, as listed, for instance, for the 4d4f P,D,F:

$4d4f^{1,3}D$:  4d nf, 5d nf, 6d nf, 7d nf, 7p nd, 5p nd, 6p nd, 7p nd, 8p nd
  (with n up to 13)
$4d4f^{1,3}P$:  all the above mentioned configurations plus 5p ns
$4d4f$  F:  all the above plus 4f ns, 5f ns, 6f ns and 7f ns.  In table 2, we have summarized some of the results of this calculation and shown the coefficients of the most important configurations, the resulting energies and the corresponding experimental energies.

Considering table 2, the agreement between theory and experiment is very reasonable albeit far from perfect. The MCHF calculation tested on other levels (like the 4d7p) gives energies within $200\ cm^{-1}$ from the experiment, which gives confidence in the above results. For the AC-Stark Shifts calculations we have used the wavefunctions obtained by MCHF but utilised the experimental energies. This calculation is to be understood as tentative and can be very much improved as more inputs are gained from the experiment. The present results are quite encouraging. Note also that states of the "5p5d" configuration have been detected here for the first time both as influencing the AC-Stark shifts and as observable resonances.

| State | Configuration | | | | | $E(cm^{-1})$ | Experiment |
|---|---|---|---|---|---|---|---|
| $5p6s\ ^3P_1$ | 5p6s | 4d4f | 4d5f | 4d7p | 4d8p | 53 236 | 52 882 (b) |
| | 0.56 | -0.33 | 0.32 | 0.54 | -0.32 | | |
| $5p6s\ ^1P_1$ | 0.63 | 0.31 | -0.33 | 0.39 | | 51 996 | 53 633 (a) |
| | 62% triplet, 38% singlet | | | | | | |
| $5p5d\ ^3D$ | 5p5d | 4d6p | 4d7p | 4d4f | | 54 627 | 53 430 (a) |
| | 0.76 | -0.33 | -0.42 | 0.28 | | | |
| $5p5d\ ^1D$ | -0.78 | +0.34 | 0.46 | -0.18 | | 54 544 | 53 223 (a) |
| | 80% triplet, 20% singlet | | | | | | |
| $4d4f\ ^3D$ | 4d4f | 4d5f | 5p5d | 4d7p | | 53 495 | 53 286 (a) |
| | -0.61 | 0.67 | -0.23 | 0.22 | | | |
| $4d4f\ ^1D$ | 0.63 | 0.67 | 0.23 | 0.15 | | 53 451 | |
| | 61% triplet, 39% singlet | | | | | | |

Table 2. Coefficients for configuration and singlet-triplet mixing.
(a) Feldmann (1982) and Petite (1986)
(b) Garton and Colding (1968) and Connerade et al. (1980).


## 5. Conclusion

Multiphoton resonance ionisation of complex atoms has proven to be a very valuable tool for spectroscopy of autoionising states (i.e., structures in the ionisation continuum) of alkaline-earth atoms, even at high laser fields, provided the resonances are not washed out by saturation. Especially electron energy analysis brings information about the spectroscopy of atoms and ions and the interactions with the laser field. In particular the singlet-triplet mixing coefficients are a direct output from angular distribution measurements. In forthcoming papers, we will extend these investigations to Ca and Mg. It is hoped that those will provide keys for understanding multiple ionisation of atoms by strong laser fields, still a very open question.


## 6. References

Agostini P and Petite G 1984 J. Phys. B: At. Mol. Phys. 17, L811.
Agostini P and Petite G 1985 J. Phys. B: At. Mol. Phys. 18, L281.
Connerade J P, Baig M A, Garton W R S and Newsom G H 1980, Proc. R. Soc. Lond. A371, 295.
Feldmann D and Welge K H 1982 J. Phys. B: At. Mol. Phys. 15, 1651 and private communication.
Garton W R S and Colding K 1982 J. Phys. B 1, 106.
L'Huillier A 1986 Comments on At. and Mol. Phys. (to be published).
Petite G and Agostini P 1986 J. Phys.: 47, 795.
Sobelman I I 1972 Introduction to the theory of Atomic Spectra, Pergamon Press NY.
Tang X 1985 Doctoral dissertation University of Southern California (Unpublished).
Yergeau F, Petite G and Agostini P 1986 J. Phys. B: in press.

*Inst. Phys. Conf. Ser. No. 84: Section 1*
*Paper presented at RIS 86, Swansea, Wales, 7–12 Sept. 1986*

# Analytical laser-enhanced ionization spectroscopy with thermionic diode detection

K. Niemax, J. Lawrenz and A. Obrebski

Institut für Spektrochemie und Angewandte Spektroskopie (ISAS), Dortmund, FR Germany

Abstract.  We have set up an arrangement of two thermionic diodes which allows the performance of very sensitive analytical laser spectroscopy. In one of the diodes (the analytical diode), small samples of pure or ultra-pure metals are evaporated electrothermally. The second diode serves as a reference for frequency-locking and power control of the lasers. Applying Doppler-free techniques with appropriate cw lasers, isotope selective analytical spectroscopy is possible. The power of the method is demonstrated by determining particular Yb isotopes in pure metal samples applying the resonant Doppler-free 2-photon spectroscopy.

## 1. Introduction

In contrast to RIS where the ionization of the resonantly excited analyte species is induced by off-resonant, powerful laser radiation, in laser-enhanced ionization spectroscopy (LEI) the ionization of the excited atoms is due to collisions with other particles. The ionization probability by collisions can be very large or even unity if the atoms are excited to high-lying states. E.g. it was shown in a cell experiment that Rb atoms in high Rydberg states (n > 20) are ionized with a probability of unity in low pressure noble gases (Niemax 1983). The other difference in our approach compared to the usual RIS-technique with high-power pulse lasers is the application of cw lasers in the power range of a few mW to a few hundred mW sufficient to saturate dipole transitions. There is the advantage of the high duty circle (residence time of the analyte in the laser field) using a cw system and, if the laser is running single mode, the possibility of applying Doppler-free techniques and excite single isotopic species of the analyte.

## 2. The Analytical Thermionic Diode

The operation of thermionic diodes as ion detectors and their application in spectroscopy is described in detail in a recent review paper (Niemax 1985). In its simplest form a thermionic diode is composed of a cylindrical anode with an inner, directly heated cathode filament.  If no bias or only a weak bias is applied, the diode current is space-charge limited.  Ions, created by laser excitation and subsequent collisional ionisation, diffuse into the space charge and lower the potential barrier for electrons from the cathode surface. As a result the diode current is increased. The amplification factor was found to be high ($10^5$ - $10^7$). It was also found that only a few ions/s are required to create an increase of the diode current which is above the noise level (Popescu et al. 1969 and Niemax 1985). Optimal operating conditions in thermionic diodes, independent on the atoms investigated, can be achieved by activating the cathode surface with alkaline earth atoms which are permanently present in the thermionic diode (Niemax and Weber 1985). A thermionic diode prepared for analytical laser spectroscopy of small solid samples like pure or ultra-pure metal is shown in Fig. 1. It is made of a stainless steel tube (anode) filled with an activating metal vapour (e.g. Ba) and an inert buffer gas (e.g. Ne). The cathode filament is made of tungsten or molybdenium. The sample, on a Re-band with a shallow deepening is brought by a probe into a pre-chamber which is then evacuated and filled with buffer gas of the same kind and pressure as inside the thermionic diode. Then the valve is opened and the probe with the sample is shifted into the diode, and the sample can be evaporated electrothermally from the Re-band.

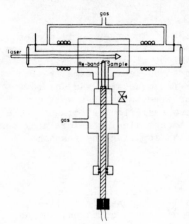

Fig. 1   Analytical thermionic diode with sample introduction systems for electrothermal evaporation

## 3. Set-Up For Isotope Selective Trace Element Detection

Fig. 2 presents schematically the basic set-up for isotope selective analytical laser spectroscopy with the thermionic diode. At least two tunable cw single mode lasers have to be used. The light beams of the two

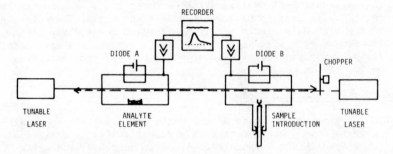

Fig. 2   Experimental set-up for isotope-selective analytical laser spectroscopy

lasers are transmitted through two thermionic diodes (diode A and B) in counter- or co-propagating directions. If they are tuned to resonances of a two-step excitation scheme and if the light of one or both lasers is

modulated, Doppler-free spectra can be obtained applying the lock-in
detection technique. Diode A filled with buffer gas, the activating element (e.g. Ba) an a larger quantity of the analyte element serves as a
reference. The lasers can be tuned and locked to a Doppler-free isotopic
component. The signal height of diode A is a measure of the frequency
and/or intensity constancy of the two lasers. If there is change of the
signal it has to be taken into account for the analyte signal which is
measured with diode B. Diode B is filled with a low pressure noble gas
(p ≈ 300 mTorr) and a cathode activating element at low number density.
The procedure of introducing samples by a probe is described above.
Because of the slow response time of the thermionic diode to ions it is
recommended to keep the modulation frequency of the laser light now
(Niemax 1985).

## 4. Excitation And Detection Of Isotopes

In our first experiments on the application of thermionic diodes in
analytical laser spectroscopy we chose Yb as analyte element (Niemax et
al. 1986). The most important reason of choosing Yb was the availability
of tunable dye lasers with the proper wavelength to excite the Yb atom
selectively. The first step in the Yb atom $6s^2$ $^1S_0$-$6s6p$ $^3P_1$ was induced by
green-yellow laser radiation (555.6 nm, dye: Rh 560), while the second
step $6s6p$ $^3P_1$-$6s6d$ $^3D_2$ needs blue laser light (457.8 nm, dye: Stilbene 3).
It should be pointed out that the final excited level is still about 10600
$cm^{-1}$ below the ionization level. Therefore the collisional ionization
probability is still expected to be very low. A part of the Doppler-free
spectrum obtained from diode A by scanning the blue laser while keeping
the green-yellow laser fixed in frequency within the Doppler profile of
the first transition is shown in Fig. 3. We see that the line of the
$^{174}$Yb isotope (natural abundance: 31.8 %) is strongly enhanced compared
with the lines of the isotopes 172 (21.8 %) and 176 (12.7 %). This is a

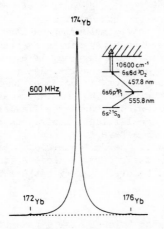

Fig. 3  Resonant Doppler-free
2-photon spectrum of the $6s^2$ $^1S_0$-
$6s6p$ $^3P_1$-$6s6d$ $^3D_2$ transition in
Yb

result of the frequency position of the first laser. Locking both lasers
to the $^{174}$Yb component the transient signal of the $^{174}$Yb trace in pure and
ultra-pure metals have been recorded (Niemax et al. 1986). E.g. in Fig. 4
we display the directly reproduced signals when 100 μg pure Eu was
introduce to the analytical diode B and evaporated electrothermally from
the Re band. The analyte signal schown corresponds to about 14 ng of
$^{174}$Yb. In these first experiments the extrapolated detection limit for Yb

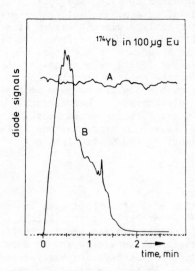

Fig. 4 Transient signal of the analytical diode B together with the signal of the laser locking diode A

isotopes was about 10 pg. With a more appropiate excitation scheme (final electronic state close to the ionization limit) one should easily enter the fg-range for the detection limit. A further improvement would be obtained keeping the sample vapour for a longer time in the laser beams. Evaporation in a narrow tube would help to be more efficient.

## 5. Calibration In the Ultra-Trace Regiem

Depending on the experimental set-up and the degree of activation of the cathode filament,the amplification factor for ion detection may change from thermionic diode to thermionic diode or from day to day. Therefore the amplification factor has to be found out by measuring calibration lines with known properties. Because there is always an activating element in the thermionic diode, it can be used for this purpose. If e.g. Ba atoms are excited to high-lying Rydberg states having a collisional ionization probability of unit and if the atomic oscillator strengths of the transitions and the Ba number density are known, the lines can serve as an 'internal standard' for trace element lines. Of course the final excited state in the trace atoms must also have a collisional ionization probability of unity and the oscillator strengths have to be known. In principle the ratio of the oscillator strengths of the Ba transition to the trace element transitions can be determined in the reference diode A, if the number density ratio has been determined before, e.g. by optical methods.

## 6. Isotope Dilution Technique

Applying Doppler-free laser spectroscopy techniques, it is possible to determine the ratio of isotopes by optical spectroscopy. The maximum ratio which can be measured is dependent on the magnitude of the isotope shift of the lines. If the shift between two isotopes with very different abundance is large the background under the component of the rare isotope due to velocity changing collisions of the more abundant isotope is small (see e.g. Weber et al. 1986a). In ordinary cases differences in the ratios of about six orders of magnitude are directly measurable. Using the frequency offset locking technique one of the lasers can be switched from one to the next isotopic component and, if necessary, the background. First experiments show that we can expect ratio data with high accuracy.

## 7. Application Of Single Mode Diode Lasers

We believe that the interest in analytical laser spectroscopy will strongly increase if the complicated and expensive $Ar^+$ (or $Kr^+$)-laser pumped dye lasers can be replaced by laser systems which are cheaper and easier to operate. In the last two years there has been a very rapid development in the field of semiconductor diode lasers. These lasers provide tunable cw single mode laser radiation with sufficient power to saturate atomic transitions. They are low-cost systems, easy to operate and with a long lifetime. The efficiency of these lasers is very high and they do not need a lot of space. Currently they are commercially available with lasing wavelengths down to about 680 nm, but in research labs wavelength in the yellow spectral range have already been obtained (Hino et al. 1986). Another direction of the development of diode lasers points towards higher output powers. With diode array lasers cw powers of several watts have already been obtained. Second-harmonic generation by diode laser radiation in non-linear crystals should enlarge the wavelength range considerably (Baumert et al. 1983). In our labs, we are already using diode lasers in analytical spectroscopy. In Fig. 5 we demonstrate the capability of these amazing devices. Here we compare resonant Doppler-free 2-photon spectra of Ga. In spectrum a and b the first transition (4P-5D)

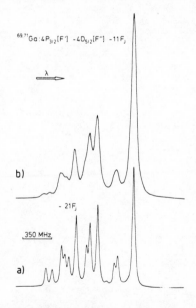

Fig. 5 Resonant Doppler-free 2-photon spectra of $^{69,71}$Ga: a) using two dye ring lasers; b) using one dye ring laser and one diode laser

is induced by frequency double radiation from a dye ring laser. But whilst in case a we used a second dye ring laser for the second step (5D-21F), in case b the second dye laser was replaced by a $ 50 diode laser. The spectra can be directly compared because the isotope shifts of the second transitions are the same within the experimental uncertainty (Weber et al. 1986b). The power of the diode laser was about 1 mW. The lower resolution in case b compared with a is due to the line width of the diode laser. If necessary, the line width could be reduced applying e.g. external cavity laser diode stabililzation with phase control of the optical feedback. Recently Tai et al. (1985) reported on a reduction of the line width of a diode laser to less than 500 kHz by a fiber-optic ring resonator. The use of more than two diode lasers would enable the performance of simultaneous multi-element analysis at reasonable costs.

## 8. Acknowledgements

This project is supported by the Deutsche Forschungsgemeinschaft under project no. Ni 185/9. The financial help is gratefully acknowledged.

## 9. References

Baumert J-C, Günter P and Mechior H 1983 Opt.Commun. 48 215
Hino I, Kawata S, Gomyo A, Kobayashi K and Suzuki T 1986 Appl.Phys.Lett. 48 557
Niemax K 1983 Appl.Phys. B 32 59
Niemax K and Weber K-H 1985 Appl.Phys. B 36 177
Niemax K 1985 Appl.Phys. B 38 147
Niemax K, Lawrenz J, Obrebski A and Weber K-H 1986 Anal.Chem. 58 1566
Popescu D, Popescu I and Richter J 1969 Z.Phys. 226 160
Tai S, Kyuma K and Nakayama T 1985 Appl.Phys.Lett. 47 439
Weber K-H, Lawrenz J and Niemax K 1986a Phys.Scr. 34 14
Weber K-H, Lawrenz J, Obrebski A and Niemax K 1986b Phys.Scr, to be published

*Inst. Phys. Conf. Ser. No. 84: Section 2*
*Paper presented at RIS 86, Swansea, Wales, 7–12 Sept. 1986*

# Application of RIMS to the study of noble gases in meteorites

Grenville Turner

Physics Department, Sheffield University, Sheffield S3 7RH

## Introduction

Being the most primitive objects available for study in the laboratory, meteorites are particularly important sources of direct information about the early history of the solar system. Recent isotopic studies involving the noble gases and a dozen or so other elements indicate that they also contain detailed information concerning the astrophysical environment which preceded the formation of the solar system. In the case of the noble gases the isotopic effects appear to be localised in obscure microscopic phases which remain to be properly identified. The small sizes of sample involved has led a number of noble gas groups including our own to look actively at the possibilities of RIMS as a way of providing the ultimate analytical sensitivity possible, namely that associated with individual atom counting.

Meteorites are broadly classified into two major groups, the more primitive (undifferentiated) chondrites, and the less primitive (differentiated) achondrites and irons. The latter are largely the products of igneous processes of melting and consequent chemical differentiation on planetary bodies, principally asteroids, although there are a handful of achondrites which may be samples of the planet Mars, and two recently discovered in Antarctica which are ejecta from the lunar surface.

The chondrites are composed largely of iron magnesium silicates, olivine $[(Mg,Fe)_2SiO_4]$ and orthopyroxene $[(Mg,Fe)SiO_3]$, metallic nickel-iron alloy, and a variety of accessory minerals. They are named after the characteristic mm-sized spherules, chondrules, of which many are composed and which appear to be melt droplets frozen while in a dispersed state. Chondrites show slight variations in total iron content and a wide variation in oxydation state, from the highly reduced E (enstatite) chondrites, in which iron occurs only as metal or in FeS, through three intermediate and numerically the largest classes (H, L, LL) referred to collectively as the ordinary chondrites, to the highly oxydised C (carbonaceous) chondrites, in which iron is largely in the silicate phase. Systematic variations are also seen in the content of volatile elements.

Chondrites are primitive in the sense that their chemistry is very similar for involatile elements to that of the sun and does not appear to have been affected by large scale melting on a planetary body. The chemistry and mineralogy of chondrites can be predicted rather convincingly by simple thermodynamic models of a cooling nebula of solar composition. In

detail though these models break down and it appears more likely that
chondrites resulted from a complex mixture of processes such as localised
melting, evaporation and condensation of material of roughly solar
composition occurring in a dispersed solar nebula (using that term in the
broadest possible sense). The effects of planetary, as distinct from
nebular, processes on chondrites appear to be restricted to thermal
metamorphism and recrystallization of a sizeable proportion of the
ordinary chondrites, and the formation of hydrous minerals by aqueous
solutions, in the case of the volatile rich C-chondrites.

Until relatively recently it was believed that the material from which the
planetary bodies formed was effectively homogenized in the gaseous state
during a nebular phase preceding the formation of solid bodies. The major
experimental evidence in support of this view was the absence of variation
in the isotopic composition of elements between the Earth and meteorites.
Variations which did exist could be understood in terms of well understood
ongoing processes such as physical or chemical fractionation, radioactive
decay, or other nuclear reactions such as those induced in meteorites by
the action of cosmic rays. The first clear indication of a pre-solar
component in meteorites which had escaped homogenization was a component
of neon referred to as Ne-E and later shown to be essentially monoisotopic
$^{22}$Ne. Since then 'isotope anomalies' resulting from a range of
nucleosynthetic and other astrophysical mechanisms have been observed in
C, N, O, Mg, Si, Ca, Ti, Cr, Sr, Ba, Nd, Sm, Xe and Kr. The daughter
products of seven extinct radionuclides ($^{26}$Al, $^{41}$Ca, $^{53}$Mn, $^{107}$Pd, $^{129}$I,
$^{146}$Sm and $^{244}$Pu) have also been identified by way of their decay products.
The presence of the latter does not of itself imply the survival of
unhomogenized pre-solar material. Provided that the parent is live in the
early solar system, chemical differentiation of parent and daughter
element during either the nebular or the planetary stage can give rise to
phases with a present day excess of the daughter isotope.

Some of the scientific problems associated with the noble gas anomalies
are described briefly below. Since separation and identification of
different components is to a large degree constrained by the techniques
used to extract the gases from meteorites some of the more commonly used
methods are first described.

Extraction Techniques

Many of the techniques for handling noble gases are well established, in
particular the use of chemical getters for sample cleaning and cryogenic
methods for separation and manipulation. The principal method used for
extracting gases from the sample involves heating in a UHV furnace at a
sequence of temperatures, typically up to 1650°C or so to ensure complete
extraction, particularly of xenon. This stepped heating method has the
advantage that different mineral sites release gas at different
temperatures and in this way a partial separation of components occurs
which gives rise to correlated variations in isotopic composition.
Stepped heating has a particular appeal on account of its simplicity and
the potential it provides for thermal diffusion calculations as a means of
interpreting observations. Although simple to apply it has limitations
when particular components are associated with minor phases and in the
last decade has been complemented by both chemical and physical separation
methods prior to the application of stepped heating.

The chemical procedures introduced by Lewis et al (1975) have been as spectacularly successful in concentrating the carrier phases of meteoritic noble gases as they are extreme. The Chicago group discovered that by dissolving (typically) 99.5% of a C-chondrite (Allende) in an HC1-HF mixture they obtained a black residue which contained on the order of 30% of the original noble gas complement. Further treatment involving the use of increasingly drastic oxydizing acid mixtures (HNO3 and HC1O4) permitted the selective removal of oxydizable minerals and further separation of different noble gas hosts. In parallel other groups, principally those in Berne and Berkeley, have developed physical separation methods involving disaggregation and size and density separations, as a method of concentrating the ill defined exotic components.

Stepped heating is normally used to extract noble gases from the concentrates, though a useful alternative is selective combustion. In this technique, developed as a means of separating and identifying the isotopic composition of different carbon phases in meteorites (Swart et al, 1983), the sample is heated at a sequence of increasing temperatures in an oxygen atmosphere, produced in the vacuum system by heating copper oxide. Since several of the exotic noble gas bearing phases in meteorites now appear to be different forms of carbon this procedure is potentially very useful in correlating isotopic anomalies in the noble gases with those in carbon.

The mechanisms by which gases are released during stepped heating and selective combustion are often obscure. More direct extraction procedures involve localised extraction from (for example) a thin section. At least two methods are being used which permit this, ion sputtering, and laser ablation. Of these ion sputtering is the most controlled and if operated in a pulsed mode most capable of matching the duty cycle of a RIMS experiment. Ablation using a pulsed laser releases gas by localised melting and is relatively simple to put into operation. The need to match the duty cycle of the extraction system to that of the RIMS laser is less compelling in the case of noble gases than for a solid or a chemically reactive gaseous species, since the gas to be analysed remains in the vacuum system and is exposed repeatedly to the ionising laser and it is only important that steps be taken to maximise the ratio of illuminated volume to the effective volume occupied by the sample gas. Laser ablation systems have been developed in a number of laboratories, including our own, for use in $^{40}Ar$-$^{39}Ar$ dating studies (McConville et al, 1985).

## Noble Gas Components

Noble gas components are conveniently organized into three broad categories; radiogenic, nucleogenic, and trapped. Radiogenic species are those produced by radioactive decay. $^{4}He$ is produced by $\alpha$-decay of U and Th, and $^{40}Ar$ by $\beta$-decay (electron capture) in $^{40}K$. Excess amounts of $^{129}Xe$ are common in meteorites from the $\beta$-decay of $^{129}I$ which was present in the early solar system but is now extinct. Spontaneous fission of $^{238}U$ gives rise to minor amounts of heavy (neutron rich) Xe and Kr isotopes which can be regarded as radiogenic. A more prominent source of fission produced Xe and Kr in meteorites is the now extinct $^{244}Pu$. Radiogenic isotopes are especially useful in establishing time scales.

An example of a recent application of our laser ablation probe is shown in figures 1 and 2. The data points represent argon extracted from individual 80μm diameter laser pits which release gas from 0.5 μg of sample. $^{39}Ar$ has been generated artificially from potassium by fast

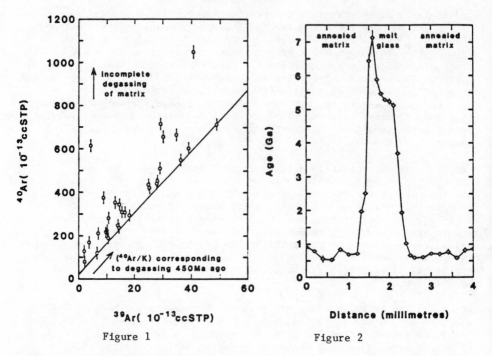

Figure 1                                    Figure 2

neutron activation and the K-Ar age of the sample is obtained from the
($^{40}$Ar/$^{39}$Ar) ratio. The meteorite is a chondrite which has been involved
in a major (asteroidal) collision and heated sufficiently to cause
extensive loss of argon. The lower bound to the points in figure 1
corresponds to an age of 450 Ma and indicates regions of the meteorite
which were totally outgassed at that time as a result of the collisional
heating and subsequent annealing. Points above the line correspond to
only partial degassing. The traverse in figure 2 crosses a region of melt
glass produced in the instant of the impact. Surprisingly, due to a
combination of rapid quenching and low diffusion coefficients the glass
lost potassium but very little argon and gives meaningless ages, older
indeed than the age of the solar system! The measurements are state of
the art for K-Ar dating and involve sample sizes of order 5 x 10$^7$ atoms
for $^{39}$Ar and 10$^9$ atoms of $^{40}$Ar. These requirements, which are orders of
magnitude larger than those for Xe for example, are limited by hydrocarbon
interferences, 5 x 10$^5$ atoms equivalent at mass 39 (and 36), adsorbed
atmospheric $^{40}$Ar blank, ~ 10$^8$ atoms, and atmospheric $^{40}$Ar build up,
~ 10$^8$ atoms/min.

Nucleogenic species are those isotopes produced in situ in meteorites,
largely as a result of bombardment by galactic cosmic rays and their
secondaries. They are more often referred to as cosmogenic or spallogenic
species for obvious reasons. The term nucleogenic can be more accurately
extended to isotopes produced within the Earth by bombardment by
(radiogenic) alpha particles and neutrons produced by ($\alpha$,n) reactions. At
shallow depth in lunar samples reactions induced by solar flare particles
are also important in producing nucleogenic isotopes. High energy
spallation reaction products are characterised by a smooth flat mass
distribution below the mass of the target and are easily identified.
Spallation products are most clearly evident at those masses where the

'natural' abundance is low, e.g. $^3$He, $^{21}$Ne, $^{38}$Ar, $^{80}$Kr and $^{124}$Xe. Abundances of spallation products have been used extensively to trace the record of meteorites as small metre sized bodies, to determine preatmospheric sizes, and to examine the variability of cosmic ray flux as a function of orbital distance and time.

The most consistently intriguing of the three categories is that referred to as the trapped component. This is essentially what is left when all the components produced by in situ processes have been accounted for. The term primordial is sometimes used synonymously though most workers prefer trapped as being more general. Trapped gases may arise in a variety of ways representing sources implanted in any of the three epochs referred to above as, planetary, nebular, and presolar. Uncovering and account for possible presolar components is an exciting feature of several recent and ongoing research programmes.

Ne-A, also called planetary neon, $(^{20}N/^{23}Ne) = 8.1$, is ubiquitous in chondrites and was probably trapped from the solar nebula. Ne-B or solar neon, $(^{20}Ne/^{21}Ne) = 12.5$, arises from the implantation of solar wind ions and is common in meteorites which have been exposed to the solar wind on the surface of a parent asteroid. Ne-S is the spallogenic component with roughly equal abundances of the three isotopes. Ne-E discovered by Black and Pepin in 1969 has been shown by Jungck and Eberhardt (1979) to be essentially monoisotopic $^{22}$Ne, with $(^{20}Ne/^{22}Ne) < 10^{-2}$ and $(^{21}Ne/^{22}Ne) < 10^{-3}$. Two carriers of Ne-S have been isolated, a low density carbonaceous phase, which releases neon below 1000°C, and a high density phase, apatite and spinel, which releases neon above 1000°C. The most recent experiments by Jungck and Eberhardt (1985) indicate that the high density phase can be made to release its neon a lower temperature by treatment with acid, possibly indicating that the Ne-S is carried in carbonaceous inclusions encased in apatite, rather than in the apatite itself. The only plausible source for Ne-S is β-decay of $^{22}$Na. The short half life of $^{22}$Na (2.6a) implies that it was synthesised and trapped on a similarly short time scale, a possible astrophysical site being in a nova. Support for this astrophysical link is provided by the observation of high (13C/12C) ratios in the Ne-E bearing residues.

Argon and krypton are probably the least interesting of the noble gases with regard to trapped components, at least on current evidence. The $(^{36}Ar/^{38}Ar)$ ratio of trapped argon from all sources shows no clear systematic variation. Trapped krypton shows small fractionation like trends when comparisons are made between meteoritic, terrestrial and solar krypton. A fission like component related to XeHL (see below) is observed and probably most interesting a discrete component produced by s-process nucleosynthesis, again related to a much larger effect in Xe described below. The significance of the s-process Kr composition is that due to a branch in the s-process neutron capture path it can be used to estimate the stellar neutron flux involved in its production.

The most challenging and rewarding of the noble gases is undoubtedly xenon. With nine isotopes the possibilities for unscrambling multicomponent mixtures is greatest. However the complexities of meteoritic xenon are such that nine is barely enough to do the job! Meteoritic xenon consists of at least six clearly separable components, with suggestions that some of these are themselves mixtures. A further fractionated xenon component is required to account for the composition of the Earth's atmosphere.

Spallogenic xenon is well characterised and affects mainly the lighter low abundance proton rich isotopes in meteorites with low concentrations of trapped gases. Two radiogenic components are present, $^{129}$Xe from the decay of extinct $^{129}$I and heavy xenon $^{131-136}$Xe from the spontaneous fission of extinct $^{244}$Pu. The fission products of $^{244}$Pu are particularly prominent in the Ca (and U) rich basaltic achondrites, and are recognizable in contributing around 6% of the $^{136}$Xe of the Earth's atmosphere. Both radiogenic components may be present from in situ decay or as part of a trapped component. At least three trapped components have been identified (figure 3). The major component variously referred to as planetary or AVCC xenon was probably trapped from the solar nebula and is held to a large degree in an unidentified, minor, acid insoluble, but oxydizable, phase referred to cryptically as Q. Strictly speaking AVCC, average carbonaceous chondrite xenon, contains all three trapped components but is so dominated by this planetary end member that the distinction is unimportant).

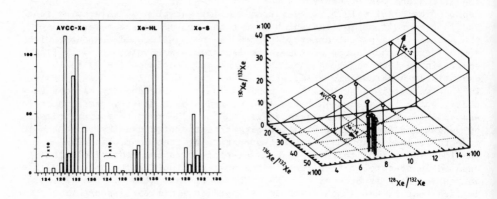

Figure 3                                    Figure 4

A second trapped component was discovered from isotope correlations more than twenty years ago (Reynolds and Turner, 1964) but its origin is still unclear. The presence of high $^{134}$Xe and $^{136}$Xe suggested a neutron rich source mechanism such as fission. On the assumption that $^{130}$Xe is absent (due to shielding at the end of a neutron rich β-decay chain by $^{130}$Te) a fission like spectrum can be calculated and led to the component being referred to as CCF (Carbonaceous Chondrite Fission) xenon. The calculation by Reynolds and Turner of the composition of CCF xenon left unanswered problems over the role of the light isotopes and in 1972 Manuel et al demonstrated a correlation between ($^{124}$Xe/$^{132}$Xe) and ($^{136}$Xe/$^{132}$Xe) in CCF xenon which cast doubts on the fission hypothesis. An alternative nucleosynthetic source for light and heavy isotope enriched CCF xenon, now renamed Xe-HL, could be a supernova. The carrier phase of Xe-HL is possibly an oxydation resistant carbon phase.

The most recently discovered xenon component (Srinivasan and Anders, 1978, Xe-S, appears to be clearly established as the product of s-process nucleosynthesis, a process which is believed to occur in red giant stars. Involving as it does quasi steady state neutron capture (Nσ ~ constant) it is characterised by high abundances of even isotopes, low abundance of odd

isotopes, and zero abundance of $^{134}$Xe and $^{136}$Xe. The measured composition
agrees well with that predicted from theory. A 3-D isotope correlation
diagram (3 component mixing plane), figure 4, illustrates clearly the
presence of all three trapped components in acid residues from the
Murchison meteorite (data from Alaerts et al, 1980, and Ott et al, 1985).

## The application of RIMS to meteoritic noble gases

Concentrations of the major isotopes of Kr and Xe in meteorites are
typically in the range $(10^{10} - 5.10^{11})$ atoms/g or $(5.10^{-14}$ to $3.10^{-11})$
atomic ratio. For $^{20}$Ne and $^{36}$Ar, concentrations are an order of magnitude
or so higher. In the minor phases containing the bulk of the trapped gas
inferred Xe concentrations may be at least as high as $10^{14}$ atoms/g or
$5.10^{-9}$ atomic ratio. Thus the sample weights required for $10^6$ atoms of Xe
range from 10 ng to 100 µg, equivalent to (spherical) grains from 20 µm to
400 µm in diameter. SEM studies indicate that the phases of interest are
micron or submicron in size. Analysis of individual grains does not
therefore seem feasible unless the gases are even more localized than
indicated above or compositions so extreme in large individual grains that
the statistical measurement precision can be relaxed.

An aspect which is worth noting in passing is the possible problem of
adsorbed atmospheric gases on small samples with a high surface to volume
ratio. A 100 µm square monolayer of terestrial atmosphere contains the
following numbers of atoms; $^{22}$Ne = $6.10^6$, $^{36}$Ar = $10^7$, $^{40}$Ar = $3.10^9$, $^{84}$Kr =
$2.10^5$, $^{132}$Xe = $7.10^3$. Clearly removal of adsorbed air is an important
aspect of experiment design associated with low level noble gas analyses.
Fortunately in the case of Xe the (unexplained) relative depletion of Xe
in the atmosphere eases the problem. Unfortunately xenon is the most
readily adsorbed of the noble gases.

The attractive features of RIMS as a technique for low level analysis are
well rehearsed. They are; chemical selectivity, which for isotopic
analysis means reduction or elimination of isobaric interferences; and,
potentially, high sensitivity. The high sensitivity itself has two
distinct aspects. On the one hand high sensitivity is important as a
means of measuring smaller samples than was previously possible. Equally
important since it opens the way to more complex 'probe type' experiments,
is the ability to reduce analysis time.

Sensitivities of conventional noble gas mass spectrometers are usually
expressed as the ratio of detector ion current to sample pressure in the
source. Optimum and limiting sensitivities for the best designed
instruments are of the order of S = $10^{-3}$ amp/torr. A more useful
parameter expressing the overall sensitivity of the mass spectrometer is
the lifetime, $\tau$, against detection which involves in addition the
instrument volume, V (typically 1 litre). It is easily shown that $\tau$ and S
are related by

$$\tau = 5.7 \times 10^{-3} \frac{V}{S} \text{ sec.}$$

where V is expressed in cm$^3$ and S in amps/torr. For the best conventional
noble gas mass spectrometers detection lifetimes for Xenon are thus of the
order of 2 hours and while samples as low as a few thousand atoms of Xenon
have been detected analysis times are unacceptably long for probe
experiments.

For RIMS the corresponding expression for $\tau$ and S are

$$\tau = \frac{Vt}{v} \quad \text{and} \quad S = 5.7 \times 10^{-3}\frac{v}{t}$$

where v is the effective volume illuminated by the laser with unit ionization probability, and t the time between laser pulses, typically 1/10 sec. For Xe a value of $v \sim 2.10^{-4}$ cm$^3$ has been demonstrated (Chen et al 1980) corresponding to $S \sim 10^{-5}$ amp/torr. As it stands this is clearly far short of the best sensitivities using electron impact sources and is a reflection of the 'duty cycle' penalty of the laser system. Several strategies are available and are being used to improve the situation and capitalise on the high intrinsic ionization probability of the laser system.

Improvements in energy output from the laser using alternative schemes is a clear option which is being investigated and should lead to improvements in v. The use of multiple reflections could in principle achieve a similar end. Since it is $\tau$ rather than S which is the key measure of sensitivity other strategies centre round decreasing the effective value of V. The 'atom buncher' of Hurst et al, 1984 produces the most dramatic reduction in V by the use of a cryogenic finger and is particularly appealing for noble gases. A further 'duty cycle' penalty arises in that the value for t in the above expression for $\tau$ refers now to the time constant for recondensation on the cold finger. Hurst et al quote values for (V/r) of order 50 and t $\sim$ 1 sec. corresponding to analysis times of the order of a minute. An alternative method to reduce V is by the use of a spectrometer with an intrinsically low volume such as an FTMS cell, or the quadruple ion trap. Both spectrometers operate in a cyclic fashion and are thus better matched to the pulsed laser source than to (low efficiency) pulsed electron impact sources. Values for V of a few cm$^3$ should be readily attainable. While the potential for ultra high sensitivity is less than that of the atom buncher, the ion trap in particular is considerably less demanding technically. It could certainly be competitive with conventional spectrometers and far less costly.

## References

Lewis R S, Srinivasan B and Anders E 1975 Science, 190 1251-1262
Black D C and Pepin R O 1969 Earth Planet. Sci. Lett. 6 395-405
Jungck M H H and Eberhardt P 1979 Meteoritics 14 439-441
Reynolds J H and Turner G 1964 J. Geophys. Res. 69 3263-3281
Manual O K, Hennecke E W and Sabu D D 1972 Nature 240 99-101
Srinivasan B and Anders E 1978 Science 201 51-56
Alaerts L, Lewis R S, Matsuda J and Anders E 1980 Geochim. Cosmochim. Acta. 44 189-209
Ott U, Yang J and Epstein S 1985 Meteoritics 20 722-723
McConville P, Kelley S and Turner G 1985 Meteoritics 20 707
Chen C H, Hurst G S and Payne M G 1980 Chem. Phys. Lett. 75 473
Hurst G S, Payne M G, Phillips R C, Dabbs J W T and Lehmann B E 1984 J. Appl. Phys. 55 1278-1284
Swart P K, Grady M M, Pillinger C T, Lewis R S and Anders E 1983 Science 220 406-410

*Inst. Phys. Conf. Ser. No. 84: Section 2*
*Paper presented at RIS 86, Swansea, Wales, 7–12 Sept. 1986*

59

# Vacuum ultraviolet light generation applications to resonance ionization spectroscopy

M.G. Payne

Chemical Physics Section

Oak Ridge National Laboratory, Oak Ridge, Tenn. 37830

## 1. Introduction

Just ten years ago it was difficult to propose efficient ionization schemes for most atoms with ionization potentials above 10 eV. This difficulty was due to the fact that the lowest dipole allowed one photon transition usually required tunable light with wavelength $\lambda < 200$ nm. The other alternative method of starting a selective ionization scheme is two-photon excitation, which requires high power densities and necessitates focusing of the laser beam in order to achieve saturating power densities over a very small volume.

The period since 1976 has included the development of dye laser pumps based on the use of the second harmonic of the Nd-YAG laser and of the XeCl excimer laser. The new, more powerful dye lasers which have resulted, make possible the use of four-wave mixing techniques which permit the generation of intense and tunable vacuum ultraviolet light (VUV) over most of the wavelength region between 110 nm and 210 nm. Outstanding contributions in this area have been made by many workers. An extensive set of references to this field of research can be found in a recent review by Jamroz and Stoicheff (1983). In this limited space we will discuss, most extensively, the important work of Wallenstein et al (See: Hilbig and Wallenstein 1981, 1982). The latter workers have devised particularly simple methods for generating VUV which is tunable, nearly continuously, from 110 nm to 210 nm. Before proceeding with our discussion of VUV generation we will discuss briefly the impact of VUV generation on Resonance Ionization Spectroscopy (RIS).

## 2. RIS Schemes Using VUV Light Sources

The VUV generation schemes due to Hilbig and Wallenstein (1982) typically result in around 10 Watt peak power in a pulse of length 6 nanoseconds. The light was generated with a bandwidth of $< 0.1 cm^{-1}$. Let's now investigate how well saturated one photon excitation can be carried out with such a source.

Suppose that the above light source is tuned to a strong transition. Strong transitions in the VUV region typically have oscillator strengths of the order of magnitude of 0.2, which translates to a dipole matrix element $D_{01} = < 0| \sum_i e\vec{r}_i |1 > \approx 0.5 e a_B$, where $a_B$ is the Bohr radius. The Rabi frequency for the transition is then $\Omega_R = D_{01}E/\hbar \approx 10^8 I^{\frac{1}{2}}$, where $\Omega_R$ is in radians/sec and I is the power density in Watt/$cm^2$. If the 10 Watt VUV beam has a diameter of 1 mm the peak power density is $\approx 1 kW/cm^2$, and correspondingly $\Omega_R \approx 3 \times 10^9 rad/sec$. This gives a power broadened width for the transition of $\approx 6 \times 10^9/sec$; which is of the same order of magnitude as both the laser bandwidth and the Doppler width of the transition for room temperature atoms radiating in the 110 nm - 210 nm region. Correspondingly, calculations show that with a pulse length of 4-8 nsec, the ground and excited states are leveled during the pulse, whether the model for the lineshape of the laser is one with two or three longnitudinal modes separated by $\approx 0.05$ $cm^{-1}$; or whether a continuous spectrum obeying the criteria for a chaotic field is assumed. Saturation is relatively complete on beam axis for atoms with lines Doppler shifted near line center (or for the discrete mode model near a mode of the light); but the efficiency is model dependent for atoms with Doppler shifted lines more than 0.05 $cm^{-1}$ from line center (or from any discrete mode). About 50% of the atoms can be ionized within a 1 $mm^2$ area if the Doppler width is $\leq 0.1$ $cm^{-1}$. If there are several hyperfine levels within the bandwidth it is probably not safe to assume that all will be ionized with the same probability. Nevertheless, for some applications, one has a relatively efficient excitation scheme which can be combined with a second dye laser to pump a second resonance, with the second excited state finally being ionized by the Nd-YAG fundamental. If the second step and the ionization are done carefully a very reasonable ionization volume results. By careful, we mean that the power density is not turned up to the point where the excited states are split, into pairs of resonances separated by twice the local value of the Rabi frequency for the second transition about the unperturbed resonance, thereby destroying the efficiency of the VUV excitation. With the excitation source described, the Rabi frequency for the second transition should be no larger than $4 \times 10^9/sec$, and the ionization rate out of the second state should be no larger than $4 \times 10^9/sec$.

## 3. Vacuum Ultraviolet Light Generation Using Four-Wave Mixing

### 3.1 Introduction to Nonlinear Optics

When intense, narrow bandwidth light enters a gas cell the interaction between individual atoms and the laser field is well represented by the Hamiltonian

$$\hat{H} = \hat{H}_0 - \hat{D} \cdot \vec{E}, \tag{1}$$

where $\hat{H}_0$ is the electronic Hamiltonian of the unperturbed atom, $\hat{D} = \sum_i e\vec{r}_i =$ electronic dipole operator, and $\vec{E}$ is the laser field described as a classical electromagnetic wave. Most of the situations where this treatment fails concern problems such as resonance fluorescence where fluctuations related to spontaneous emission play a dominant role, or where high Rydberg states are involved, thereby invalidating the multipole expansion involved in arriving at the $-\hat{D} \cdot \vec{E}$ form of the interaction.

If the laser is tuned to a wavelength which is well away from any one-, two-, or three-photon resonance and the laser power is $\leq 10^{12}$ $W/cm^2$ ,the effects of the laser fields are usually described accurately by time dependent perturbation theory. We expand the time dependent state vector of an atom at $\vec{R}$ in terms of the complete set of eigenvectors of $\hat{H}_0$:

$$|\Psi(\vec{R},t)> = \sum_{\epsilon_n, \mu} a(\vec{R}, \epsilon_n, \mu, t) e^{-i\omega_n t}|\epsilon_n, \mu>, \tag{2}$$

where $\hat{H}_0|\epsilon_n, \mu> = \epsilon_n|\epsilon_n, \mu>$, and $\epsilon_n = \hbar\omega_n$. The set of quantum numbers $\mu$ may be regarded as angular momentum quantum numbers, and the sum over $\epsilon_n$ is to be interpreted as a sum over the discrete eigenenergies and an integral over the continuous part of the eigenenergy spectrum. The proper interpretation of $a(\vec{R}, \epsilon_n, \mu, t)$ is as a probability amplitude for an atom at $\vec{R}$ being in state $|\epsilon_n, \mu>$ at time t. In terms of time dependent perturbation theory we find (See for instance: Payne et al 1981.)

$$a(\vec{R}, \epsilon_n, \mu, t) =< \epsilon_n, \mu|\hat{S}|\epsilon_0, \mu_0 >, \tag{3}$$

where $|\epsilon_0, \mu_0 >$ is the atomic ground state and

$$\hat{S} = \hat{1} + \sum_{k=1} \hat{S}_k, \tag{4a}$$

with,

$$\hat{S}_k = \int_{-\infty}^{t} dt_1 \cdots \int_{-\infty}^{t_{k-1}} dt_k \hat{V}_I(t_1) \cdots \hat{V}_I(t_k), \tag{4b}$$

and,

$$\hat{V}_I(t) = \exp(i\hat{\mathcal{H}}_0 t/\hbar)(-\hat{D} \cdot \vec{E}) \exp(-i\hat{\mathcal{H}}_0 t/\hbar). \tag{5}$$

More will be said about the actual evaluation of the perturbation theory result for $|\Psi(\vec{R}, t) >$ later.

In a neutral gas the laser beam produces induced dipoles oscillating at the laser frequencies and at the odd harmonic frequencies of the lasers. With more than one laser beam overlapping there are also dipole oscillations at frequencies such as $2\omega_{L1} - \omega_{L2}$ and $2\omega_{L1} + \omega_{L2}$, etc. The source of the dipoles is the atoms in the medium as they respond to the intense laser fields. A concentration of oscillating dipoles is equivalent to a current density $\vec{J} = \frac{\partial \vec{P}}{\partial t}$. With only a current density of this value present Maxwell's equations can be manipulated to obtain the following wave equation for the $\vec{E}$ field.

$$\nabla^2 \vec{E} - \frac{1}{c^2} \frac{\partial^2 \vec{E}}{\partial t^2} = \frac{4\pi}{c^2} \frac{\partial^2 \vec{P}}{\partial t^2}. \tag{6}$$

Actually, there is also a charge density associated with the polarizability $\rho = -\nabla \cdot \vec{P}$. However, if the laser beams are parallel and either unfocused, or focused with long focal length lenses, then $\rho$ produces negligible effects due to the propagation vector and the $\vec{E}$ field being orthogonal.

We see that if we could evaluate $\vec{P}$ without ambiguity for rather general forms of the $\vec{E}$ field, then the problem of calculating the fields generated in the medium would involve the solution of Eq (6) and the evaluation of some quantities related to atomic response. It is not difficult to write a formula for the polarizability. It is just the atomic concentration times the time dependent expectation value of $\hat{D}$. Thus,

$$\vec{P}(\vec{R}, t) = N < \Psi(\vec{R}, t)|\hat{D}|\Psi(\vec{R}, t) >, \tag{7}$$

$$= N \sum_{\epsilon_n, \mu} \sum_{\epsilon_m, \mu_1} < \epsilon_n, \mu|\hat{D}|\epsilon_m, \mu_1 > a^*(\vec{R}, \epsilon_n, \mu, t) a(\vec{R}, \epsilon_m, \mu_1, t) e^{-i(\omega_m - \omega_n)t}.$$

We first consider the contributions to $\vec{P}$ which are linear in $\hat{V}_I$. Within the framework of perturbation theory the population of the atomic groundstate remains undepleted, so that $|a(\vec{R}, \epsilon_0, \mu_0, t)| \approx 1$, and for all excited states $|a(\vec{R}, \epsilon_n, \mu, t)| << 1$. As a consequence of Eqs. (2)-(7) the contribution to $\vec{P}(\vec{R}, t)$ which is linear in $\hat{V}_I$ leads to a separate term for each laser beam $\cdots$ and for any coherent beams generated in the medium. For a particular beam at frequency $\omega$ we find

$$\vec{P}_\omega^{(1)}(\vec{R}, t) = \chi_1(\omega)\vec{E}_L(\vec{R}, t), \tag{8}$$

where $\chi_1(\omega)$ is

$$\chi_1(\omega) = \frac{Ne^2}{m} \sum_{\epsilon_n,\mu} \frac{f_{n,\mu;0,\mu_0}}{(\omega_n - \omega_0)^2 - \omega^2}, \tag{9}$$

with $f_{n,\mu;0,\mu_0}$ being the oscillator strength for the transition from the ground state to the state $|\epsilon_n, \mu>$. The next non-zero contributions to $\vec{P}(\vec{R}, t)$ are of third order in $\hat{V}_I$.

If there are laser beams which coincide in space and time at frequencies $\omega_{L1}$ and $\omega_{L2}$, then there will be third order terms in $\vec{P}(\vec{R}, t)$ at frequencies $3\omega_{L1}$, $3\omega_{L2}$, $2\omega_{L1} \pm \omega_{L2}$, and $2\omega_{L2} \pm \omega_{L1}$, as well as terms which are linear in the laser field amplitudes at the same frequencies. The field amplitudes at the sum and difference frequencies have their origin in the third order in $\hat{V}_I$ contributions to $\vec{P}(\vec{R}, t)$, and serve as source terms at these frequencies in Eq. (6). Any term at a frequency $\omega$ together with any nonlinear source terms at the same frequency, can be considered to obey a separate equation like Eq. (6), which becomes

$$\nabla^2 \vec{E}_\omega - \frac{1}{v^2} \frac{\partial^2 \vec{E}_\omega}{\partial t^2} = \frac{4\pi}{c^2} \frac{\partial^2 \vec{P}_\omega^{NL}}{\partial t^2}, \tag{10}$$

where, $v^{-2} = c^{-2}[1 + 4\pi\chi_1(\omega)/c^2] = n^2(\omega)/c^2$ and $n(\omega)$ is the index of refraction of the medium at frequency $\omega$. In arriving at Eq. (10) the term at $\omega$ which is linear in the corresponding field has been brought to the left hand side of the equation to give a modified phase velocity for the generated wave. In the case under consideration, the new frequencies generated in the medium are weak, and of third order in the laser field amplitudes (or perhaps second order in one field and linear in the other), and consequentially they do not appear in $\vec{P}_\omega^{NL}(\vec{R}, t)$. However, if we considered the fifth order contributions to $\vec{P}_\omega^{NL}$ the third order fields generated, in product with two laser fields, can be just as important (or more so) than terms that are directly fifth order in the laser fields.

In principle the $\vec{P}_\omega^{NL}(\vec{R}, t)$ for a particular new frequency generated in the medium can be calculated. It is just a matter of evaluating Eq. (3) (using $\hat{S}_3$, as given by Eq. (4b)) by inserting unit operators in the form $\hat{1} = \sum_{\epsilon_n,\mu} |\epsilon_n, \mu><\epsilon_n, \mu|$ between pairs of $\hat{V}_I$ and using the fact that $\exp(\pm i\hat{\mathcal{H}}_0 t/\hbar)$ operating on these bras and kets can be evaluated immediately (they are eigenstates of $\hat{\mathcal{H}}_0$). The resulting time integrals are all of the type $\int_{-\infty}^{t_k} d\tau Q(\tau)e^{iv\tau} \approx -iQ(t_k)e^{ivt_k}/v$; where in evaluating Eq. (12) the fact that $|v| >> |d(ln(Q(\tau))/dt|$ has been used. Finally, one must include the all coherent fields in the $\vec{E}$ *in* $-\hat{D} \cdot \vec{E}$ and pick out all products which lead to the frequency under consideration. Inserting the resulting terms into Eq. (7) then leads to a very messy expression for $\vec{P}_\omega^{NL}(\vec{R}, t)$ in terms of a nonlinear susceptibility for the atomic species times a product of three laser fields. The nonlinear susceptibility depends only on the concentration and upon the properties of the atoms, but involves the sum of several terms that are themselves sums over three sets of intermediate states. The sums involve *"resonant"* denominators which give enhancements to the susceptibility whenever there is a near coincidence with a one-, two-, or three-photon resonance with the combinations of laser amplitudes involved in generating the frequency in question.

The above description of the nonlinear susceptibility approach to nonlinear optics can be "fleshed out" by filling in the intermediate steps to yield a complete derivation of Eq. (10); which will then be considered to be the fundamental equation for non-resonant VUV generation in an atomic gas medium.

### 3.2 Sum Mixing With Unfocused Beams

Solutions to Eq. (10) bring in another aspect of nonlinear mixing processes which has a rather simple origin. The aspect in question has to do with arranging for complete constructive interference of the dipole waves generated at the position of each atom in the laser beams. The achievement of this constructive interference is referred to as *phase matching*.

In order to give the reader a better physical feel for phase matching consider two unfocused, concentric, plane polarized (with the same plane of polarization) laser beams of beam diameter d entering a gas cell of length l. If $(l\lambda)/(\pi d^2) \ll 1$, where $\lambda$ is the wavelength of either beam, then diffraction effects are unimportant during passage through the cell and we can consider the effects of the beams on the atoms in the beams as if the laser beams are plane waves propagating in the x direction. Suppose also that the laser at angular frequency $\omega_{L1}$ is tuned on, or very near, a dipole allowed two-photon resonance, between the ground state $|0>$ and an excited state $|1>$ and that $2\omega_{L1} + \omega_{L2}$ is close to three photon resonance with state $|2>$. Assuming that the laser at $\omega_{L1}$ is too weak to produce saturation effects in the two-photon resonance, one can show that the amplitude for being in $|2>$ is

$$a(\vec{R}, 2, t) = \frac{\Omega_2(0 \to 1)}{\Gamma_1} \frac{\Omega_1(1 \to 2)}{\delta_2} e^{i\delta_2 t}, \tag{11}$$

where $\Omega_2(0 \to 1)$ is the Rabi frequency for the $|0> \rightleftharpoons |1>$ transition, $\Omega_1(1 \to 2)$ is the Rabi frequency for the $|1> \rightleftharpoons |2>$ transition, $\delta_2 = 2\omega_{L1} + \omega_{L2} - \omega_2 + \omega_0$, and $\Gamma_1$ is an effective level width for the two-photon transition. The laser at $\omega_{L1}$ is assumed to be detuned from resonance by just enough so that the VUV signal has started to drop due to a decrease in nonlinear susceptibility. The drop coincides with the detuning becoming larger than effective level width.

The simplicity of Eq. (11) is due to the detuning from two- and three-photon resonances being so small that a dominant resonance reduces two of the three sums over intermediate states to single terms. We can use Eq. (11) to obtain the nonlinear polarizability

$$\vec{P}_{2\omega_{L1}+\omega_{L2}}(x, t) = \vec{e}_1 N |D_{20}| \frac{\Omega_2(0 \to 1)}{\Gamma_1} \frac{\Omega_1(1 \to 2)}{\delta_2} e^{i\phi(x,t)} + c.c \tag{12}$$

where, $\phi(x,t) = (2\omega_{L1} + \omega_{L2})t - (2k(\omega_{L1}) + k(\omega_{L2}))x + \phi_0(t)$. Above, $\vec{e}_1$ is a unit vector parallel to the electric vector of the laser beams. Notice that if the two laser beams were circularly polarized in the same sense that they could only drive $\Delta J = 3$ transitions with three-photon processes, and that for such transitions $D_{02}$ is zero. With the latter state of polarization there can be no nonlinear generation.

Using the plane wave feature of the laser beams, Eq. (10) becomes independent of the y and z coordinates and can be solved.

$$E_{VUV}(x,t) = -\frac{2\pi v}{c^2} \int_0^x \frac{\partial P_{2\omega_{L1}+\omega_{L2}}}{\partial t}\left(x_1, t - \frac{|x - x_1|}{v}\right) dx_1, \tag{13}$$

where, $v = (2\omega_{L1} + \omega_{L2})/k(2\omega_{L1} + \omega_{L2})$.

If we use Eq. (12) in Eq. (13) the resulting integral is elementary and we find for the flux of VUV photons,

$$\mathcal{F}_{\omega_s} = \kappa N \left|\frac{\Omega_2(0 \to 1)\Omega_1(1 \to 2)}{\Gamma_1 \delta_2}\right|^2 \left\{\frac{1 - 2e^{-\beta x}\cos(\Delta k x) + e^{-2\beta x}}{\Delta k^2 + \beta^2}\right\}, \tag{14}$$

where $\omega_s = 2\omega_{L1} + \omega_{L2}$, $\kappa = (2\pi\omega_s N|D_{02}|^2)/(\hbar c)$, $\Delta k = k(\omega_s) - 2k(\omega_{L1}) - k(\omega_{L2})$, and the absorption of the medium has been introduced through $\beta$, which is half of the absorption coefficient. If the laser fields are Gaussian and drop to $1/e$ at a distance r from beam center, we find for the total number of photons generated with $\Delta k = 0$: $N_{\omega_s} = \{(\pi r^2 \kappa N\tau)/3\}|(\Omega_2\Omega_1)/(\delta_2\Gamma_1)|^2(1 - e^{-\beta x})^2/\beta^2$. It is very clear from the expression for the VUV photon flux that a peak occurs when $\Delta k = 0$. With unfocused laser beams the condition $\Delta k = 0$ makes the phase velocity of the generated wave just equal to the effective phase velocity of the nonlinear polarizability. This is the condition for constructive interference; and indeed we see that with $\Delta k = 0$ the photon flux depends on the atomic concentration as $N^2$. In order to achieve $\Delta k = 0$ at a chosen wavelength a gas mixture is ordinarily used, with the "active" gas being chosen so that it is negatively dispersive (i.e. $\delta_2 > 0$) and a "buffer" gas is added which is positively dispersive. In the latter circumstance, if the ratio of concentrations is chosen properly the condition $\Delta k = 0$ is achieved. By keeping the ratio of concentrations fixed while increasing the concentration of the "active" gas the value of $N^2$ can be made very large while still preserving phase matching. This cannot be continued indefinitely due to the fact that $\beta$ generally increases either as the square of the concentration one of the gases, or as a product of concentrations of the two components. The latter fact ultimately leads to a decreasing VUV output as N is increased.

The most impressive demonstrations of VUV generation have been based of the sum generation method described above (See: Tomkins and Mahon 1981, and Hurst et al 1985). The main drawback to this scheme is that it frequently requires doubling or doubling plus sum mixing to get the beam at $\omega_{L1}$. $\omega_{L2}$ is frequently in the infrared, and may require a stimulated Raman process for its generation. It is for this reason that we now consider a simpler scheme which requires only one dye laser and its second harmonic.

### 3.3 Four-Wave Sum and Difference Mixing in Xenon and Krypton

The work which we will now describe (Hilbig and Wallenstein 1982) was carried out with just two "active" gases: xenon and krypton. In the regions where xenon was negatively dispersive either argon or krypton was used as the buffer gas; while in the negatively dispersive regions of krypton the argon served as a buffer gas. The region of the spectrum between 110 nm and 130 nm was investigated by using a dye laser tuned in the region between 550 nm and 650 nm. Part of the dye laser output was

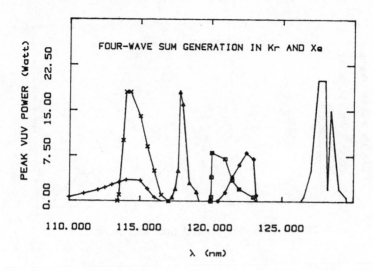

Figure 1. Four-wave sum generation in Xe and Kr (Hilbig and Wallenstein 1982)

frequency doubled to give $\omega_{L1}$ in the wavelength region 275 nm $< \lambda <$ 325 nm, while $\omega_{L2}$ was the remainder of the dye laser beam. The process $2\omega_{L1} + \omega_{L2}$ then yielded VUV light at 1/5 the wavelength of the dye laser. By restricting the wavelength of the light used in this way one gives up the large nonlinear susceptibility which results from having a two-photon resonance, and tries to gain back the efficiency by making use of the excellent dye laser output available when pumping is done by the second harmonic of a Nd-YAG laser. By focusing the laser beams tightly, and by using a buffer gas in order to phase match at the desired wavelength with high concentrations of the active gas excellent results were obtained. Figure 1 shows the output generated in this way.

In order to generate light at a sum frequency the active gas must be negatively disper- sive. However, in the case of difference four-wave mixing (i.e. generating $2\omega_{L1} - \omega_{L2}$) the requirement with focused beams is that:$(\Delta k = k(2\omega_{L1} - \omega_{L2}) - 2k(\omega_{L1}) + k(\omega_{L2}))$ $\Delta k$ times the length of the high intensity region (i.e. the confocal parameter) should not be large compared with unity. The latter quantity can be either positive or neg- ative, with optimum phase matching occuring for $\Delta k = 0$. To generate light between 160 nm and 210 nm Hilbig and Wallenstein used xenon as the active gas. A dye laser in the wavelength region 550 nm to 650 nm was frequency doubled and the doubled output was used as $\omega_{L1}$; with either the fundamental of the dye laser or the funda- mental of the Nd-YAG being used as $\omega_{L2}$. All of the generation was carried out in a positively dispersive region far from the lowest xenon resonance, so that relatively high xenon pressures could be used without producing a large $\Delta k$. The output achieved by this process is shown in Figure 2.

It is clear that eventhough the methods described in this subsection often yield less

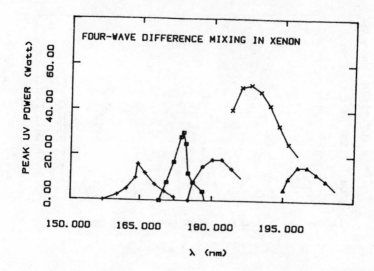

Figure 2. Four-wave difference mixing in Xe (Hilbig and Wallenstein 1982.)

power than a method which uses a two-photon resonant enhancement, that the output can still be useful for spectroscopy and sensitive detection methods.

## Acknowledgments

This research is sponsored by the Office of Health and Environmental Research, U.S. Department of Energy under contract DE-AC05-84OR21400 with Martin Marietta Energy Systems, Inc.

## References

Hilbig R and Wallenstein R 1981 IEEE J. Quantum Electron. <u>QE-17</u> 1566
Hilbig R and Wallenstein R 1982 Appl. Opt. <u>21</u> 913
Hurst G S, Payne M G, Kramer S D, Chen C H, Phillips R C,
    Allman S L, Alton G D,Dabbs J W T, Willis R D, and Lehman B E
    1985 Rep. Prog. Phys. <u>48</u> 1333-70
Jamroz W and Stoicheff B P 1983 *Progress in Optics vol. XX* ed E. Wolf
    (New York: North-Holland) pp 327-80
Payne M G, Chen C H, Hurst G S, and Foltz G M 1981
    *Advances in Atomic and Molecular Physics* vol <u>17</u> ed D. Bates
    and B. Bederson (New York: Acedemic Press) pp 229-74
Tomkins F S and Mahon R 1981 Opt. Lett. <u>6</u> 179

*Inst. Phys. Conf. Ser. No. 84: Section 2*
*Paper presented at RIS 86, Swansea, Wales, 7–12 Sept. 1986*

67

# Resonance ionization spectroscopy with xenon

K. Schneider

Max-Planck-Institut für Kernphysik, 6900 Heidelberg, FRG

The measurement of very small quantities of rare gases is of great importance in modern physics. Examples for Xe are the detection of double beta decay of $^{128}$Te and $^{130}$Te (1), the search for extinct radioactivities in meteorites (2), or the Xe-Xe dating method (3), all depending on the accurate measurement of Xe isotope ratios. Presently, efforts are under way to combine conventional mass spectrometry with resonance ionization spectroscopy (RIS) (4) to reduce the quantity of rare gases necessary for measurements of isotope ratios. At MPIK Heidelberg a laser system for RIS with Xe has been built and test measurements with a proportional counter have been carried out. A special time of flight mass spectrometer has been constructed and combined with a RIS-ion source.

For resonance ionization of Xe two photon absorption (5) has been used to excite the 6p-niveaus directly from the 5p-ground state. Table 1 shows the energy levels and wavelengths necessary for two photon excitation.

Table 1: 6p-niveaus of Xe core: $5p^5(^2P^o_{3/2})6p$, JL-coupling

| Configuration | J | Energy (eV) | Wavelength (nm) |
|---|---|---|---|
| 6p(0 1/2) | 1 | 9.58 | - |
| 6p(2 1/2) | 2 | 9.68 | 256.02 |
| 6p(2 1/2) | 3 | 9.72 | - |
| 6p(1 1/2) | 1 | 9.78 | - |
| 6p(1 1/2) | 2 | 9.82 | 252.48 |
| 6p(0 1/2) | 0 | 9.93 | 249.63 |

The core state for all these two photon resonances is $5p^5(^2P^o_{3/2})$ with an ionization limit of 12.127 eV. The excited Xe atoms are ionized by the absorption of an additional photon.

The laser system (Fig.1) used for this experiment consists of an excimer laser pumped dye laser. The excimer laser EMG 201 MSC (Lambda Physik) operates at the wavelength of XeCl (308 nm), with a pulse energy of 400 mJ and a pulse length of 28 nsec. The maximum repetition rate is 80 Hz. The tuning range of the dye laser (FL2000 Lambda Physik) is 490–530 nm (Dye: coumarin 307). At the exit of the dye laser a KPB-frequency doubling crystal partly converts the output beam into UV-radiation tunable from 245-260 nm. This wavelength range covers all three 6p-resonance of Xe. The maximum pulse energy in the UV is 0.1 mJ with a pulse duration of about 10 nsec. The laser beam is focussed in the respective detection system by a quartz lens with 150 mm focal length.

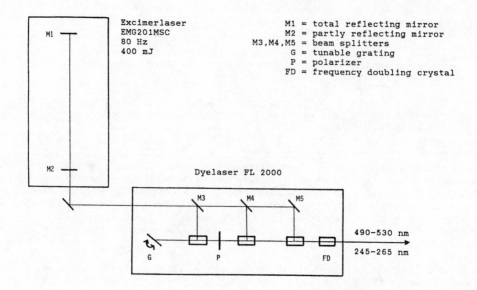

Figure 1: Laser system for production of pulsed UV-radiation in the range of 250 nm.

First test measurements have been carried out using a proportional counter (Fig.2) for the detection of xenon ions. Due to the gas amplification of the proportional counter as few as twenty ions could still be detected.

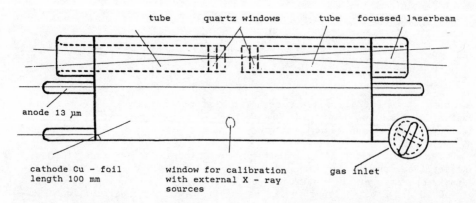

Figure 2: Proportional counter for detection of Xe ions produced by RIS.

Figure 3 shows the Xe ionization signal versus wavelength with constant pulse energy of the laser. From this diagram it is possible to estimate the relative strengths of the 6p-resonances since the pulse energy of the laser system has been kept constant (Table 2).

The signal ratio in resonance/off resonance exceeds 1000. This is important since it allows to measure the signal-to-background ratio of a RIS-ion source during a real measurement of small amounts of Xe by tuning the wavelength in and off resonance.

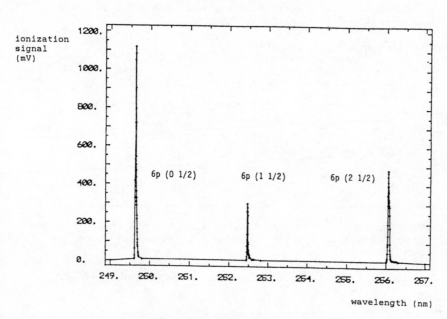

Figure 3: Xenon ionization signal vs. wavelength.

The counter is equipped with a glass window used for calibration with an external X-ray source. By calibration the unknown factors of gas amplification and electronic amplification can be eliminated:

$$N_{Xe} = \frac{E_R \ S_{Xe}}{U_I \ S_R} \quad \text{(Eq. 1)} \ , \quad V_{eff} = \frac{N_{Xe} \ P_N}{N_A \ P_p} \ V_{mol} \quad \text{(Eq. 2)}$$

$N_{Xe}$ = number of Xe ions detected
$S_{Xe}$, $S_R$ = signals of Xe ions or X-ray source
$E_R$ = energy of X-rays
$U_I$ = mean ionization potential of counting gas
$V_{eff}$ = effective ionization volume
$P_p$ = partial pressure of xenon in the counting gas
$N_A$ = Avogadro constant $6.023 \ 10^{23} \ \text{mol}^{-1}$
$V_{mol}$ = Molvolume $224 \ 10 \ \text{cm}^3$
$P_N$ = normal pressure 1013.25 mbar

With Equation 1 the number of detected Xe ions $N_{Xe}$ can be calculated. $N_{Xe}$ does not take into account recombination processes and losses of primary produced photoelectrons due to electronegative components in the counting gas. The effective ionization volume (corresponding to one hundred percent ionization rather than to the real volume of ion production) can be computed from the Xe-partial pressure in the counter. The calculated values for a measurement with 800 mbar Xe are shown in Table 2.

Table 2:  Measurement of effective ionization volumes and the relative strength of the 6p-resonances of Xe.

pulse energy=0.1 mJ     $p_{Xe}$=800 mbar
signal of $^{109}$Cd-source = 60 mV (used for calibration)

| 6p-niveaus | (0 1/2) | (1 1/2) | (2 1/2) |
|---|---|---|---|
| $S_{Xe}$ (mV) | 1120 | 300 | 480 |
| $N_{Xe}$ | $2.2\ 10^4$ | $6.0\ 10^3$ | $9.6\ 10^3$ |
| $V_{eff}$(cm$^3$) | $1.1\ 10^{-15}$ | $2.9\ 10^{-16}$ | $4.6\ 10^{-16}$ |
| relative signal of 6p-niveaus | 1 | 1/3.7 | 1/2.3 |
| theoretical expectation (6) | 1 | 1/6 | 1/2 |

The highest ionization signal is obtained at 249.63 nm, the wavelength of the 6p(0 1/2)-resonance of Xe. Hence, a RIS-ion source should be operated at this wavelength.

The properties of a RIS-ion source can be expressed by its duty cycle:

$$P = f\ N_{Xe}\ \frac{V_{eff}}{V_{spectr.}}  \quad (Eq.3)$$

$N_{Xe}$ = number of Xe ions of the sample
$f$    = repetition rate of the laser system
$V_{eff}$= effective ionization volume (volume which corresponds to 100% ionization)
$V_{spectr}$ = total volume of the spectrometer
$P$    = number of Xe ions produced per second

Equation 3 applies to measurements in the static mode, whereby the sample gas is distributed homogeneously in the whole spectrometer volume. Obviously, the spectrometer volume should be as small as possible. For this purpose, a special time of flight mass spectrometer with resolution M/ $\Delta$M up to 300 ($\Delta$ M at 10% of peak maximum) has been developed (Fig. 4).

The spectrometer consists of three potential plates (P1, P2, P3) forming homogeneous electric fields in the regions I and II. Ions produced near P1 are accelerated to P2, cross P2 through a grid and then slow down in region II until they are stopped near P3. Subsequently, they are again accelerated towards P2 and slow down in region I. This process can be repeated several times such that ions oscillate in the spectrometer volume. After n oscillations, the electrical potential of P1 is lowered while the ions are moving in region II.

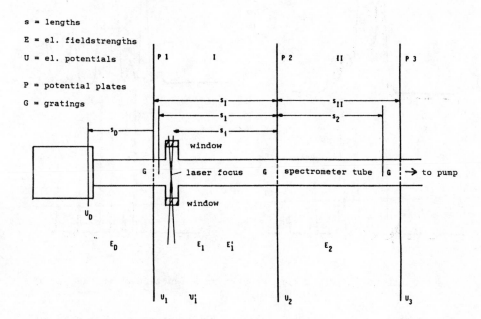

Figure 4: Oscillator time of flight mass spectrometer.

Ions crossing region I no longer loose their total energy but cross P1 with a residual velocity and can be detected in D. The total flight time can be written as a sum of five single times:

$$T_{tot} = T_{i2} + (2n-1) \, T_{12} + T_{12}' + 2n \, T_{23} + T_{1D} \qquad (Eq.4)$$

$T_{i2}$ = flight time from ionization point to P2
$T_{12}, T_3$ = flight time from P2 to the points of zero velocity
$T_{12}'$ = flight time from P2 to P1 with reduced field strength
$T_{1D}$ = flight time from P1 to the detector
$n$ = number of oscillations

Besides an analytical treatment of Equation 5, the results of a computer simulation shall be discussed. The flight times of an ensemble of $10^6$ Xe ions of natural iostopic abundance have been computed. A thermal velocity distribution (T=300 K) and a spatial distribution (ds=0.5 mm) have been chosen as initial parameters. Figure 5 shows four flight time spectra with different time scales for better comparison of the mass peaks. Dimensions of the oscillator time of flight mass spectrometer:
$s_I = s_{II} = 0.1$ m, $s_D = 0.05$ m, $s_1 = 0.09$ m
$E_1 = E_2 = 2000$ V/m, $E_D = 40\ 000$ V/m, $E_1' = 1750$ V/m.

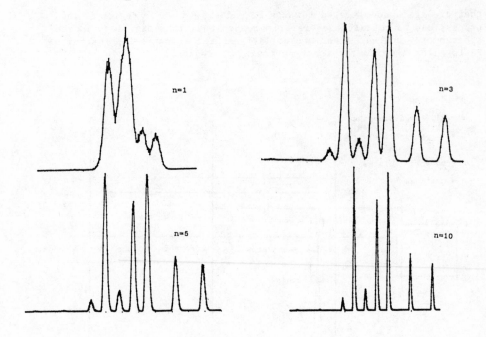

Figure 5: Flight time spectra as a function of the number of
oscillations n.

With a test module of an OTOF a time of flight spectrum of Xe after one
oscillation has been measured and compared with a simulated spectrum
with identical parameters.

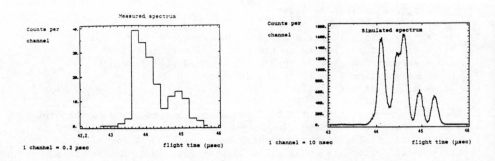

Figure 6: Comparison of a measured Xe spectrum with a
simulated spectrum.

Measurements at different wavelength in and off resonance of the 6p(0 1/2)-niveau gave a signal-to-background ratio of 50 to 1.

The partial pressures of Xe and of residual gases were both $10^{-6}$ mbar.

The major difficulty is presently the very small ionization volume, that is the insufficient production rate of a RIS-ion source for Xe ions. As reported in (7) and in some other works (5) this restriction can be overcome.

The results of our work show that a combination of a RIS-ion source and an oscillator time of flight mass spectrometer OTOF has the potential of becoming a powerful tool for measurements of minute amounts of rare gases.

## References

(1)  T. Kirsten, H. Richter, E. Jessberger, Rejection of Evidence for Nonzero Neutrino Restmass from Double Beta Decay. Phys.Rev.Lett. **50**, 474-477, 1983.

(2)  T. Kirsten, Time and the Solar System. In: **Origin of the Solar System**. Ed. S.F. Dermott, Wiley&Sons Ltd., London (1978), p.267.

(3)  J. Shukoljukov, T. Kirsten and E.K. Jessberger, The Xe-Xe Spectrum Technique, A New Dating Method. Earth Planet.Sci.Lett. **24** (1974) 271-281.

(4)  G.S. Hurst, M.G. Payne, S.D. Kramer and J.P. Young, Resonance Ionization Spectroscopy and One-Atom Detection. Rev.Mod.Phys. **51** (1979) 7667.

(5)  C.H. Chen, G.S. Hurst and M.G. Payne, Direct Counting of Xe Atoms. Chem.Phys.Lett., **75**, 3 (1980) 473.

(6)  M.G. Payne, C.H. Chen, G.S. Hurst and G.W. Foltz, Applications of Resonance Ionization in Atomic and Molecular Physics. Adv. Atomic Molec.Phys. **17** (1981) 229.

(7)  R.T. Hawkins, H. Egger, J. Bokor and C.K. Rhodes, A Tunable Ultrahigh Spectral Brightness KrF Excimer Laser Source. Appl.Phys.Lett. **36** (1980) 391.

*Inst. Phys. Conf. Ser. No. 84: Section 2*
*Paper presented at RIS 86, Swansea, Wales, 7–12 Sept. 1986*

# Noble gas atom counting using RIS and TOF mass spectrometry II: first results

N   Thonnard, R D  Willis, M C  Wright, and W A  Davis

Atom Sciences, Inc, 114 Ridgeway Center, Oak Ridge, Tennessee  37830

B E  Lehmann

Physics Institute, University of Bern, CH-3012 Bern, Switzerland

Abstract.  A dedicated mass spectrometer for noble gas atom counting has been designed and assembled.  Combining a cryogenic cold finger with pulsed atom desorption and resonance ionization in a simple time-of-flight mass spectrometer provides high sensitivity and adequate mass resolution.  We have measured a concentration of 0.4 $^{78}$Kr atoms per cm$^3$ with no interference from the $10^8$ atoms and molecules per cm$^3$ in the residual mass spectrometer vacuum.

## 1.  Introduction

The noble gases are ideal tracers in environmental studies.  Due to their chemical inertness, they are retained in materials only as a solute and are transported through the environment without significant interactions. For hydrological, oceanographic and atmospheric studies, the radioisotopes $^{39}$Ar, $^{81}$Kr and $^{85}$Kr are particularly useful (Lehmann and Loosli, 1984). However, the long half-life (10.8 to 210,000 yrs.) and very low atmospheric concentration (1 part in $10^{16}$ to $10^{18}$) make these radionuclides very difficult to detect with decay counting methods.  The extreme sensitivity and selectivity of RIS opens the possibility for routine measurement of these radioisotopes.  Improving on the method of Hurst et al. (1985), we are implementing all the apparatus and procedures necessary to determine the concentration of the Ar, Kr, and Xe isotopes from environmental and other samples.  We present here some first results.

## 2.  Enrichment

Many of the environmentally interesting radionuclides, in addition to being very rare, occur in the presence of the major isotopes of the same element which are more abundant by factors of $10^{11}$ to $10^{15}$.  It is therefore necessary to follow the enrichment procedures described earlier (Willis and Thonnard, 1984; Lehmann et al., these proceedings) to reduce the level of the interfering isotopes to within a factor of $10^3$ to $10^2$ of the level of the selected radioisotope.  These enrichments must proceed with minimal loss of the sample and be quantitative.

The first step of enrichment is presently performed with a 6" velocity filter (Lehmann et al., these proceedings).  Shown in Figure 1 is the second enrichment system.  An aluminized foil or silicon chip containing implanted noble gas atoms from the first enrichment step is mounted on the rotatable target wheel.  A 532 nm  pulse from a doubled Nd:YAG laser heats the target and releases the trapped noble gas by either vaporizing the aluminum or annealing the silicon. The electron impact ion source and quadrupole mass spectrometer are adjusted to tansmit the isotope of interest, which is implanted into a clean piece of silicon at 10 keV. Typical values for throughput and abundance sensitivity are 8 ma torr$^{-1}$ and $10^4$ respectively.

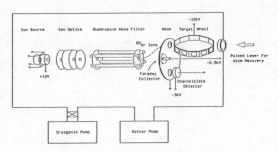

Fig. 1 Second enrichment system

Even though the losses in the various enrichment steps have been measured, quantitation can be improved considerably by utilizing a variation on isotope dilution. Rather than adding an isotope "spike", we instead collect, in addition to the isotope of interest, one of the major isotopes in the sample for a small fraction of the enrichment time. For example, if the goal were to determine the age of a water sample utilizing $^{81}$Kr (i.e., the time since the water sample was last equilibrated with the atmosphere, the source of $^{81}$Kr), then for each enrichment we collect $^{81}$Kr for 99.9% of the time and $^{83}$Kr for 0.1% of the time. After 3 enrichments, the $^{83}$Kr/$^{81}$Kr ratio is reduced by a factor of $10^9$ ($10^3$x$10^3$x$10^3$), and hence, for "modern" water with an initial $^{83}$Kr/$^{81}$Kr ratio of $2.2$x$10^{11}$, the ratio after enrichment would be 220. Therefore, if the measured ratio is 440, for instance, half of the $^{81}$Kr atoms have decayed implying a $^{81}$Kr "age" of $2$x$10^5$ years.

Table 1

Enrichment and Recovery of Krypton Isotopes from Sample F43

| Kr | Frac. | Gas in Sample | 6" VF Proc'sd | Recov'd | 2nd Step (QP) Expect. | Recov'd | 3rd Step (QP) Start | Expect. | RIS-TOF Recov'd |
|---|---|---|---|---|---|---|---|---|---|
| (1) | (2) | (3) | (4) | (5) | (6) | (7) | (8) | (9) | (10) |
| 78 | .0035 | 2.82E14 | 6.95E13 | 1.4E09 | --- | 4.5E08 | 4.1E08 | 3.4E04 | 3.4E04 |
| 80 | .0225 | 1.81E15 | 4.47E14 | 1.85E10 | 1.94E05 | 2.2E08 | 2.2E08 | 1.7E04 | 2.5E03 |
| 81 | --- | 4.28E04 | --- | --- | --- | --- | --- | --- | --- |
| 82 | .116 | 9.36E15 | 2.30E15 | (4.55E12) | 6.04E07 | 9.7E08 | 9.4E08 | 7.4E04 | 3.3E04 |
| 83 | .115 | 9.28E15 | 2.28E15 | 1.1E12 | 2.90E07 | 5.1E08 | 5.2E08 | 3.8E04 | 2.5E04 |
| 84 | .570 | 4.60E16 | 1.13E16 | 2.3E13 | 5.99E08 | 4.2E09 | 4.6E09 | 3.2E05 | 6.7E05 |
| 85 | --- | 3.93E06 | (9.7E05) | (9.7E05) | (3.25E05) | (3.25E05) | (2.4E05) | (9.96E04) | 2.92E04 |
| 86 | .173 | 1.40E16 | 3.45E15 | 3.37E12 | (1.23E10) | (1.23E10) | 2.8E09 | (8.95E06) | 2.80E06 |

Notes to Columns: (2), fraction of the major krypton isotopes in air sample; (3), number of krypton atoms introduced into first enrichment system; (4)-(9),, $^{85}$Kr values in parenthesis are calculated from system parameters; (5), $^{82}$Kr value from calibration procedure; (6)-(9), $^{86}$Kr value from calibration.

In Table 1 we follow the processing of a $^{85}$Kr test sample. We started with 0.003 cm$^{-3}$ of krypton having a $^{85}$Kr activity of 161 dpm cm$^{-3}$ (3.7x background), which is equivalent to the krypton from 3 liters of air or 33 liters of water. Comparing column (3), the number of krypton atoms introduced into the first enrichment system with column (4), the number processed by the 6" velocity filter system, we note that 25% of the sample has been processed. This fraction should be less than 33% to minimize errors due to differential isotope losses. The number of krypton atoms recovered after release in the quadrupole system for the next enrichment step, column (5), gives the actual enrichment factor when compared to column (4) ranging from 490 for $^{84}$Kr to >$10^4$ for $^{78}$Kr and $^{80}$Kr. The expected number of atoms in the sample after the second and third enrichment steps, columns (6) and (9), are calculated from the measured abundance sensitivity ($9.5 \times 10^3$) of the quadrupole and the measured recovery (42%) after having processed 73% of the sample. Note lower limit of $\sim 2 \times 10^8$ in the number of krypton atoms recovered, column (7), which is caused by outgassing from contaminated surfaces in the quadrupole and limits the effective enrichment for the third step. Using resonance ionization with time-of-flight mass spectrometry (RIS-TOF) to determine the remaining krypton after the third enrichment gives the values in column (10), which are reasonably close to the expected result, column (9).

As we are relying on an isotope ratio measurement to determine the concentration of the $^{81}$Kr or $^{85}$Kr isotopes in the original sample after several isotope enrichment steps, it is vital that each system process all isotopes with equal losses. From preliminary tests, this seems to be the case for the first enrichment system (6" velocity filter) and the RIS-TOF final counting system but have encountered a number of difficulties in characterizing the quadrupole system. The results of the quadrupole tests are summarized in Figure 2.

Starting with $\sim 2 \times 10^{13}$ krypton atoms in the quadrupole system, we have operated with the quadrupole tuned to mass 83 for a period equivalent to 1.3 e-folds (typically 60 to 90 minutes), in which one would expect 73% of the sample to be processed. Various combinations of subsystems were turned off (i.e. source ON, poles OFF, etc.) to isolate the different loss mechanisms. If there were no losses, one would expect the curve labeled GAS, the residual gas in the system after enrichment, to be at 100% for all masses except mass 83,

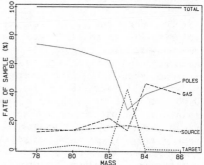

Fig. 2 Quadrupole enrichment

which would be 27%. Conversely, the curve labeled TARGET, the amount of krypton recovered after annealing the target, would be at 0% for all masses except mass 83, which would be 73%. This is not the case and only 42% of the $^{83}$Kr is recovered, which would not in itself be a serious problem (other than needing a larger sample to start with) if all losses were mass independent, as is the case for the curve labeled SOURCE. Unfortunately, a very large fraction of the sample is lost to the poles of the quadrupole, with the losses being greater for masses below resonance than for those above resonance. Therefore, the isotopic composition of the residual gas is constantly changing during the enrichment process, leaving a skewed distribution that varies by more than a factor of 4 at the conclusion of the enrichment. We have attempted to compensate for this behavior, but because the losses are variable and seem to depend on detailed tuning of the quadrupole and its previous sample processing history, significant errors are introduce in the isotope ratios. This is still useful for determining very old ages (i.e., more than one half-life) but limits the usefulness when accurate values are required, such as for relatively young samples.

The errors introduced by the quadrupole enrichment are compounded by the need for two consecutive enrichment steps with that system. The measured abundance sensitivity of the quadrupole mass filter is significantly greater than $10^4$; frequently even $10^5$. Therefore, if the first enrichment (velocity filter) could be improved sufficiently to achieve $10^9$ enrichment with only one additional step (quadrupole), then the errors could potentially be improved by a factor of 4 and the sample losses reduced by a factor of 2.5. We are in the process of implementing a larger velocity filter for the first enrichment step. Shown in Figure 3 are mass plots for the present 6" velocity filter system and a larger, 15" velocity filter. Data for both plots were collected through a 1 mm aperture which transmits only 2/3 of the ions focused on the image plane (hence the peaked profiles). This is the largest aperture that can be used with the low (2 mm/amu) dispersion of the 6" system and still achieve useful abundance sensitivity. The much higher dispersion (12 mm/amu) of the 15" system, having relatively narrower peaks, allows collection of all the ions by using a 2 mm aperture while still achieving an abundance sensitivity of $\sim 10^5$.

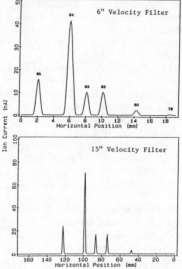

For the near future, we expect the 15" velocity filter to eliminate the need for two quadrupole enrichments, increasing accuracy and sample utilization. The minimum practical gas loading for the first enrichment step is presently 0.003 cm$^{-3}$ of Kr, which is required to maintain the operating pressure in the plasma ion source (Lehmann et al, these proceedings). Reduction in gas recirculation system volume could reduce this by perhaps an additional factor of 5. Looking further into the future, we hope to replace the quadrupole with a different mass spectrometer, perhaps mounted in tandem with the first enrichment step. This should improve accuracy and reduce sample losses to permit routine measurement of $^{81}$Kr and $^{85}$Kr concentrations in 1 liter air or 10 liter water samples at the few percent error level.

Fig. 3 Performance comparison

## 3. RIS-TOF Atom Counting System

The final atom counting using resonance ionization and time-of-flight mass spectrometry is shown schematically in Figure 4. The RIS scheme and laser set-up are essentially the same as described by Kramer et al. (1983). With careful attention to UHV practices for the four-wave mixing cell, a critical component in the generation of the VUV

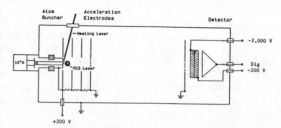

Fig. 4 RIS-TOF atom counting system

radiation required to reach the first excited state of krypton, a single filling of the gas cell remains stable for many months, making VUV generation for krypton ionization routine. Our present laser system allows saturation of all three steps in the Kr RIS process for a 2 mm diameter cylinder, yielding good sensitivity. The time-of-flight mass spectrometer utilizes the time delay between desorption of the krypton from the cold finger and the ionization, and with the proper choice of accelerating potential (Thonnard et al., 1984) achieves good abundance sensitivity at low extraction potentials.

The ionization selectivity of the RIS process is shown in Figure 5. For this mass spectrum, the system pressure was $2 \times 10^{-8}$ torr, including $6 \times 10^8$ Ar $cm^{-3}$ and $1 \times 10^8$ CO $cm^{-3}$, while the $^{84}$Kr concentration was only $6 \times 10^3$ $cm^{-3}$. Note that the CO ionization efficiency is only $10^{-5}$ relative to Kr and that no Ar signal is seen, indicating an element ionization discrimination $>10^7$. The sensitivity of the RIS-TOF system is shown in Figure 6. The spectrum was recorded by averaging 3000 laser shots (5 minutes during a period in which the average krypton content of the 6.6 liter volume mass spectrometer was $8 \times 10^5$ atoms. The 2800 $^{78}$Kr atoms in the system are clearly detected, indicating a sensitivity for $^{81}$Kr of ~1000 atoms. Note that the $^{78}$Kr concentration was 0.4 $cm^{-3}$, while the $2 \times 10^{-9}$ torr pressure in the chamber implies $10^8$ atoms and molecules per $cm^3$.

The formal resolution of the TOF mass spectrometer, $m/\Delta m$ is ~200, but the mass peaks are asymmetric, with a very steep leading edge and a long high mass tail. The cause of this shape has not yet been elucidated. Fortunately, the shape of the individual peaks is very stable and seems to be independent of most system operating parameters. It is therefore possible to numerically fit a universal profile shape to each mass peak by only adjusting the FWHM width, peak position and peak amplitude, allowing accurate deconvolution of the individual mass peaks (Figure 7).

## 4. Results and Future Outlook

To date, we have processed five $^{85}$Kr samples and one $^{81}$Kr sample through all the steps described here. All of the $^{85}$Kr samples, two of which started with normal atmospheric background concentration, were detected. The one $^{81}$Kr sample, which due to losses in the various enrichment steps was expected to have a $^{81}$Kr content at or somewhat below our present RIS-TOF sensitivity limit, was not detected.

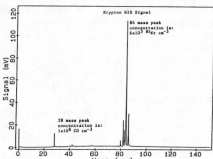

Fig. 5 RIS selectivity

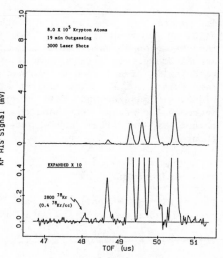

Fig. 6 RIS-TOF sensitivity

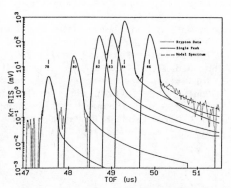

Fig.7 Modeled fit of TOF data

We show in Figure 8 the mass spectrum of a 0.003 cm³ krypton sample that was enriched in $^{85}$Kr, the numerical results being given in Table 1. The peculiar isotope signature in the figure is due to the enrichment process, with the large $^{86}$Kr peak being the calibration. The net $^{85}$Kr signal was obtained by utilizing the $^{86}$Kr peak as a "standard" shape and fitting it to the data, as was done in Figure 7. Even though the final $^{85}$Kr signal was only 1/3 of what had been expected from the known system performance, the measured $^{85}$Kr/$^{86}$Kr ratio, 0.0104, was within 6% of the expected result, 0.0111. Though this close agreement is somewhat fortuitous, most results have been within 30%, indicating the power of the calibration method.

In conclusion, we feel the method will allow measurements of the $^{81}$Kr and $^{85}$Kr concentration in reasonably sized environmental samples. Improvements that are technically feasible and planned for the near future will allow smaller samples by reducing losses and will significantly increase accuracy. We hope to extend the method to the other noble gases in the future.

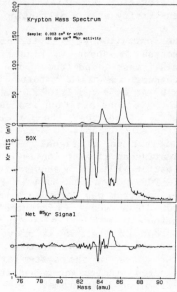

Fig. 8 RIS-TOF mass spectrum for a natural $^{85}$Kr sample processed through all enrichment systems

Acknowledgement:

This work was supported in part by the U.S. Department of Energy SBIR program under contract No. DE-ACO5-86ER80404.

References:

Hurst G S, Payne M G, Kramer S D, Chen C H, Phillips R C, Allman S L, Alton G D, Dabbs J W T, Willis R D and Lehmann B E 1985 Rep. Prog. Phys. 48 1333
Kramer S D, Chen C H, Payne M G, Hurst G S and Lehmann B E 1983 Appl. Opt. 22 3271
Lehmann B E and Loosli H H 1984 Resonance Ionization Spectroscopy 1984 ed G S Hurst and M G Payne, Inst. Phys. Conf. Ser. 71 (Bristol: The Institute of Physics) pp 219-26
Thonnard N, Payne M G, Wright M C and Schmitt H W 1984 Resonance Ionization Spectroscopy 1984 ed G S Hurst and M G Payne, Inst. Phys. Conf. Ser. 71 (Bristol: The Institute of Physics) pp 227-34
Willis R D and Thonnard N 1984 Resonance Ionization Spectroscopy 1984 ed G S Hurst and M G Payne, Inst. Phys. Conf. Ser. 71 (Bristol: The Institute of Physics) pp 213-218

*Inst. Phys. Conf. Ser. No. 84: Section 2*
*Paper presented at RIS 86, Swansea, Wales, 7–12 Sept. 1986*

81

# The challenge of sorting out 1000 atoms of krypton-81 from $10^{25}$ molecules of water

B E Lehmann*, D F Rauber*, N Thonnard** and R D Willis**

*Physics Institute, University of Bern, Switzerland
**Atom Sciences Inc, Oak Ridge, Tennessee, USA

## 1. Introduction

Resonance Ionization Spectroscopy (RIS) has proven its capability of detecting individual atoms of almost every element in the periodic table and in combination with various mass filters, a technique is now available to count single atoms with isotopic selectivity (Hurst et al. 1985). In many practical applications, however, this ultimate sensitivity and selectivity can develop its full power only if the sample to be analyzed has been brought into a suitable form to be accepted by the final detecting system. Elaborate sample processing steps are sometimes necessary prior to the use of laser resonance ionization spectroscopy. For the case of $^{81}Kr$ dating in hydrology the task is to find 1000 atoms in one litre of water. Obviously one cannot simply pour water into a RIS-detector system or fire RIS-lasers through a bottle of water.

## 2. Krypton-81 Dating in Environmental Sciences

$^{81}Kr$ is produced in the earth's atmosphere by cosmic-rays in spallation reactions from the stable isotopes $^{82}Kr$, $^{83}Kr$, $^{84}Kr$, and $^{86}Kr$ and by neutron-capture of $^{80}Kr$. Its atmospheric equilibrium activity is extremely low (1.7 x $10^{-6}$ Bq per $m^3$ of air) making activity measurements in environmental applications by low level decay counting techniques impossible.

With a halflife of 210,000 years, $^{81}Kr$ is an excellent isotope to study underground water circulation in the time range of 50,000 to 1,000,000 years or to date very old ice from deep drilling projects in Greenland or in Antarctica. One of the main advantages of this isotope in environmental studies is its inert character. This not only ensures that concentrations in a natural archive are not changed by chemical reactions over extended periods of time but it also enables to work with very small samples of gas. The sorting procedure described in this text is certainly only possible with a noble gas. On the other hand, RIS for noble gases of course requires much more elaborate laser systems as compared to e.g. work with alkali-atoms.

While $^{81}Kr$ dating of groundwater by low level counting techniques would require degassing of more than $10^5$ litres of water, we can now work with water samples of typically 20 to 100 litres using RIS-counting of $^{81}Kr$ atoms (Lehmann 1985).

3. Sample Processing

We currently use five different systems in processing hydrological samples for $^{81}$Kr analysis:

| System | Purpose | Input | Output |
|--------|---------|-------|--------|
| 1 | Degassing of water | 20 to 100 lt of water | 0.5 to 2 lt of gas |
| 2 | Separation of krypton | 0.5 to 2 lt of gas | $10^{-3}$ to $10^{-2}$ cc of krypton |
| 3 | 1st step of isotope enrichment | $(1-5) \times 10^{-3}$ cc of Kr gas | 10,000 $^{81}$Kr atoms Kr/$^{81}$Kr ratio reduced by >1000 |
| 4 | 2nd (and possibly 3rd) step of isotope enrichment | pre-enriched sample implanted in Al target | Kr sample with Kr/$^{81}$Kr ratio reduced by $10^4$ in each step |
| 5 | RIS-counting of $^{81}$Kr atoms | $^{81}$Kr atoms and remaining stable Kr atoms implanted in Si target | result |

In this paper we describe the use of systems 1 to 3. The initial sorting problem of finding 1000 atoms of $^{81}$Kr among $3 \times 10^{25}$ molecules of water is thereby reduced to detecting 1000 atoms of $^{81}$Kr in $10^{11}$ atoms of $^{82}$Kr.

3.1 Degassing of water

For degassing water samples under field conditions in remote areas a system has to be used which is as simple and robust as possible but still guarantees that the groundwater samples do not come into contact with atmospheric air and get contaminated with modern krypton. The set-up is schematically shown in Figure 1:

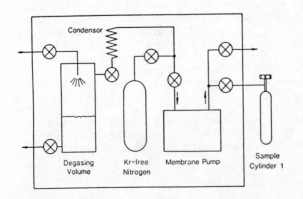

Fig. 1

System for degassing groundwater samples in the field.

A two-stage membrane pump is used both to evacuate the system prior to sampling and for compressing the gas into a sample cylinder. Water is degassing inside an evacuated stainless steel cylinder of a volume of 8 litres. A copper tube coil is used as a condensor for water vapour. Since the ultimate pressure that can be reached with such a simple pump is only about 50 mbar, it is necessary to flush the system with Kr-free nitrogen prior to sampling. Nitrogen gas is also used to press the water out of the degassing volume after each batch of 8 litres that has been degassed. This procedure eliminates the need for an extra water pump to empty the degassing volume. The final gas sample, compressed to approximately 10 bars into a 0.5 lt sample cylinder, is a mixture of gas that was dissolved from groundwater and nitrogen used in this method. It is therefore essential that high-purity Kr-free nitrogen is used.

## 3.2 Separation of krypton

Air-saturated water at 10°C contains $9.2 \times 10^{-5}$ cc of krypton. The next step in the sorting work is to get rid of gases such as $N_2$, $O_2$, $CO_2$, Ar, $H_2O$ vapour, and sometimes $CH_4$, $H_2S$ or other components found in gases extracted from natural deep groundwater.

We use a system with major components as outlined in Figure 2:

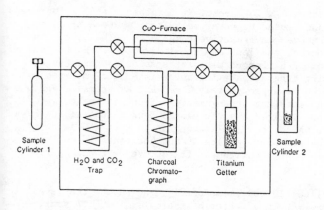

**Fig. 2**

System for separating krypton from extracted gases.

A first trap, consisting of a Cu-coil at liquid nitrogen temperature, removes $H_2O$ and $CO_2$. A simple charcoal-filled second Cu-coil held at -80°C traps Kr-atoms but passes most of the $O_2$, $N_2$ and Ar gas. Since $CH_4$ has almost the same cryogenic properties as Kr, the two cannot be separated in this step. If necessary, $CH_4$ is removed by circulating the gas through a hot CuO-filled quartz tube. A hot Ti-getter removes remaining traces of chemically reactive gases. To bring the Ar to Kr ratio down to the required level of at least 5:1 a second smaller chromatographic step is often used. At the end of this unit a Kr-gas sample is trapped in a small $LN_2$-cooled charcoal-filled sample container.

The sample now contains approximately $(1-5) \times 10^{-3}$ cc of Kr.

3.3 First step of isotope enrichment

It is necessary to reduce the ratio of total Kr to $^{81}$Kr which is 1.9 x 10$^{12}$ in a modern sample by several orders of magnitude, before a sample can be analyzed in a RIS mass-spectrometer system. The first 3 orders of magnitude are accomplished in the system, outlined in Figure 3:

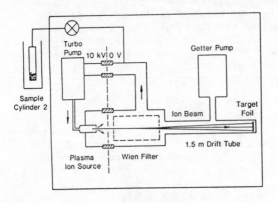

<u>Fig. 3</u>

System for first step of isotope enrichment procedure

In a standard run we introduce 3 x 10$^{-3}$ cc of Kr together with 3 x 10$^{-3}$ cc of Ar into the evacuated and closed high vacuum system, where a Zr-alloy getter pump continuously removes any reactive components. The Kr gas is ionized in a plasma ion source, and ions are accelerated by 10 kV through a Wien velocity filter. Ions of different mass are spatially separated as they pass down a 1.5 m drift tube and are implanted into an aluminized kapton foil. The dispersion is approximately 2 mm, and Kr atoms of various isotopes are sorted with an aperture arrangement in front of the target foil. Kr atoms that are not ionized are circulated back to the source by a small turbo-molecular pump which is also floating at 10 kV. A beam stabilization system centers the ion beam vertically and horizontally onto a target aperture to maintain a high enrichment factor over extended periods of time.

In a run of 6 hours we collect 10,000 atoms of $^{81}$Kr and reduce the level of interfering isotopes at mass-82 by a factor of typically 1000. A small piece of foil (8 x 8 mm) containing 10,000 $^{81}$Kr atoms, approximately 10$^{12}$ atoms of $^{82}$Kr and the remaining lower levels of other stable Kr isotopes is finally extracted from the system and shipped for further processing to systems 4 and 5. These steps are described in a successive paper in this session.

<u>References</u>

Hurst G S, Payne M G, Kramer S D, Chen C H, Phillips R C, Allman S L,
    Alton G D, Dabbs J W T, Willis R D and Lehmann B E 1985 Rep. Prog. Phys.
    <u>48</u> 10 1333
Lehmann B E, Oeschger H, Loosli H H, Hurst G S, Allman S L, Chen C H,
    Kramer S D, Payne M G, Phillips R C, Willis R D and Thonnard, N 1985
    J. Geophys. Res. <u>90</u> B13 11547

*Inst. Phys. Conf. Ser. No. 84: Section 2*
*Paper presented at RIS 86, Swansea, Wales, 7–12 Sept. 1986*

# Ultrasensitive laser isotope analysis of krypton in an ion storage ring

R.E.Bonanno, J.J.Snyder, T.B.Lucatorto, P.H.Debenham, and C.W.Clark

National Bureau of Standards, Gaithersburg, MD 20899, USA

ABSTRACT. We are developing a new instrument for ultrasensitive isotope analysis that combines magnetic mass selection, resonant charge exchange, and laser reionization. For krypton, this technique is expected to achieve isotope abundance sensitivities better than $10^{-12}$.

## I. INTRODUCTION

At the present time, few techniques are available for measuring isotopes with abundances lower than $10^{-10}$. Conventional magnetic mass spectrometry is limited to approximately $10^{-9}$, principally by isobaric and molecular interferences (i.e. different atomic or molecular species which have the same mass). For radioactive isotopes, counting techniques are available but become impractical as concentrations decrease and as half-lives increase (e.g. $^{129}I$, $^{81}Kr$, $^{90}Sr$). High energy mass spectrometry has been successful for measuring abundances as low as $10^{-15}$, for a limited number of low mass isotopes such as $^{10}Be$ and $^{14}C$ (Mast et al. 1980). However, these techniques do not exhibit the same capabilities for heavier isotopes and the cost of these instruments is often prohibitive for use in routine analysis. Exploitation of the optical isotope shift, allowing the selective excitation and/or ionization of a specific isotope over all others, has led to the development of several laser-based techniques (Lucatorto et al. 1984, Miller et al. 1985 and Cannon and Whitaker 1985). These methods are appealing due to their generality and relatively low cost. We are proposing a technique which combines, in a novel way, magnetic mass spectrometry with laser photoionization (Snyder et al. 1985). We expect this technique to be useful for ultrasensitive analysis of rare isotopes which are otherwise difficult to measure. In this paper we describe the technique and discuss its advantages and possible limitations. We will also describe preliminary experiments which are currently underway in our laboratory to determine the overall feasibility of the technique. As an example, we describe the applicability of this method to measurement of the environmentally important radioactive isotopes of krypton. The production of krypton-85 is primarily a result of fission processes and has an isotopic abundance of $10^{-12}$. Krypton-81 is produced by cosmic ray interactions in the upper atmosphere and has an abundance of approximately $10^{-11}$.

## II. DESCRIPTION OF THE TECHNIQUE

In our proposed instrument, sample ions are continuously injected into a small, racetrack shaped ring where they are stored for several orbits as the undesired isotopes are removed. This multistaged separation technique has three distinct elements (see figure 1). First, the ions are separated by two 180 degree bending magnets. After this initial stage of enrichment, the ions are neutralized in a charge exchange cell. A laser reionizes the ion beam before it passes back into the magnet region for the second time to be further concentrated in the desired isotope. After n orbits the desired degree of enrichment is attained and the ions are directed to a particle multiplier. An essential feature of the charge exchange stage is that the cell is electrically isolated and maintained at a potential of approximately 75 volts. Accordingly, the ion beam is decelerated each time it enters the cell. As illustrated in figure 1, this leads to a physically discrete orbit for each pass through the system. All orbits share a common path through the straight section containing the charge exchange cell. However, the trajectories through the second straight section are separated as a result of each successive deceleration. The amount of separation is proportional to the difference in the kinetic energy of ions in different orbits. Thus, the primary function of the charge exchange cell is to switch ions, after each stage of enrichment, into the next lower orbit. The end result is that ions of the desired species follow a decaying spiral path until they are intercepted by the detector. Orbit separation reduces contamination of the highly purified ions entering the detector by ions which have not yet been enriched to the same degree.

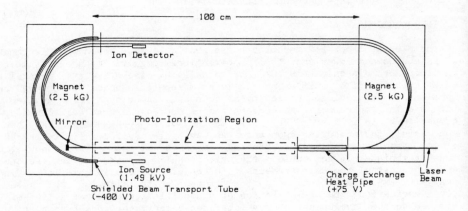

Fig. 1. Schematic of the ion storage ring configured for $^{81}$Kr measurements

The instrument just described has several important advantages for ultrasensitive isotopic measurements. First, a high degree of isotopic selectivity is obtained each time the beam passes through the magnets. However, in contrast to conventional magnetic mass spectrometers, the total enrichment is equal to the n$^{th}$ power of the single-pass selectivity, where n is the number of orbits. In addition, the use of multiple orbits reduces contamination due to elastic scattering processes, a factor which limits the sensitivity of conventional single-pass instruments. If the charge exchange process is resonant or near-resonant, isobaric and molecular interferences are greatly reduced since only isotopes of the correct element will have a high probability for charge exchange. The reionization stage is also elementally selective, and serves to eliminate isobars since only those atoms reionized can remain in the ring. The reionization step can add additional isotopic selectivity to the overall process. However, in the case of krypton the optical isotope shifts are small and the selectivity attained in the magnets alone should be more than adequate.

With the instrument configured for $^{81}$Kr, the storage ring parameters would have the following values. Ions initially accelerated to 1.5 keV travel at

7 x 10⁶ cm/sec. A uniform magnetic field of 2500 gauss produces an initial orbit diameter of 20 cm. Each time the beam is incrementally decelerated by 75 volts the orbit diameter is reduced by 1 cm. The orbits produced by adjacent mass isotopes will be separated by about 0.1 cm from the desired isotope (e.g. $^{82}$Kr vs $^{81}$Kr). An appropriately positioned aperture plate after the exit aperture of the first magnet will allow only the desired isotope to pass.

For each isotope to be measured, the charge exchange gas and laser ionization scheme must be chosen carefully so as to maximize the efficiency of these processes and therefore maximize the overall transmission of the instrument for the desired isotope. In the case of krypton, the charge exchange gas is rubidium. The reason is apparent in figure 2, which is a partial energy level diagram of krypton and rubidium, drawn relative to a common ionization potential. There is a large cross section for transfer of an electron from the ground state of rubidium into the first excited state of krypton because of the near resonance between these two states. This first excited level consists of two closely spaced states, one of which is metastable (5s;J=2) with respect to the ground state ($\tau > 1$ s). Atoms emerging from the charge exchange region in this state are easily ionized via a two-photon transition through a resonant intermediate p state to an ns or nd autoionizing level. This represents a significant advantage over other laser-based techniques for ultrasensitive analysis of krypton which are based on ionization from the ground state, which lies 14 eV below the ionization threshold. Atoms which are left in the non-metastable state (5s;J=1) will decay rapidly ($\tau = 4.4$ ns) to the ground state before they reach the ionization region and must therefore be considered a loss. The cross section for charge exchange has not yet been measured experimentally, but has been calculated. At the krypton ion velocity of 7 x 10⁶ cm/sec, the cross section is calculated to be 7 x 10$^{-15}$ cm² (Ice and Olson 1975). With this we estimate the atom density of rubidium required to completely neutralize the krypton ion beam within a 10 cm path length to be ≈10¹⁴ cm⁻³.

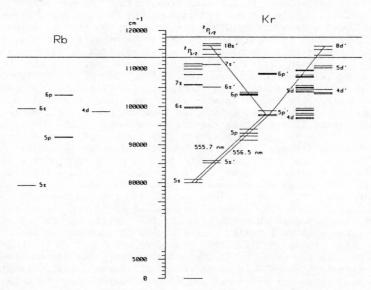

Fig. 2. Partial energy level diagram of krypton and rubidium. Two possible photoionization schemes are shown

Ideally, since ions are continuously injected into the ring we would like to continuously reionize them. We have estimated the cross section for the transition from an intermediate 5p' state to an autoionizing level to be $10^{-15}cm^2$. In order to saturate this transition (a requirement if we are to maximize the overall efficiency) a cw laser which can generate $\approx 1$ kW at 555 nm is necessary. Unfortunately, this requirement is not easily met without employing difficult intracavity ionization techniques. A more practical approach is to use a high repetition rate copper-vapor laser to pump a dye laser. The power generated by this system is more than adequate for saturating the transition. The drawback of this approach is that only a segment of the total number of isotopes in the ring are utilized, those that are in the ionization region when the laser fires. In addition, the repetition rate of the laser must be synchronized with the velocity of the ions. Then each time this "bunch" passes back into the ionization region, regardless of the orbit number, the laser will fire thereby allowing these isotopes to continue in the ring. We estimate that about 20% of the total number of isotopes injected into the ring will be reionized. It is important to note that this will not affect the overall enrichment or sensitivity of the technique, since the detector can be gated and synchronized with the arrival of each successive ion bunch. The only negative effects of this approach are a loss in overall sample utilization and an increase in the total time required to detect a given number of isotopes.

III. **ACHIEVABLE ENRICHMENT**

The fundamental objective is to maximize transmission through the instrument for the desired isotope (i.e. that which we wish to measure) and to minimize transmission through the instrument of the undesired species (all others). For krypton, the net enrichment of the sample in the desired isotope for each pass through the ring will depend entirely on the isotopic selectivity of the two bending magnets. A conservative estimate of $10^2$ for the selectivity per magnet gives a factor of $10^4$ per orbit and $10^{12}$ after three orbits. At the same time, after three orbits we must account for losses in the desired isotope due to less than perfect transmission. These losses are incurred primarily in the charge exchange cell where, in addition to populating the desirable metastable state, the state which is radiatively coupled to the ground state is also populated. As mentioned in section II, only those atoms which are in the metastable state can be reionized and therefore continue through the system. Although energy defect and statistical considerations indicate a propensity for charge transfer into the metastable state, we assume that only 50% of the total number of krypton ions injected into the system will be left in the metastable state after exiting the charge exchange cell. Then, assuming saturation of the ionizing transition, the maximum achievable single-orbit transmission efficiency is 50%, or about 10% after three orbits. As is the case for the net enrichment, the overall efficiency of the enrichment process after n orbits will be the $n^{th}$ power of the single-orbit efficiency. For example, consider a naturally occurring krypton sample containing one part in $10^{11}$ $^{81}Kr$. Given an ion source current of 2 microamperes, in a 100 second counting time $10^{15}$ total ions, principally $^{84}Kr$, will be injected into the ring. Of the total number of ions injected, there will be $10^4$ $^{81}Kr$. Assuming the maximum enrichment, theoretically, after three orbits the undesired isotopes in the beam will be reduced by a factor of $10^{12}$. If the net efficiency for the desired isotope is 10%, then $10^3$ $^{81}Kr$ should be detected. Of course this unrealistically assumes that there are no losses due to less than perfect beam handling or other processes which can degrade the net efficiency, such as charge exchange collisions with background gas. However, it should be possible to design the instrument so that these losses are kept to about 10%.

## IV. PRESENT EXPERIMENTS

In order to determine the overall feasibility of the proposed technique, we are currently conducting experiments designed to measure the single-pass laser ionization efficiency and the effectiveness of the charge exchange process in populating the desired metastable state in Kr. Figure 3 is a schematic of the present experimental set-up. The major components of the linear beamline are an ion source, two translatable current detectors (Faraday cups) and a rubidium heatpipe. The other components of the apparatus include: ion beam steering and deflection units and ion beam profile diagnostics. The laser system consists of a 5 kHz repetition rate, 10 W copper-vapor laser which pumps a single-stage dye laser. The laser beam enters longitudinally through an optical window mounted at the end of the apparatus. The plasma ion source is capable of producing several microamperes of singly ionized krypton. Measurement of the beam current before and after the rubidium heatpipe allows determination of the charge exchange efficiency. We have recently demonstrated 100% neutralization of a 0.5 microampere $Kr^+$ beam. Experiments to measure the reionization of the beam are now in progress. The degree of reionization is measured by time-resolved detection of ions produced after the laser is fired. If the two-photon ionization transition can be saturated, then the branching ratio for charge exchange into the two excited 5s states can be determined. Future plans include extending these experiments to measure the absolute cross section for the $Kr^+$-Rb charge exchange process.

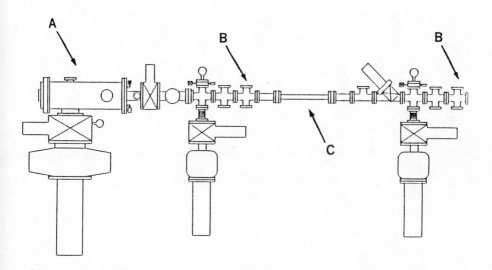

Fig. 3. Present experimental apparatus, A: ion source; B: Faraday cups; C: rubidium charge exchange heatpipe

## V. SUMMARY

We have proposed a novel technique for the ultrasensitive isotopic analysis of krypton. The method combines magnetic mass selection, near-resonant charge exchange neutralization, and laser reionization of a sample ion beam in a racetrack shaped storage ring. With this approach, measurement of isotope abundance sensitivities in the range of $10^{-12}$ - $10^{-14}$ appears to be technically feasible. We believe that this is a general technique with the potential to make a direct impact in the field of ultrasensitive isotope analysis. In this paper we have described the prospective application of this technique to the analysis of the radioactive isotopes of krypton. In a previous paper, applicability of the method was discussed for the measurement of $^{90}$Sr (Snyder et al. 1985). The future development of this instrument will depend principally on the outcome of experiments currently underway.

REFERENCES

Cannon B.D. and Whitaker T.J. 1985 Appl. Phys. B 38 57
Ice G.E. and Olson R.E. 1975 Phys. Rev. A 11 111
Lucatorto T.B., Clark C.W., and Moore L.J. 1984 Opt. Commun. 48 406
Mast T.S., Muller R.A., and Tans P.P. Proceedings of the Conference on the Ancient Sun, ed. P.O.Pepsin, J.A.Eddy, and R.B.Merril (New York:Pergamon) p.191
Miller C.M., Engleman R., and Keller R.A. 1985 J. Opt. Soc. Am. 2 1503
Snyder J.J., Lucatorto T.B., Debenham P.H., and Geltman S. 1985 J. Opt. Soc. Am. B 2 1497

*Inst. Phys. Conf. Ser. No. 84: Section 3*
*Paper presented at RIS 86, Swansea, Wales, 7–12 Sept. 1986*

# Resonance ionization mass spectrometry

J.C. Travis, J.D. Fassett, and T.B. Lucatorto
U.S. National Bureau of Standards
Gaithersburg, MD. 20899, USA

## 1. Introduction

A variety of mass spectrometric tools of inorganic analytical chemistry share a common deficiency which may be serious in certain applications: the physical mechanisms employed to produce free atomic ions for mass analysis are inefficient, non-uniform, and non-selective. Specifically, the neutral free atom yield is typically at least $10^2$ greater than the atomic ion yield; different elements ionize to different extents; and elements having isotopes of the same nominal mass (isobars) are simultaneously ionized. These shortcomings are directly addressed by the use of resonance ionization spectroscopy (RIS) (Hurst et al 1979) for the selective ionization of atomic neutrals in these devices.

Resonance ionization mass spectrometry (RIMS) has experienced an impressive rate of growth and success, witnessed by the 42 elements reported since about 1980: Al, Am, B, Ba, Be, C, Co, Cr, Cu, Dy, Er, Fe, Ga, Hf, I, In, K, Li, Lu, Mg, Mo, Na, Nb, Nd, Ni, Np, Os, Pb, Pu, Re, Sm, Sn, Sr, Ta, Tc, Th, Ti, U, V, W, Y, Zr (Auschwitz et al 1985; Cannon et al 1985; Donohue et al 1985c; Harrison et al 1984a,b; Parks et al 1985a; Travis et al 1984). The time frame used for the present overview of RIMS, 1984 – present, overlaps the time period assessed by Young et al (1984a) at RIS II, including nine contributions to the RIS II proceedings (Hurst and Payne 1984), and early 1984 works referred to therein. The application of RIS to rare gas isotope analysis by repeated enrichment is arbitrarily excluded, as a separate topic of the RIS conferences. The period surveyed includes two general reviews of RIMS (Nogar et al 1985c; Fassett et al 1985) and several more restricted reviews/tutorials (Moore 1984; Nogar et al 1984b; Parks et al 1985a,b; Whitaker 1986).

## 2. Provision of Free Atoms

As of RIS II, the RIMS literature was largely dominated by thermal vaporization (Donohue et al 1984b; Nogar et al 1984a; Travis et al 1984). Nanograms to micrograms of sample may be dried or electroplated (Downey et al 1984b; Peuser et al 1985) onto electrically-heated filaments of Re, Ta, W, etc., for vaporization in the mass spectrometer. Samples are sometimes adsorbed onto resin beads which are then mounted on the filament (Donohue et al 1985c; Fassett and Walker 1986). Codeposition or overcoating with some form of graphite is often used to effect the reduction of metals deposited as salts, thus favoring the volatilization of free atoms over molecular forms (Fassett et al 1984b). Uranium codeposition (Miller et al 1985) and Re overplating (Kroenert et al 1985) have been employed as diffusive barriers. Thermal vaporization from enclosed furnaces has been utilized (Auschwitz and Lacmann 1985; Bekov 1984; Cannon 1985; Andreev et al 1986b), as well as from filaments.

Ion beam sputtering for direct solid sampling was also discussed at RIS II (Fairbank et al 1984a; Parks et al 1984b; Winograd 1984). The sputter-initiated resonance ionization mass spectrometer (SIRIS) developed by Atom Sciences (Schmitt 1985) utilizes high energy, mass selected $Ar^+$ ions, pulsed by electrostatic beam deflection. The 50-μA, 1-μs duration ion pulse produces about $10^9$ sputtered atoms, at an estimated sputter yield of ~5 (Parks 1985b), appropriately timed to the RIS laser pulse. Kimock et al

(1984a,b) employ a hollow cathode source of $Ar^+$ ions, yielding 2 µA at 5 kV, which may also be pulsed by ion deflection. Available beam energies from <1 keV to ~30 keV have been used to study the metastable state distribution of sputtered atoms. An ion microprobe mass analyzer (IMMA), delivering a 21-nA beam current in a 2-µm spot, has been modified by Donohue et al (1985a,b) for RIMS analysis of particles. Several groups (Winograd 1984; Becker 1985; Pellin et al 1984) stress the use of low dose (~nA beam current) ion sputtering for surface studies.

The glow discharge represents a very different approach to ion sputtering of solid samples (Harrison et al 1984a,b; 1986; Savickas et al 1984). This approach requires differential pumping since Ar pressures of 0.1 - 1 Torr in the source chamber are required to sustain 1-3 mA glow discharges at 300 - 4500 V. Samples serving as the cathode may be metals machined into pins or salts pressed with graphite into rods. The discharge may be pulsed for improved sample utilization with pulsed RIS lasers.

Pulsed laser ablation/desorption, has seen continued development (Becker and Gillen 1985; Williams et al 1984) owing largely to the efficiency of sample utilization for pulsed RIS analysis. Thermodynamic properties necessary to understand and optimize the efficient use of this source with pulsed ionization have been studied by Beekman and Callcott (1984), Hurst et al (1984), and Nogar et al (1984b).

## 3. Resonance Ionization

The relatively long (1-µs) pulse length and broad (~3-cm$^{-1}$) bandwidth of the flashlamp-pumped dye laser (Donohue et al 1984b; Downey et al 1984a) are ideally suited to the cross section requirements of most RIS excitation/ionization schemes (Hurst et al 1979). The use of short-pulse (5-30 ns) dye lasers may require sacrificing photo-ionization yield to avoid unwanted background ionization and power-induced spectral distortions (Fassett and Walker 1986). Alternatively, separate lasers may be employed for excitation and ionization (Beekman and Callcott 1984; Peuser et al 1985; Williams et al 1984; Young et al 1984b). Short-pulse dye-pumping lasers employed in RIMS include Nd:YAG (Kimock et al 1984 a,b), excimer (Savickas et al 1984), $N_2$ (Young et al 1984b), and Cu vapor (Pueser et al 1985). With moderate focusing, continuous wave (cw) lasers are capable of saturating resonant transitions, but the excited state photo-ionization rate is sufficiently low that small ion yields (<10$^{-3}$) are normally achieved during the transit time of the free atom across the laser interaction volume (Donohue et al 1984b; Nogar et al 1984a,b). For many applications, the low ionization efficiency is more than adequately compensated by the gain in duty factor. Single mode cw lasers provide the highest available spectral resolution (~1-MHz) for isotopically selective RIMS (Nogar et al 1984a; Miller et al 1985a,b). Cannon et al (1985) have combined high resolution and high ionization yield by stepwise exciting Ba with two cw dye lasers, and ionizing by means of a cw Nd:YAG laser at 400 W/cm$^2$.

Single-laser ionization schemes, corresponding to variations on types I, II, and V RIS (Hurst et al 1979), were used for the majority of the elements observed by RIS II (Travis et al 1984). Since then, special attention has been accorded to the single-laser, "three-photon, doubly resonant" scheme first employed for RIMS by Donohue et al (1984a). This variant of type V RIS reduces the density of observed transitions in the optical spectra of actinides (Donohue et al 1984a,b,c; 1985c; Downey et al 1985a; Kroenert et al) or lanthanides (Smith et al 1985) to reduce the risk of spectral interference among isobars. Efficient 3-photon ionization is restricted to combinations of near- and exact-coincidences at the one- and two-photon levels. Nogar et al (1985b) have characterized the ionization of Ta by this scheme, having first encountered undesired background resonances from their Ta filament in the course of another experiment (Nogar et al 1985a). Two- and three-color schemes (type III RIS) have been employed for greater selectivity and sensitivity (Cannon et al 1985; Nogar et al 1985a; Peuser et al 1985).

A recurring theme in recent RIMS is the possibility of enhancing the "abundance sensitivity" of mass spectrometers -- the ability to measure extreme ($~10^{12}$) isotope ratios -- by means of isotopically selective resonance ionization (Whitaker 1986). A single mode, cw dye laser has been used to measure isotope shifts and hyperfine splittings in Lu, as well as to demonstrate the measurement of an isotope ratio of $~10^6$ (Miller et al 1985a,b; Nogar et al 1984a). Stepwise excitation with two single-mode cw lasers was used for isotopic studies of Ba, including a determination of a selectivity of 800 for the isotope pair $^{137}$Ba:$^{138}$Ba (Cannon et al 1985). Counter-propagation of the two exciting beams was used to achieve partial cancellation of Doppler broadening. Full cancellation of Doppler broadening can be achieved by utilizing two-photon excitation steps (Clark et al 1984; Lucatorto et al 1984). Such type V RIS schemes have been demonstrated for Be, C, I, and Mg (Travis et al 1984; Clark et al 1985; Bonanno et al 1985), with Doppler-free isotope shift measurements reported for Be (Wen et al 1986; Johnson et al 1986) using a high-resolution pulsed-dye-amplifier (PDA), which achieves $~150$ MHz bandwidths by the pulsed amplification of single-mode cw dye laser radiation. Isotope shift measurements are reported in U with an etalon-narrowed flashlamp-pumped dye laser (Donohue et al 1984). Doubly-resonant, stepwise excitation studies of Pu have utilized Cu-vapor-laser-pumped dye heads operating broad-band, but with the capability for isotopic resolution (Peuser et al 1985). Fedoseyev et al (1984) place their apparatus "on-line" with an isotope separator to measure shifts of short-lived isotopes.

Several alternatives are utilized for maximizing the ionization yield of photo-excited atoms. Tunable ionization lasers may be used to exploit the superior cross section of autoionizing states in the continuum (Cannon et al 1985; Clark et al 1985; Peuser et al 1985). Laser-enhanced electron impact ionization has been employed by Bushaw et al (1986), and collisional ionization of laser-excited atoms in the glow discharge is postulated by Harrison et al (1984a,b) and Savickas et al (1984). Field ionization, preferred by Bekov et al (1985), utilizes three resonant excitation lasers to populate excited states sufficiently near the ionization limit.

## 4. Mass Discrimination

Three common mass dispersive instruments are encountered in RIMS. Quadrupole mass spectrometers (Savickas et al 1984) are reasonably efficient and inexpensive, and can be operated in a low resolution mode for simultaneous display of a number of isotopes (Cannon et al 1985). Time-of-flight mass spectrometry is natural for pulsed-laser-RIMS, inexpensive, efficient, and yields a complete mass spectrum with each laser shot (Frey et al 1985; Thonnard et al 1984). Magnetic sector mass spectrometers provide high mass resolution, with good stability and throughput. Magnetic sectors are employed on thermal-ionization-derived systems (Travis et al 1984), as well as some sputtering instruments (Donohue et al 1985a; Parks et al 1984a,b,1985a,b). For pulsed-laser RIMS, time-of-flight information can be combined with magnetic dispersion for improved mass resolution and abundance sensitivity (Fassett et al 1986).

## 5. Analytical Performance

Real sample analyses have been performed for iron in human serum and in water, using isotope dilution (Fassett et al 1984c,d), and for ruthenium in sea water (Bekov et al 1984). Fassett et al (1984c,d) quoted typical precision and accuracy of $~2$-3%. Isobaric discrimination is a key to accurate isotope ratios of Tc in the presence of Mo and Ru (Downey et al 1984) and Pu in the presence of U and Am (Peuser et al 1985; Donohue et al 1984a). Donohue et al (1984b) reported on $~2$-3% isotope ratio biases resulting from isotope shifts in U. RIMS accuracy may also be compromised by the appearance of molecular analyte species (Kroenert et al 1985) and/or the observation of anomalous (high) metastable populations (Nogar et al 1985a; Winograd 1984; Young et al 1984b). The observation of different saturated ion yields for different ionization pathways (Kimock et al 1984b) and the metastable population anomaly could both relate to photofragmentation yields of analyte molecules.

The sensitivity of RIMS is limited by  the analyte free atom fraction, ionization yield, geometrical and temporal overlap of the atom plume and laser focal volume, state distribution of the free atom population, mass spectrometer throughput, ion detection efficiency, and background noise. Fassett et al (1984b) have computed geometric distributions of atoms emitted from a filament.  Keller and Nogar (1984) estimate a 5% geometric efficiency for a typical thermal filament cw RIMS experiment, and propose gasdynamic focussing of the sample atoms to achieve a ~20-fold improvement in geometry, and associated improvement in other efficiency factors. Kimock et al (1984b) derive the spatial distribution function for ion beam sputtered atoms, and compute combined spatial/temporal overlap functions for the case of pulsed sputtering.  Pulsed laser desorption may be used for improved temporal efficiency (Beekman et al 1984; Williams et al 1984). Nogar et al (1985b) have measured the temporal profile of laser-ablated Ta and determined a temporal overlap efficiency of about 10% for optimum laser timing.  Hurst et al (1984) show theoretical calculations of transient laser heating of metal substrate surfaces and of spatial distributions for Kr atoms desorbed from such (low temperature) surfaces.  Alpha counting of $^{239}$Pu has been used to determine  an overall system efficiency of $10^{-7}$ for a 6.5 kHz Cu-vapor-pumped system with near unit temporal effeciency (Kroenert et al 1985; Peuser et al 1985).  Improvement by at least two orders of magnitude was predicted for improved overlap geometry ($10^{-4}$) and higher laser power or efficient use of autoionizing states.

Measured limits-of-detection (LODs) at the 2-ppb level in silicon using SIRIS compare favorably with the $10^{-10}$ design criterion, considering the 1-mA ion beam current (Parks et al 1985a,b).  Fairbank et al (1984,1985) have established upper limits of $10^{-10}$ for the presence of H-quarks in Nb, and $10^{-11}$ for superheavy Li atoms in Li metal.  Peuser et al (1985) predict the extension of their $10^8$-atom limit for Pu to $10^6$ atoms of Pu and other actinides.  For extreme isotope ratios, $10^6$ atoms of the lesser isotope may correspond to samples of tens-of-micrograms.  Charge exchange considerations limit the utility of large samples and high volatilization rates (Lucatorto et al 1985).

## 6. Development Directions

Several instrumental trends are driven by convenience, generality, efficiency, increased financial backing, and practical experience. Laser-pumped dye heads are replacing flashlamp-pumped systems, primarily for the ability of a single laser to pump multiple dye heads.  Resulting deficiencies in ionization yield are being addressed by an array of alternative options, including the use of high-power, fixed-frequency lasers, autoionizing levels, electron impact, collisional ionization, and field ionization. Sample utilization efficiency is being increasingly addressed by the use of cw RIS or pulsed sample vaporization.  Direct sampling of solids by laser ablation, ion beam sputtering, and glow-discharge sputtering  now seems to be practiced as widely as thermal vaporization RIMS.  The trend in mass spectrometers favors TOF, driven by the natural affinity to pulsed lasers, new high-resolution designs, relative simplicity and affordability, recent advances and descending prices  in commercial transient-digitizer/signal averager technology and the advantage of simultaneous detection of all masses.

Trends in methodology are marked by a high interest in optical isotope selectivity, theoretical modeling of relevant processes, increasing complexity and sophistication of excitation/ionization schemes, and diagnostic studies of atom yields, state distributions, matrix effects, efficiencies and fragmentation processes. Achievement of necessary performance standards is apparently on schedule for accurate, ultra-trace detection applications in solid samples and chemically-processed ecological and clinical samples in the reasonably-near future.  Though early in the development cycle as yet, isotopically selective RIMS has shown great promise in proof-of-principle demonstrations.  The relative maturity of RIMS instrumentation and methodology evidenced in this review promises increasing importance of applications-oriented research at future RIS conferences.

# REFERENCES

Andreev S V, Mishin V I and Letokhov V S 1986 Opt. Commun. 57 317
Auschwitz B and Lacmann K 1985 Chem. Phys. Lett. 113 230
Becker C H and Gillen K T 1985 J. Opt. Soc. Am. B B2 1438
Beekman D W and Callcott T A 1984 Resonance Ionization Spectroscopy 1984 ed
  G S Hurst and M G Payne (Bristol:Inst. of Phys.) pp 143-50
Bekov G I, Letokhov V S, Radaev V N, Baturin G N, Egorov A S, Kursky A N,
  and Narseyev V A 1984 Nature 312 748
Bonanno R E, Clark C W and Lucatorto T B 1986 Phys. Rev. A A34 (Sept)
Bushaw B A, Cannon B D, Gerke G K and Whitaker T J 1986 Opt. Lett. 11 422
Cannon B D, Bushaw B A and Whitaker T J 1985 J. Opt. Soc. Am. B B2 1542
Clark C W, Fassett J D, Lucatorto T B and Moore L J 1984 Resonance Ioniza-
  tion Spectroscopy 1984 ed G S Hurst and M G Payne (Bristol:Inst. of
  Phys.) pp  107-17
Clark C W, Fassett J D, Lucatorto T B, Moore L J and Smith W W 1985 J. Opt.
  Soc. Am. B B2 891
Donohue D L, Smith D H, Young J P, McKown H S and Pritchard C A 1984a Anal.
  Chem. 56 379
Donohue D L, Smith D H, Young J P and Ramsey J M 1984b Resonance Ionization
  Spectroscopy 1984 ed G S Hurst and M G Payne (Bristol:Inst. Phys.) pp 83-9
Donohue D L, Young J P and Smith D H 1984c Analytical Spectroscopy ed W S
  Lyon  (Amsterdam:Elsevier) pp 143-8
Donohue D L, Christie W H, Goeringer D E and McKown H S 1985a Proc. Int.
  Conf.  Lasers 295
Donohue D L, Christie W H, Goeringer D E and McKown H S 1985b Anal. Chem.
  57  1193
Donohue D L, Young J P and Smith D H 1985c App. Spectrosc. 39 93
Downey S W, Nogar N S and Miller C M 1984a Anal. Chem 56 827
Downey S W, Nogar N S and Miller C M 1984b Int. J. Mass Spectrom. Ion
  Processes 61 337
Fairbank W M Jr., Hurst G S, Parks J E and Paice C 1984  Resonance Ioniza-
  tion  Spectroscopy 1984 ed G S Hurst and M G Payne (Bristol:Inst. Phys.)
  pp 287-96
Fairbank W M Jr., Perger W F, Riis E, Hurst G S and Parks J E 1985 Laser
  Spectroscopy 7 (Springer Ser. Opt. Sci. 49) ed T W Hansch and Y R Shen
  (Berlin:Springer-Verlag) pp 53-4
Fassett J D, Moore L J, Shideler R W and Travis J C 1984a Anal. Chem. 56
  203-6
Fassett J D, Travis J C and Moore L J 1984b Applications of Lasers to
  Chemical   Diagnostics (Proc. SPIE 482) ed A B Harvey (Bellingham
  WA:SPIE) pp 36-43
Fassett J D, Moore L J and Travis J C 1984c Analytical Spectroscopy (Anal.
  Chem. Symp. Ser.19) ed W S Lyon (Amsterdam: Elsevier) pp 137-42
Fassett J D, Powell L J and Moore L J 1984d Anal. Chem. 56 2228-33
Fassett J D, Moore L J, Travis J C and DeVoe J R 1985 Science 230 262-7
Fassett J D, Zeininger H J and Moore L J 1986 Int. J. Mass Spectrom. Ion
  Processes 69 285
Fassett J D and Walker R J 1986, These proceedings
Fedoseyev V N, Letokhov V S, Mishin V I, Alkhazov G D, Barzakh A E, Denisov
  V P, Dernyatin A G and Ivanov V S 1984 Opt. Commun. 52 24
Frey R, Weiss G, Kaminske H and Schlag E W 1985 Z. Naturforsch. A 40A 1349-
  50
Harrison W W, Savickas P J, Hess K R and Marcus R K 1984a Resonance
  Ionization   Spectroscopy 1984 ed G S Hurst and M G Payne (Bristol:Inst.
  Phys.) pp 119-25
Harrison W W, Savickas P J, Marcus R K and Hess K R 1984b Analytical
  Spectroscopy (Anal. Chem. Symp. Ser. 19) ed W S Lyon (Amsterdam:Elsevier)
  pp  173-77
Harrison W W, Hess K R, Marcus R K and King F L 1986 Anal. Chem. 58 341A-
  56A
Hurst G S, Kramer S D, Payne M G and Young J P 1979 Rev. Mod. Phys. 51 767
Hurst G S, Payne M G, Phillips R C, Dabbs J W T and Lehmann B E 1984a J.
  Appl.  Phys. 55 1278-84
Hurst G S and Payne M G 1984b Resonance Ionization Spectroscopy 1984
  (Bristol:Inst. Phys.)

Johnson B C, Wen J, Travis J C, Fassett J D, Bonanno R E and Lucatorto T B 1986 Bull. Am. Phys. Soc. 31 946

Keller R A and Nogar N S 1984 Appl. Opt. 23 2146-51

Kimock F M, Baxter J P, Pappas D L, Kobrin P H and Winograd N 1984a Analytical Spectroscopy (Anal. Chem. Symp. Ser. 19) ed W S Lyon (Amsterdam:Elsevier) pp    179-84

Kimock F M, Baxter J P, Pappas D L, Korbin P H and Winograd N 1984b Anal. Chem. 56 2782

Kroenert U, Bonn J, Kluge H J, Ruster W, Wallmeroth K, Peuser P and Trautmann    N 1985 Appl. Phys. B B38 65

Lucatorto T B, Clark C W and Moore L J 1984 Opt. Commun. 48 406

Miller C M, Engleman R Jr. and Keller R A 1985a J. Opt. Soc. Am. B B2 1503

Miller C M, Engleman R Jr. and Keller R A 1985b Laser Spectroscopy 7 (Springer    Ser. Opt. Sci. 49) ed T W Hansch and Y R Shen (Berlin:Springer-Verlag) pp    385-6

Moore L J 1984 Stable Isotopes in Nutrition (ACS Symp. Ser. 258) ed J R Turnland and P E Johnson (Washington:Am. Chem. Soc.) pp 1-26

Moore L J, Fassett J D and Travis J C 1984 Anal. Chem. 56 2770

Moore L J, Fassett J D, Travis J C, Lucatorto T B and Clark C W 1985 J. Opt.    Soc. Am. B. B2 1561

Nogar N S, Downey S W and Miller C M 1984a Resonance Ionization Spectroscopy    1984 ed G S Hurst and M G Payne (Bristol:Inst. Phys.) pp 91-5

Nogar N S, Downey S W, Keller R A and Miller C M 1984b Analytical Spectroscopy    (Anal. Chem. Symp. Ser. 19) ed W S Lyon (Amsterdam:Elsevier) pp 155-60

Nogar N S and Keller R A 1985 Anal. Chem. 57 2992

Nogar N S, Downey S W and Miller C M 1985a Anal. Chem. 57 1144

Nogar N S, Estler R C and Miller C M 1985b Anal. Chem. 57 2441

Nogar N S, Downey S W and Miller C M 1985c Spectroscopy 1 54

Parks J E, Schmitt H W, Hurst G S and Fairbank W M Jr. 1984a Analytical Spectroscopy (ACS Symp. Ser. 19) ed W S Lyon (Amsterdam:Elsevier) pp 149-54

Parks J E, Schmitt H W, Hurst G S and Fairbank W M Jr. 1984b Resonance Ionization Spectroscopy 1984 ed G S Hurst and M G Payne (Bristol:Inst. Phys.) pp 167-74

Parks J E, Beekman D W, Schmitt H W and Taylor E H 1985a Nucl. Instrum. Methods Phys. Res. B10-11 280

Parks J E, Beekman D W, Schmitt H W and Spaar M T 1985b Applied Materials Characterization (Mater. Res. Soc. Symp. Proc. 48) ed W Katz and P Williams    (Pittsburgh: Mat. Res. Soc.) pp 309-17

Pellin M J, Young C E, Calaway W F and Gruen D M 1984 Surface Sci. 144 619

Peuser P, Hermann H, Rimke H, Sattelberger P, Trautmann N, Ruster W, Ames F, Kluge H-J, Kroenert U and Otten E-W 1985 Appl. Phys. B B38 249

Savickas P J, Hess K R, Marcus R K and Harrison W W 1984 Anal. Chem. 56 817

Schmitt H W 1985 Nucl. Sci. Eng. 90 442

Smith D H, Donohue D L and Young J P 1985 Int. J. Mass Spectrom. Ion Processes    65 287

Thonnard N, Payne M G, Wright M C and Schmitt H W 1984 Resonance Ionization Spectroscopy 1984 ed G S Hurst and M G Payne (Bristol:Inst. Phys.) pp 227-34

Travis J C, Fassett J D and Moore L J 1984 Resonance Ionization Spectroscopy    1984 ed G S Hurst and M G Payne (Bristol:Inst. Phys.) pp 97-106

Wen J, Johnson B C, Travis J C, Fassett J D, Bonanno R E and Lucatorto T B 1986 Bull. Am. Phys. Soc. 31 798

Whitaker T 1986 Lasers & Applications V (August) 67

Williams M W, Beekman D W, Swan J B and Arakawa E T 1984 Anal. Chem. 56 1348

Winograd N 1984 Resonance Ionization Spectroscopy 1984 ed G S Hurst and M G Payne (Bristol:Inst. Phys.) pp 161-6

Young J P, Donohue D L and Smith D H 1984a Resonance Ionization Spectroscopy 1984 ed G S Hurst and M G Payne (Bristol:Inst. Phys.) pp 127-33

Young J P, Donohue D L, and Smith D L 1984b Int. J. Mass Spectrom. Ion Proc.    56 307

*Inst. Phys. Conf. Ser. No. 84: Section 3*
*Paper presented at RIS 86, Swansea, Wales, 7–12 Sept. 1986*

# The laser resonance ionization spectrometer

G.I. Bekov*, Yu.A. Kudryavtsev*, I. Auterinen**, and J. Likonen**

* Institute of Spectroscopy, USSR Academy of Sciences, SU-142092 Troitsk, Moscow Region, USSR

** Technical Research Centre of Finland, Reactor Laboratory, Otakaari 3 A, SF-02150 Espoo, Finland

## 1. Introduction

The laser stepwise resonance atomic photoionization technique suggested and implemented experimentally 15 years ago (Ambartsumyan 1971) has recently come to be actively used in trace element analysis (Bekov 1983). This has been favoured by the extremely high sensitivity and selectivity of the technique in the detection of neutral atoms. The high detection sensitivity level (up to $10^{-10}$ % (Bekov 1985)) already reached by the investigators and the wide variety of the objects analyzed and the elements determined enable one to speak now of the emergence of a new substantial analysis method – the laser resonance photoionization analytical spectroscopy.

The present paper describes a laser resonance photoionization spectrometer for ultrasensitive trace element analysis designed and constructed at the Institute of Spectroscopy of the USSR Academy of Sciences and the Technical Research Centre of Finland within the framework of the LARISA joint scientific project. The spectrometer depends for its operation on the thermal evaporation and atomization in a vacuum ($p \lesssim 10^{-4}$ Pa) of the substance to be analyzed, followed by the stepwise laser resonance excitation and field ionization of the atoms released, the ions thus produced being detected by an electron multiplier after time-of-flight mass separation.

Figure 1 shows a diagram of the laser resonance photoionization analytical spectrometer. The instrument may be divided into several independent parts: a laser system, a vacuum-analytical system, and a registration system.

## 2. The Laser System

The laser system of the spectrometer depends essentially on the adopted stepwise photoionization scheme for the atoms to be detected. A most universal and easy-to-realize scheme is ionization via the Rydberg states (Bekov 1983). It is also most suitable for analytical applications. The energy fluencies for saturation of the single photon resonant transitions are low, for the first step $E_p \lesssim 10$ $\mu J/cm^2$ and for the following steps $E_p \lesssim 1$ $mJ/cm^2$ with a laser bandwidth of $\Delta\nu \approx 0.2$ $cm^{-1}$. As a result,

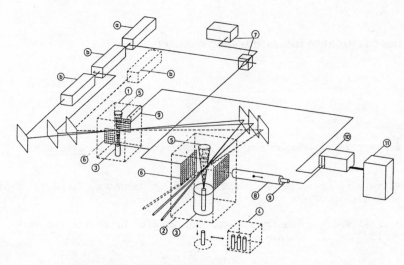

Fig. 1 General scheme of the laser resonance ionization spectrometer.
a) Excimer laser (XeCl, 308 nm).
b) Dye laser. The laser beams are normally expanded by Galilean telescopes
   and combined by dichroid mirrors.
1) Reference chamber for wave length control of the dye lasers.
2) Analytical chamber for analytical measurements.
3) Atomizer oven where a graphite crucible is heated resistively by AC
   current.
4) Transport box for the graphite crucibles.
5) Low divergent atomic/molecular beam from the narrow channel of the
   crucible.
6) Field ionization plates.
7) Spark gap system for generation of the high voltage pulses.
8) Time-of-flight tube for mass separation of the ions.
9) Electron multiplier for ion detection.
10) Registration electronics for the ion signal: gated integrators or a
    transient recorder.
11) Laboratory computer.

the nonresonant ion background inevitable in direct analytical experiments
will be two or three orders of magnitude lower than with stepwise
ionization on a transition to the continuum. The narrow bandwidths (0.2 to
0.04 cm$^{-1}$) combined with low amplified spontaneous emission (ASE)
background and moderate intensities in the unfocused laser beams provide
high elemental and in many cases even isotopic selectivity. The level of
the usually low nonselective background caused by multiphoton ionization of
atoms and molecules is measured by detuning one of the narrow laser lines
from the absorption line.

The laser system best suited to multistep resonance atomic photoionization
through the Rydberg states comprises a combination of two or three tunable
dye lasers pumped simultaneously by a single high-power UV laser (in our
case, a XeCl excimer laser, $E_p \approx 30 - 150$ mJ, $\lambda = 308$ nm, Fig. 1). The
tuning range (217 - 970 nm) of such a system covers the atomic transition
wavelengths of more than 80 % of the elements in the periodic table.

The long-term frequency and fluence stabilities of the system make it possible to attain a high absolute sensitivity of atomic photoionization detection in lengthy analytical experiments. In order to illuminate every atom in the atomic beam by the lasers the pulse repetition frequency should be over 10 kHz. However, with the excimer laser it is possible to go only up to 200 Hz, which gives a maximum duty cycle of 2 %.

## 3. The Analytical Chamber

A schematic drawing of the analytical chamber can be seen in Figure 2. The analysis of a sample proceeds as follows. For sample evaporation and atomization an electrically heated pyrolytic graphite crucible is used (see Fig. 3). The sample (solid or dissolved, weight ca. 1 – 100 mg) is introduced into the crucible and the liquid solvents and water are evaporated by gentle heating (< 100 $^0$C) under a light bulb. All this should be done in a clean-room environment to avoid contamination from dust particles of normal laboratory air, which would render the analysis of e.g. aluminium below the 1 µg level impossible (Zilliacus 1986).

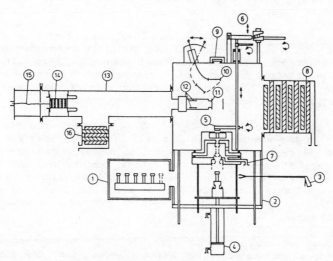

Fig. 2 Schematic drawing of the analytical chamber. Details are not to scale, the side of the cubic chamber is 23 cm.
 1) Transport box for the graphite crucibles.
 2) Overpressurized cover box for the sample exchange system.
 3) Manual manipulator for the crucibles.
 4) Pneumatic piston to move the lower electrode of the atomizer oven.
 5) Vacuum-tight valve to separate the atomizer from the ionization chamber.
 6) Manipulators for the above valve.
 7) Roughing line for the atomizer.
 8) Turbomolecular pump for the main chamber.
 9) Window for measuring the oven temperature with an optical pyrometer.
10) Beam dump for the atomic beam used for protection of the window.
11) Field ionization plates.
12) Constant field plates to steer the ions to the detector.
13) Time-of-flight tube.
14) Ion detector; 'venetian blind' -type electron multiplier.
15) Signal lead from the anode of the electron multiplier.
16) Turbomolecular pump for the time-of-flight tube.

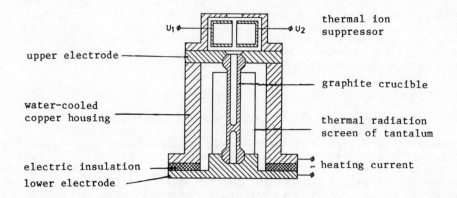

Fig. 3 The atomizer with a resistively heated pyrolytic graphite crucible (RWO675/PYC, Ringsdorff, FRG). The length of the crucible is 42 mm and the channel diameter 3 mm.

The crucibles are transported to the laser laboratory in a specially designed box which can be mounted with an air tight connection to the over-pressurized cover box under the analytical chamber. The crucible is positioned to the atomizer by means of a manipulator. A vacuum-tight valve separates the atomizer from the ionization chamber thus enabling rapid sample exchange without need of airing the main vacuum.

The crucible can be heated up to 2500 $^0$C by AC current. The heating procedure depends on the sample type. Typically it begins with gentle heating to evaporate the rest of the solvents and to outgas porous materials. Then the temperature is increased to an ashing temperature and further on stepwise to higher temperatures keeping the analytical signal on an appropriate level until the signal goes to zero. The inner wall temperature of the crucible is measured with an optical pyrometer through the window above the atomizer.

The atomic and molecular vapour from the hot, narrow channel of the crucible forms an atomic-molecular beam. Thermal ions and electrons are prevented from getting into the detection volume by a static electric field in an ion supressor over the atomizer. As a result, the residual thermal ion background at 2500 $^0$C is no more than an ion per second. The collinear laser beams (diameter 2 to 5 mm) and the atomic beam cross perpendicularly with good spatial overlap determining the ionization volume. To utilize the second-stage laser radiation more effectively and to improve the sensitivity of the analysis, in some cases a multipass system installed inside the vacuum chamber has been used.

The selectively excited atoms are ionized by an electric field pulse (up to 20 kV/cm, duration $\tau_p \approx$ 20 ns, rise time $\tau_r <$ 1 ns) generated with a laser triggered spark-gap (Fig. 1). The electric field pulse arrives along the coaxial transmission line to the electrodes some 70 ns after the laser pulses, so the excitation takes place without interference from external electric fields. The amplitude of the pulse is adjusted to be well above the critical value for the field ionization of the Rydberg-levels (Bekov 1978) $F_c = 3.21 \cdot 10^8$ $(n^*)^{-4}$ V/cm. For a typically used effective principal quantum number $n^* \approx$ 15 this gives a value of 6.3 kV/cm. The excited atoms are ionized instantaneously on the rising edge of the electric field pulse.

## 4. Ion Detection

The ions resulting from the ionization of the Rydberg atoms acquire in the electric field pulse a velocity component normal to the thermal atomic velocity in the beam. This component is two or three orders of magnitude greater than the average atomic velocity in the beam, and so the ions move almost at right angles to the atomic beam axis. Having passed through the electrode screen, the ions enter the deflection system of a time-of-flight mass separator and after being mass-separated, they are detected by the electron multiplier (venetian blind type, effective area 16x16 mm$^2$) with a near-unity efficiency. When the ion optical system has been optimally aligned the ion collection efficiency has been measured to be over 70 %. The specialty of the equal impulse acceleration is a linear time to mass dependence, in our case $\tau_{flight} \approx M \cdot 0.5$ μs (M is the mass of the ion in atomic mass units).

The time-of-flight mass separator is used above all to resolve the analyzed element from the background caused by molecular and atomic ions. Due to the mass selective ion detection it is also possible to monitor the isotopic selectivity of the ionization and to carry out isotope analysis.

## 5. The Reference Chamber

The reference vacuum chamber is of a design similar to the analytical one. The reference atomic beam produced as a result of vaporization of a weight of the pure element to be detected or of some compound of this element is used first for the exact tuning of the laser radiation to resonate with the atomic transitions selected. Later on, the ion signal from this beam is used in each measurement cycle to normalize the main analytical signal. This procedure makes it possible to eliminate the effect of the dye laser frequency departures (both fast fluctuations and slow drift) on the integral analytical signal, which is especially important in lengthy analytical experiments. Finally, the reference chamber can sometimes be used to analyze samples with high concentrations of the element.

## 6. Signal Registration

The signal from the ion detector is amplified, monitored by means of an oscilloscope, and processed with e.g. a gated integrator, transient recorder or multichannel amplitude analyzer. A minicomputer is used to normalize the analytical signal using the reference signal, to sum up the total signal from a sample, and to calculate the analytical result based on standard sample calibration.

## 7. The Vacuum System

The vacuum in the system is maintained by turbomolecular or oil diffusion pumps. For sealing rubber, mainly Viton, rings are used. The recidual pressure in the system is $10^{-4}$ Pa ($10^{-6}$ torr). During evaporation the pressure in the ionization chamber is not allowed to rise over $10^{-3}$ Pa ($10^{-5}$ torr); pumping speeds between 200 and 850 1/s are used. The pressure in the TOF-tube is kept at $10^{-4}$ Pa ($10^{-6}$ torr) by a separate pump (ca. 150 1/s) to guarantee collisionless flight of the ions and to reduce the noise in the ion detector. For an easy and secure operation of the vacuum system the valve actuations at start up, shut down and sample change are effected by a micro switch and timer-controlled relay unit.

## 8. Discussion

It should be noted in conclusion that the block arrangement of the instrument is very promising, as it enables one to improve constantly its individual units.

The applied thermal atomization in a graphite crucible in a vacuum has been proved to be feasible for the direct analysis of both liquid (Bekov 1983b) and solid (Bekov 1984 and Bekov 1985) samples without any noticeable matrix effect.

The analysis capacity of the system is some 10 samples in one hour. This, in connection with to the capability for direct analysis of many types of samples without any need for sample pretreatment makes it feasible to analyse a large number of samples in a reasonable time.

## Acknowledgements

The authors are grateful to Dr. O.N. Kompanets for his active participation in the construction of the spectrometer, to Prof. R.Salomaa for many useful discussions as well as to Profs. E. Byckling, P.Hiismäki and V.S. Letokhov for their support in this work.

The work of the Finnish LARISA-team has been supported by the Finnish Academy and the Helsinki University of Technology, Department of Technical Physics.

## References

Ambartsumyan R V, Kalinin V P, Letokhov V S 1971 Pis'ma Zh. Eksp. Teor. Fiz. (Russian) 13 305 , JETP Lett. 13 217
Bekov G I, Letokhov V S, Mishin V I 1978 JETP 27 p. 52
Bekov G I, Letokhov V S 1983a Appl.Phys.B30 161
Bekov G I, Yegorov A S, Letokhov V S, Radaev V N 1983b Nature 301 410
Bekov G I, Letokhov V S, Radaev V N, Baturin G N, Egorov A S, Kursky A N, Narseyev N A 1984 Nature 312 748
Bekov G I, Letokhov V S, Radaev V N 1985 J. Opt. Soc. Am. B2 1554
Zilliacus R, Lakomaa E-L, Auterinen I, Likonen J 1986 Proceedings of this conference

*Inst. Phys. Conf. Ser. No. 84: Section 3*
*Paper presented at RIS 86, Swansea, Wales, 7–12 Sept. 1986*

103

# High efficiency, high abundance sensitivity RIMS by double-resonance excitation with cw lasers

B A Bushaw, B D Cannon, G K Gerke and T J Whitaker

Pacific Northwest Laboratory, Richland, Washington 99352

Abstract.   Two single-frequency cw dye lasers and a medium-power infrared cw laser have produced efficient, highly selective ionization of barium isotopes in the source region of a quadrupole mass spectrometer. Laser ionization efficiencies of >5% have been observed while maintaining isotopic selectivities of ~$10^5$.

## 1. Introduction

Resonance ionization spectroscopy (RIS, Hurst 1979) has been a field dominated by pulsed dye laser excitation sources. While pulsed lasers have been remarkably successful in removing isobaric interferences and improving sensitivity, the resolution of isotopes has generally been left to mass spectrometers (RIMS). The resolution of single mode lasers can provide additional isotopic selectivity beyond that of the mass spectrometer. High-resolution pulsed dye lasers have been used (Bushaw 1981) and suggested (Lucatorto 1984) for achieving isotopic selectivity, but the transform-limited bandwidths and loss of resolution due to saturation effects may limit these systems to applications that require the generation of hard ultraviolet radiation or intensity sufficient to drive nonresonant two-photon transitions. Bound-bound one-photon transitions can usually be saturated with powers available from continuous-wave dye lasers and thus in the past few years they have received increasing attention for use in RIS. First experiments (Miller 1983) used a broad-band dye laser and concentrated upon improving sensitivity by removing duty cycle constraints. More recently, single-frequency cw dye lasers have been used in high resolution (Doppler limited, Miller 1985), and ultra-high resolution (Doppler free, Cannon 1985a) studies to provide additional isotopic selectivity in the ionization process. With the availability of frequency-doubled ring dye lasers, these cw techniques may be extended to nearly 80% of the elements, and with the efficient generation of metastable states (Cannon 1985b) may be applied to noble gases as well.

We have continued our investigations of continuous-wave double-resonance ionization mass spectroscopy (cw-DRIMS) because it offers several advantages:

1) Efficiencies may approach unity using medium-power infrared lasers for the ionization step. This may be done without significant distortion of the resonance lineshapes and the infrared radiation produces no detectable background ionization.

*This work was sponsored by the U S Department of Energy's Office of Health and Environmental Research under contract DE-AC06-76-RLO-1830.

2) Selectivity for the two resonance steps may be cascaded, improving isotopic selectivity beyond the limit of the natural Lorentzian tails that are encountered in single resonances.

3) Doppler broadening of a divergent atomic source may be substantially reduced, often to less than a few percent, by counterpropagating the two dye-laser beams.

We have used barium as a model element to test these measurement techniques for several reasons: it has a wide range of natural isotopic abundances (.1 to 70%), it has both even and odd isotopes and hence hyperfine structure for $J \neq 0$ states, it lies in the region of the periodic table where isotopic shifts are smallest (Whitaker 1986), and the spectroscopy of its Rydberg states is well characterized (Aymar 1984). Figure 1 is an energy level diagram for the neutral barium atom showing the double-resonance excitation processes that have been studied. While S, P, and D states in both the singlet and triplet manifolds have been observed, the $6snd\ ^1D_2$ Rydberg series has been examined most extensively. The states in this series exhibit the highest ionization efficiencies and larger hyperfine splittings for the 135 and 137 isotopes provide greater selectivity.

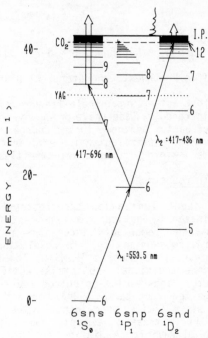

**Fig. 1** Energy levels of Barium showing double-resonance excitation pathways studied.

## 2. Experimental

The experimental apparatus used in these studies is shown in Figure 2. An effusive barium atomic beam is created by heating solid metal in an oven at 400 - 600 °C. The atomic beam is collimated to 14 mrad divergence (full angle) by an aperture at the entrance to the quadrupole mass spectrometer's crossed beam ionizer head (Extranuclear model 041-2). Within the ionizer head the atomic beam perpendicularly intersects three overlapping laser beams: two counterpropagating single-frequency dye laser beams and a medium power cw infrared beam. The first resonance ($6s^2\ ^1S_0 \rightarrow 6s6p\ ^1P_1$) at 553.5 nm is excited by a standing-wave dye laser (CR 599-21). The second resonance ($6s6p\ ^1P_1 \rightarrow$ Rydberg) is excited with a ring dye laser (CR 699-21). Both dye lasers are stabilized by fringe offset locking to a stabilized single frequency He:Ne laser. This technique has been described previously (Bushaw 1986) and both dye lasers may be locked at any frequency (or scanned) throughout the visible spectrum with a long term stability of better than 0.5 MHz/hr. The third laser, used for the photoionization step, may be either a 100 watt cw Nd:YAG or a 25 watt cw $CO_2$ laser. Results using the Nd:YAG laser to ionize doubly excited atoms have been described previously (Cannon 1985a), and this paper will concentrate upon the improvements in sensitivity and selectivity that can be achieved using a $CO_2$ laser to ionize higher lying Rydberg states.

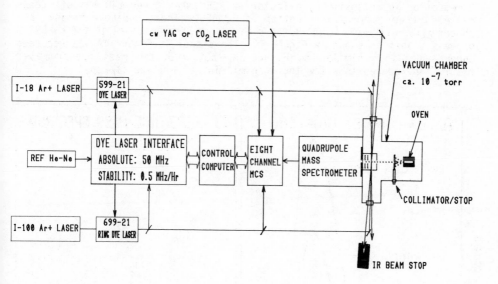

**Fig. 2** Experimental Apparatus for continuous-wave double-resonance ionization mass spectroscopy (cw-DRIMS).

## 3. Results and Discussion

In Figure 3A the mass spectrometer was tuned to mass 138 and barium was ionized via double resonance excitation through the $6s6p\ ^1P_1$ and $6s19d\ ^1D_2$ states. The first dye laser (~500 microwatts) was tuned to the Ba-138 resonance. The second dye laser (~15 milliwatts) was then scanned across the second resonance. When 1.7 watts from a cw $CO_2$ laser was used to enhance the final photoionization step (upper trace) the sensitivity improves by a factor of 30 over photoionization by the dye lasers alone (lower trace). The absolute efficiency for laser ionization is 6%, some $2.5 \times 10^3$ times greater than achieved with electron impact ionization (8.5eV, .5 mA), as shown in the reference spectrum of Figure 3B. Greater sensitivity (roughly a factor of 3) may be gained by driving the two resonance steps closer to saturation; however, the lineshape function degrades and some selectivity is lost. Another factor of three in $CO_2$ laser power could be used to advantage in driving the photoionization step to saturation, but increases beyond this will degrade lineshape due to depletion broadening of the Rydberg state. The low background of cw-DRIMS may also be seen in Figure 3A. With a 500 MHz detuning the count rate has dropped to a few counts per second (cps). With a detuning of greater than 2 GHz (.07 cm$^{-1}$) the count rate falls to less than .05 cps.

The observed lineshape function for the cw-DRIMS signal is not Lorentzian, but rather involves a convolution of the two resonance levels. The width is controlled by the shorter lived of the two states, while the longer lived controls the intensity in the tails. In Figure 3A the observed FWHM is ~30 MHz, reflecting the somewhat saturated (19 MHz natural width) first intermediate $^1P_1$ resonance level. With a Lorentzian of this width we would expect a reduction in intensity by a factor of 250 when tuned 240 MHz off resonance, yet we observe a reduction of 10,000. These sub-Lorentzian lineshapes allow extremely high isotopic abundance sensitivities in the cw-DRIMS process. This is shown in Figure 4. The signal in Figure 4A was recorded under conditions similar to Figure 3A, but with the mass spectrometer resolution reduced to allow the transmission of all the isotopes.

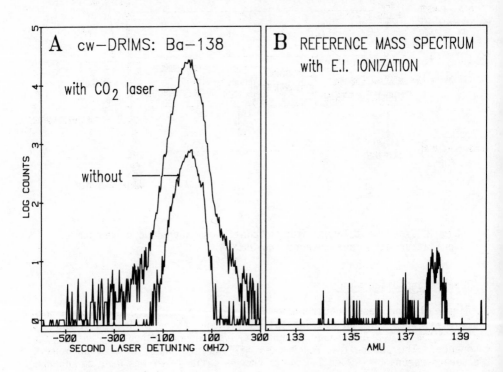

**Fig. 3** Sensitivity of cw-DRIMS. A: Signal optimized for laser ionization of 138-Ba, second dye laser scanned to determine on- and off-resonance count rates. Upper trace is with $CO_2$ laser photoionization, lower with ionization by the dye lasers. B: reference electron impact ionization mass spectrum of the atomic beam under the same conditions as A. All data was recorded with an integration period of 0.1 sec per channel.

Figure 4B shows the mass spectrum of ions produced when the lasers are tuned to resonance for Ba-137 via the 6s6p $^1P_1$ F=5/2 hyperfine and 6s15d $^1D_2$ F=7/2 hyperfine intermediate levels (the peak labeled 137-7/2 in A). The peak at mass 137 is $10^4$ greater than that at mass 138. When relative isotopic abundances are considered, this means that the laser ionization process is ionizing Ba-137 $8\times10^4$ times more efficiently than Ba-138.

Similarly, the selectivity for 137 with respect to 135 is $1.2 \times 10^3$. The other (even) isotopes, which are not observed experimentally, should be suppressed by an amount similar to Ba-138 as the spectral shifts are similar. The structureless baseline of a few counts is not due to laser ionization at these masses, but rather demonstrates the baseline limitations of quadrupole mass spectrometry: the baseline is due to laser ionized Ba-137 atoms which have undergone collisions (gas and/or wall) during their transit through the quadrupole filter.

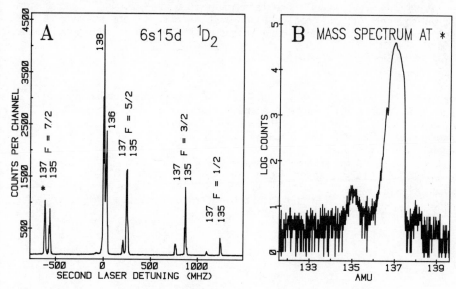

**Fig 4**. Selectivity of cw-DRIMS. A. Spectrum of the 6s15d $^1D_2$ Rydberg state achieved by scanning the second dye laser showing all isotopes (no mass resolution). B. Mass spectrum of the ions produced when both lasers are fixed at the double resonance frequencies for the first peak in A (Ba-137, F=5/2 first resonance, F=7/2 second resonance).

## 4. Conclusions

We have demonstrated that high resolution cw-dye lasers may be used to ionize specific isotopes with high abundance sensitivity. For barium, with small isotopic splittings, selectivities of almost $10^5$ have been demonstrated. For other elements having larger splittings this should improve even further. By using a modest power ($\sim 100$ W/cm$^2$) infrared laser for the photoionization step ionization efficiencies of greater than 5% have been observed without degrading the selectivity of the double resonance excitation to the Rydberg level. Background ionization is essentially zero and when combined with the high efficiency leads to extremely low detection limits: our background equivalent count rate corresponds to an atomic beam flux of 4 atoms per second or a density of 8 atoms per liter.

For atomic spectroscopy cw-DRIMS is an extremely powerful tool. The sensitivity has allowed the study of weak transitions such as $^1P_1 \rightarrow {}^3P_{0,1,2}$ in the second resonance. The sensitivity combined with the mass discrimination allows the recording of "clean" optical spectra of individual isotopes even though they may be minor components of a complex mixture. The choice of resonance enhancement in the first (known) transition also assists in making upper level hyperfine assignments. Oscillator strengths may be determined for either of the resonance transitions by running one dye laser above saturation intensity and scanning the second laser as a weak probe to determine the induced AC Stark splitting. When the dye lasers have equal offsetting detunings to produce a nonresonant simultaneous two-photon excitation, the lineshape is nearly a pure Lorentzian and transforms as the lifetime of the upper state.

Our future work will apply these techniques to close-coupled filament atomization sources for the analysis of small discrete samples and to test Doppler reduction in the double-resonance excitation for a divergent source.

## 5. References

Aymar M  1984  Physics Reports 110 163, and references cited therein
Bushaw B A, Cannon B D, Gerke G K and Whitaker T J 1986 Opt. Lett. 11 422
Bushaw B A and Whitaker T J 1981 J. Chem. Phys. 74 11
Cannon B D, Bushaw B A and Whitaker T J 1985a J. Opt. Soc. Am. B 2 1542
Cannon B D and Whitaker T J 1985b Appl. Phys. B 38 57
Hurst G S, Payne M G, Kramer S D and Young J P 1979 Rev. Mod. Phys. 51 767
Lucatorto T B, Clark C W and Moore L J 1984 Opt. Comm. 48 406
Miller C M and Nogar N S 1983 Anal. Chem. 55 481
Miller C M, Engleman R Jr and Keller R A 1985 J. Opt. Soc. Am. B 2 1503
Whitaker T J 1986 Lasers and Applications 5 67

*Inst. Phys. Conf. Ser. No. 84: Section 3*
*Paper presented at RIS 86, Swansea, Wales, 7–12 Sept. 1986*

# Resonance ionization mass spectrometry at Los Alamos National Laboratory

C. M. Miller, N. S. Nogar, E. C. Apel, and S. W. Downey

Los Alamos National Laboratory, Los Alamos, New Mexico 87545

## 1. Introduction

Numerous analytical problems at Los Alamos require the ability to measure mass spectrometrically large isotope ratios with small samples. Such analyses are often plagued by the presence of isobaric interferences, especially with low abundance isotopes. Although extensive chemical efforts, both in sample preparation and in the source of the mass spectrometer, have been able to successfully eliminate these isobaric interferences in some instances, there are still many cases in which this treatment is not adequate. Resonance ionization mass spectrometry (RIMS) is another approach which we have pursued for such highly selective and sensitive analyses.

Two approaches to RIMS will be discussed here. The first is the use of continuous-wave (cw) dye lasers as the ionization source. Continuous excitation effectively matches the ionization to the evaporation of sample from a resistively heated filament, providing for efficient use of the sample. In addition, cw ionization permits pulse counting detection, necessary for the measurement of large isotope ratios on small samples. In the cases of lutetium and technetium discussed below, average ionization rates with cw ionization are much larger than with pulsed ionization.

The second approach to RIMS is the use of multiphoton resonances in the pulsed laser excitation of atoms. The potential advantages derived from the use of n-photon resonances ($n \geq 2$) include minimal laser hardware, since the fundamental output of a single dye laser is sufficient to effect ionization in a large number of elements. The possibility also exists for Doppler-free excitation, which could be used to increase the selectivity of the ionization process. Experiments with 2+1 (photons to resonance plus photons to ionize) RIMS schemes for several elements will be discussed.

## 2. Experimental

Analyses of Lu and Tc have been discussed previously (Nogar et al 1984, Downey et al 1984). Briefly, experiments were performed in a magnetic sector mass spectrometer, equipped with both current integrating and pulse counting electronics. Lu samples were prepared by evaporating Lu solutions onto Re source filaments along with uranium binder. Resonance ionization was accomplished using a simple 1+1 scheme at 452 nm ($^2D_{3/2}$ $5d6s^2$ $\longrightarrow$ $^2D^\circ_{3/2}$ $5d6s6p$). Technetium samples were prepared by electroplating Tc from solution onto Re filaments. The Tc ionization scheme utilized two tunable dye lasers for a 1+1+1 scheme. The first laser excited the $^6S_{5/2}$ $4d^55s^2$ $\longrightarrow$ $^6P^\circ_{7/2}$ $4d^55s5p$ transition at 430 nm, while the second excited the $^6P^\circ_{7/2}$ $4d^55s5p$ $\longrightarrow$ $^6S_{5/2}$ $4d^55s6s$ transition at 612 nm. From the last state, absorption of an additional photon at 430 nm caused ionization. Argon-ion laser pumped cw dye lasers provided the appropriate wavelengths in these cases, with powers of 300 to 500 mW at a bandwidth ~1.3 cm$^{-1}$). A simple optical train directed the laser outputs into the source of the mass spectrometer, focused to beam diameters estimated to be 100 to 200 $\mu$m. In the case of Tc, the two colors were combined on a dichroic mirror; alignment of the two beams was observed to be critical to the production of ions.

The apparatus for pulsed laser excitation has also been described before (Nogar et al 1985, Downey et al 1984). Briefly, samples were prepared by drying solutions onto metal filaments which were inserted into the source region of a time-of-flight mass spectrometer equipped with pulse counting electronics. The requisite wavelengths were generated by an XeCl excimer laser pumped dye laser, with a bandwidth of ~0.5 cm$^{-1}$ and energies of 0.1 to 4 mJ, depending on wavelength. The diameter of the laser beam in the source region of the spectrometer was 0.5 to 1.2 mm.

3.  Results and Discussion

Lutetium. Resonance ionization of Lu demonstrates the applicability of RIMS for the measurement of large isotope ratios with small samples. A $^{173}$Lu/$^{175}$Lu isotope ratio measurement of 4.4 x 10$^{-7}$ was accomplished using a sample of only 60 ng, in the presence of orders of magnitude more isobaric interference from Yb (Nogar et al 1984). This dynamic range approaches our analytical needs for the Lu analysis. Efforts are underway to extend the sensitivity and dynamic range possible with RIMS, including isotopically-selective ionization, the use of autoionization structure, and intracavity cw dye laser excitation.

Technetium. Two-color cw resonance ionization of Tc was sucessfully demonstrated on samples as small as 10 ng. The estimated ionization efficiency, however, was low (~10$^{-8}$), consistent with previous observations (Downey et al 1984). This suggests the continued presence of a non-ionizing molecular species evaporating from the source filament. Assuming 1 part in 10$^3$ of the Tc leaves the filament in an analyzable atomic form (Downey et al 1984) and a 1% geometric overlap with the laser, one can estimate an ionization efficiency for atoms interacting with the lasers of ~10$^{-3}$. This reasonably high value is in agreement with a rate equations calculation and indicates the basic applicability of the RIMS scheme, provided that the atomization problems can be overcome. Saturation studies (see Figure 1) showed a strong

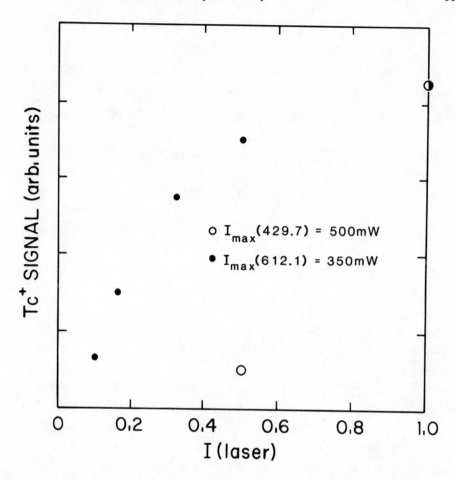

Fig. 1. Saturation curve for the two-color ionization of Tc. Note the strong power dependence of the 430 nm laser, which is used for both a bound-bound and a bound-free transition, and the evidence of saturation at high power levels of the 612 nm laser, which promotes only a bound-bound transition.

power dependence of the 420 nm laser, and a significantly smaller dependence at 612 nm. As anticipated, this reflects the necessity for two photons at 420 nm, one for the ionization step with a substantially lower cross section, whereas the bound-bound transition at 620 nm can be essentially saturated.

Multiphoton Resonances. A summary of observed elements and transitions for pulsed laser 2+1 ionization is shown in Table 1. Ionization potentials range from 6.1 to 9.2 eV. In most cases, useful transitions were predictable from published energy level tables (Moore 1949-1958). Usually, only resonant transitions

Table 1

| Element | Ionization Potential (eV) | Ground State Configuration | Term | Excited State Configuration | Term | Energy Difference[a] (cm$^{-1}$) |
|---------|---------------------------|----------------------------|------|------------------------------|------|-----------------------------------|
| Au | 9.2 | (5d$^{10}$6s) | $^2S_{1/2}$ | (5d$^{10}$7s) | $^2S_{1/2}$ | 54485 |
| Bi | 7.3 | (6p$^3$) | $^4S_{3/2}°$ | ((6p$^2$)$^3$P$_0$7p) | J=3/2 | 41125 |
| Ca | 6.1 | (4s$^2$) | $^1S_0$ | (4s4d) | $^1D_2$ | 37298 |
| Cu | 7.7 | (3d$^{10}$4s) | $^2S_{1/2}$ | (3d$^{10}$5s) | $^2S_{1/2}$ | 43137 |
| Mg | 7.6 | (3s$^2$) | $^1S_0$ | (3s6s) | $^1S_0$ | 56187 |
|    |     | (3s$^2$) | $^1S_0$ | (3s5d) | $^1D_0$ | 56308 |
| Ta | 7.9 | (5d$^3$6s$^2$) | $^4F_{3/2}$ | (5d$^3$6s7s) | J=3/2 | 44096 |
|    |     | (5d$^3$6s$^2$) | $^4F_{3/2}$ | (5d$^3$6s7s) | $^4F_{3/2}$ | 43964 |
| Y | 6.4 | (4d5s$^2$) | $^2D_{3/2}$ | (4d5s6s) | $^2D_{3/2}$ | 36421 |
|   |     | (4d5s$^2$) | $^2D_{3/2}$ | (4d5s6s) | $^2D_{5/2}$ | 36431[b] |
|   |     | (4d5s$^2$) | $^2D_{3/2}$ | (4d5s6s) | $^2D_{5/2}$ | 35901 |
| Zr | 6.8 | (4d5s$^2$) | $^3F_2$ | (4d$^2$5s6s) | $^3F_2$ | 37460 |

[a]From Moore 1949-1958.
[b]Low intensity transition.

indicative of the free atom were observed, although in some instances (Au, Y) transitions attributable to molecular ions and atoms produced from a molecular precursor were also observed. In cases where more than one transition was observed, or predicted to be observable, spin-allowed transitions were usually seen to be more intense than were spin-forbidden transitions, in agreement with observations for single-photon processes.

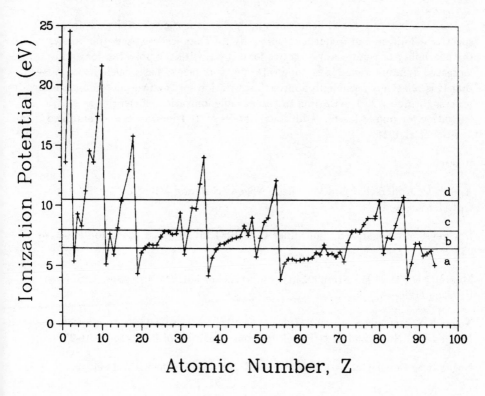

Fig. 2. Ionization potentials and simple resonance ionization schemes for the elements. Those elements falling in region "a" can be ionized with two photons from a single dye laser. Those in region "b" may require the addition of "pump" laser photons to achieve ionization. Region "c" may require frequency doubling or the utilization of n-photon resonances, while region "d" elements will usually require special techniques for ionization.

Several general observations can be made regarding the utility of the 2+1 ionization process for analytical RIMS. First, the use of three photons allows a relatively broad coverage using a single dye laser (see Figure 2). For example, using the dye coumarin 540, with a tuning range of 525 to 575 nm, atoms with ionization potentials up to 7.3 eV can be ionized. In addition, the energy range of accessible intermediate states doubles, relative to 1+1 processes, for a single dye. The use of dyes operating further to the blue and in the ultraviolet allows ionization , without frequency doubling, of atoms with ionization potentials up to 11.4 eV. Second, the resonant excitation process is not dramatically oversaturated, as is often the case with one photon resonant excitation.

This allows better spectral definition and higher isobaric selectivity. Third, the spectral definition and frequent opportunity for Doppler-free excitation allows the possibility of isotopically-selective ionization, with the potential for greatly increased dynamic range in isotope ratio measurements. Last, calculations show that it is relatively feasible to approach saturation for the two-photon absorption process, bringing 2+1 ionization to comparable ionization efficiencies as 1+1 ionization for pulsed lasers. Additional details of 2+1 ionization are contained in Apel et al 1986.

## References

Apel E C, Anderson J E, Estler R C, Nogqar N S, and Miller C M 1986 App. Opt. (submitted)

Downey S W, Nogar N S, and Miller C M 1984 Int. J. Mass Spec. Ion Proc. 61 337

Moore C E 1949-1958 Atomic Energy Levels, Vol I-III (Washington: U.S. Govt. Printing Office)

Nogar N S, Downey S W, and Miller C M 1984 Resonance Ionization Spectrscopy 1984 ed G S Hurst and M G Payne (Boston: Institute of Physics) pp 91-95

Nogar N S, Downey S W, and Miller C M 1985 Anal. Chem. 57 1144.

*Inst. Phys. Conf. Ser. No. 84: Section 3*
*Paper presented at RIS 86, Swansea, Wales, 7–12 Sept. 1986*

115

# Ultratrace elemental and isotopic analysis of osmium and rhenium using resonance ionization mass spectrometry and thermal vaporization

J. D. Fassett and R. J. Walker

Center for Analytical Chemistry, National Bureau of Standards, Gaithersburg, MD 20899

Abstract. Aspects of atomization and optical spectroscopy are described for resonance ionization of the elements osmium and rhenium. Picogram sensitivities have been developed with 1-5 % precisions and accuracies in isotopic ratio measurement.

## 1. Introduction

Resonance ionization spectroscopy coupled with mass spectrometry (RIMS), has held considerable promise in analytical chemistry for several years (Fassett et al., 1985). There have been many demonstrations of sensitive and selective measurement, but the application of the technique to solve an important measurement problem has yet to be accomplished. Significant progress is being made on the application of RIMS/RIS to a number of exciting problems, many examples of which have been discussed at this and preceding RIS conferences. The analytical difficulties have resided in sampling and quantitation, and the demand for high selectivity to achieve sensitive measurement from complex matrices.

We report on the development of a RIMS procedure for the isotopic analysis of Os and Re. The procedure relies on the efficient chemical separation of these elements from the matrix. Quantitation is achieved by isotope dilution mass spectrometry. This procedure will be applied to a problem of extreme interest in isotope geology. By studying the ingrowth of $^{187}$Os from the decay of $^{187}$Re (half-life: 4.35 x 10$^{10}$ years) in geologic materials, we hope to elucidate a number of unsettled issues in the chemical evolution of the earth and solar system.

The Os/Re problem is being attacked by a diverse array of mass spectrometric techniques besides RIMS. Secondary ion mass spectrometry (ion probe) has been the leading technique in the isotopic analysis of Os/Re for many years (Allegre and Luck, 1980; Luck and Allegre, 1982, 1984). More recently, the laser microprobe mass analyzer (LAMMA) (Englert and Herpers, 1980; Simons, 1983), inductively-coupled plasma mass spectrometer (ICP-MS) (Lindner, et al., 1986), and tandem accelerator mass spectrometer (TAMS) (Fehn, et al., 1986) have been applied to Os/Re measurement. Whether any of these techniques will have the sensitivity and selectivity to determine the sub-part-per-billion concentration of Re and Os in most geologic materials remains to be seen. We feel that RIMS has a natural advantage over these other techniques because of its selectivity and potential sensitivity. We hope that with experience this advantage will be overwhelming.

We have demonstrated the use of our RIMS system to measure picogram quantities of Os/Re with precisions and accuracies of 1-5 % (Walker and Fassett, 1986). Here, we detail the steps taken to develop this highly sensitive procedure and examine the systematics of atomization and optical spectroscopy.

## 2. Experimental

The basic RIMS system has been described (Fassett et al., 1983a,b). In summary, a 10-Hz, Nd:YAG-pumped, tunable dye laser is frequency doubled, producing 1-3 mJ of UV radiation. We have studied the wavelength ranges 278-284 nm and 297-302 nm in this experiment. The laser radiation is directed through the source of a magnetic sector mass spectrometer, ionizing atoms vaporized from a hot filament. A thermal pulser (Fassett et al., 1984a) was used for most Os measurements. An electron multiplier and transient digitizer detection and quantification system is new and has been recently described (Fassett et al., 1986).

## 3. Results and Discussion

The RIMS procedure for Os evolved from these elemental properties:
1. Os is readily oxidized and the oxide is volatile.
2. Os metal is extremely refractory (m.p. $> 3,000^\circ$ C).
3. Os ionization potential is high, $70,450$ cm$^{-1}$ (8.7 eV).
4. Os has 6 energy levels below $10,000$ cm$^{-1}$.

In contrast, the RIMS procedure for Re is much more straightforward: the ionization potential is less, the chemistry more controllable, and there are no energy levels between the ground state and $10,000$ cm$^{-1}$.

### 3.1.1 Atomization

Thermal vaporization, as used in thermal ionization mass spectrometry, has provided the basis for development and utilization of resistively heated filament sources as atom reservoirs for resonance ionization (Travis, et al., 1984). Because of its high ionization potential and ready oxidation, no successful thermal ionization procedure for Os has ever been developed. Conversely, the direct elemental atomization of Os from a filament is highly efficient. The sample loading procedure developed for Os consists of drying osmium chloride solution onto anion exchange resin beads which are then placed on a filament, fixed with collodion, and covered with graphite. This procedure results in the ultimate reduction of Os when slowly heated in nitrogen and in the vacuum of the mass spectrometer.

Osmium is an element that illustrates the importance of sample loading on atomization, in a manner similar to what has been discussed before (Fassett et al., 1984b) for vanadium. We observed in initial experiments that there existed conditions of low temperature ($<700^\circ$C) over which resonance ionization signals would occur. In these cases, there existed large off-resonance backgrounds. We believe that if reduction of Os fails, Os is released as either the tetraoxide or a chloride molecular species, which is then dissociated and ionized by the laser. In contrast to V, where a laser-produced VO ion signal was very large, we did not observe a molecular ion peak. We did not attempt to develop and use the low-temperature ionization behavior of osmium in an analytical procedure, since the ionization was highly uncontrolled and dependent on both the vacuum in the mass spectrometer and the unmeasurable temperature of the filament. More importantly, selectivity was also reduced. However, future investigation of this low-temperature behavior could be both valuable and enlightening.

When properly loaded, osmium was emitted from tantalum filaments at temperatures in excess of 1900° C. We have thermally pulsed the filaments for osmium atomization. This pulsing was done not only to better match the duty cycles of atomization and ionization (Fassett et al., 1984a) but also to prolong filament life. At the temperatures required for Os atomization, the filament lifetime is limited. Pulsing the miniature filament to 3000°C from a baseline of 2000°C, ten times a second allows the filament to survive for many hours.

### 3.1.2 Resonance Ionization Spectroscopy

The general basis of our RIMS instrument is the use of a simple, one-color, two-photon (UV) scheme where the first photon absorbed populates an excited state of the element and absorption of the second photon promotes ionization. Thus, the resonant level must be more than half way to the ionization potential. For Os with an ionization potential of $70,450 \text{ cm}^{-1}$, this represents a level greater than $35,225 \text{ cm}^{-1}$ or less than 283.9 nm. The first allowed transition is at $35,616 \text{ cm}^{-1}$ or 280.8 nm which is readily accessed using the dye R6G. For Re, with an ionization potential of $63,530 \text{ cm}^{-1}$, the resonant level must be greater than $31,765 \text{ cm}^{-1}$ or less than 314.8 nm. In the R6G dye range, the transition at $35,267.9 \text{ cm}^{-1}$, or 283.5 nm, is appropriate.

Os has both a high vaporization temperature and relatively low-lying energy levels that are predictably, significantly populated (Fassett et al., 1983b). The calculated populations of the 7 lowest lying levels are summarized in Table I. Also included in Table I are transitions observed from these levels in the 278–284 nm range of R6G.

TABLE I.  Metastable Excited State Populations of Osmium.

| Level cm-1 | Populations 2000°C | Populations 3000°C | Transitions Observed, 278–284nm # | Rel. Int. |
|---|---|---|---|---|
| 0 | 0.82 | 0.66 | 1 | 10 |
| 2740 | 0.08 | 0.11 | 5 | 3.3–0.6 |
| 4159 | 0.045 | 0.08 | 4 | 1.4–0.05 |
| 5144 | 0.039 | 0.08 | 1 | 0.2 |
| 5766 | 0.007 | 0.018 | | |
| 6093 | 0.0019 | 0.005 | | |
| 8743 | 0.0032 | 0.014 | | |

These thermally-populated low-lying levels can be used in resonance ionization schemes and, thus, wavelengths above 280.8 nm exhibit two-photon resonance ionization. We scanned the wavelength range of the combination of dyes R610/R640, 296–302 nm, and observed a significant number of resonant ionization transitions (Figure 1). The relative intensities and assignments are summarized in Table II.

We note that the three most intense transitions are simple 2-photon transitions which originate from the $4159 \text{ cm}^{-1}$ level. The observed transition which originated from the 2740 level does not have enough energy to reach the ionization potential by a simple 2-photon process alone, yet does have significant probability for ionization. We also note lines originating from transitions that are a factor of 1000 less populated than the ground state. These transitions are expected, as we have pointed out previously for Re (Fassett et al., 1983b), and point to the sensitivity of the technique. The importance of carefully evaluating the selectivity of the resonance ionization spectroscopy by examining the experimental spectra cannot be overestimated. The most intense line at 297.2 nm is comparable to the ground-state-originating transition at 280.8 nm in the R6G dye range. Re conveniently has a ground-state-originating transition at 297.7 nm, and thus, both dye systems can be used to accomplish RIMS of Os/Re.

TABLE II.   Osmium Resonant Ionization Lines Observed, 296–302 nm.

| Line nm | Initial | Energy, cm-1 Resonant | Two-Photon | Rel.Int. |
|---|---|---|---|---|
| 297.2 | 4159 | 37808 | 71457 | 10 |
| 296.2 | 4159 | 37922 | 71684 | 1.7 |
| 296.3 | 4159 | 37908 | 71658 | 0.93 |
| 296.5 | 5766 | 39494 | 73221 | 0.44 |
| 302.0 | 5766 | 38876 | 71986 | 0.26 |
| 298.3 | 11378 | 44893 | 78408 | 0.18 |
| 301.4 | 2740 | 35920 | 69099 | 0.11 |
| 297.8 | 8743 | 42317 | 75891 | 0.07 |
|  |  | 42310 | 75877 |  |
| 299.8 | 10166 | 43516 | 76866 | 0.04 |
| 300.4 | 11378 | 44663 | 77948 | 0.02 |

When one has the flexibility for using a particular analytical wavelength in resonance ionization, the choice often is made by secondary considerations. We have done the majority of work for Os/Re using the 297.2 nm wavelength because other experiments were being done on our RIMS instrument which required this dye range. An important secondary consideration exists, however, which results from the hydrocarbon background in the mass spectrometer. This laser-produced hydrocarbon background varies from mass to mass and is not sharply wavelength specific. However, there is approximately a factor of 3 lower background at 298 nm than 283 nm under the conditions of these experiments. This lower background favors low-level ion counting for Os using the lower energy photons at 297.2 nm.

The origin of the hydrocarbon background is open to speculation, whether introduced with the sample or inherent to the mass spectrometer (e.g. pump oil). We do note that our loading procedure introduces a large amount of carbon-containing species, resin beads, collodion, and graphite, and that the severe temperature used promotes the outgassing and dispersal of these materials in the source of the mass spectrometer. We note that the absolute magnitude of this background is small, only a few counts per minute, and is only a problem that must be coped with in the most sensitive measurements.

The fact that the hydrocarbon background is not wavelength dependent is to our advantage. The wavelength specificity of the laser ionization can be used to measure the on-mass signal of the hydrocarbon, off-resonance from the element of interest. This contribution to the Re and Os signals can be subtracted to provide an accurate, corrected, elemental isotopic ratio.

We have investigated the effect of focusing our laser on the selectivity and sensitivity of Os resonance ionization. We have noted previously that, in some cases, both sensitivity and selectivity is reduced when the laser is focused (e.g. vanadium (Fassett et al., 1984b) and iron (Fassett et al., 1984c)). When the laser is focused, the geometric overlap of the laser with the atom reservoir is reduced, which can more than compensate for increased ionization efficiency. Furthermore, we have observed severe broadening of resonant transitions and large off-resonance backgrounds, which presumably arise from direct 2-photon ionization. We also scanned the region between 296-302 nm for Os with the laser focused, under exactly the same conditions as with the laser unfocused (Figure 1). Besides the increase in the non-resonance background, there appeared a very large number of discrete transitions (31) in this wavelength range. The unfocused, optimum wavelength at 297.2 nm is within a factor of two as intense as 12 of these lines. The most intense, optimum wavelengths when focused are 299.35 and 299.2 nm.

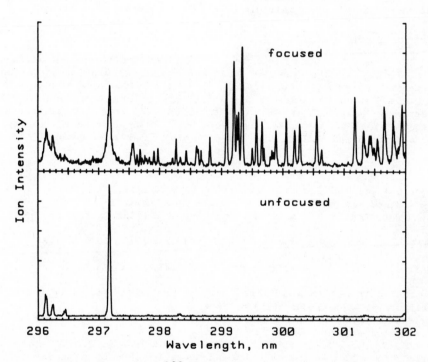

Figure 1. RIMS spectra of $^{192}$Os taken at 2000°C, using a focused and unfocused laser beam with all other conditions the same. A 25-cm lens was used to focus the laser beam (unfocused diameter, roughly 3 mm).

The origin of these lines which appear when the laser is focused is difficult to explain; however, the electronic structure of Os between 60,000 cm$^{-1}$ and the ionization limit has not been elucidated. A three-photon scheme, 2-photon resonance plus one photon for ionization, originating from the ground state is hypothesized. The two-photon transistion would access an untabulated level in this region below the ionization potential. This scheme has been used to ionize Be, C, and I in our laboratory, and can be highly efficient (Moore et al., 1985). Whereas the two-photon transition at 297.2 nm is probably saturated, these hypothesized three-photon transitions are probably not, and thus could be enhanced with increasing laser power. We have not made any laser power studies of the spectrum for Os, but have noted increased ratios (up to 3:1) of the intensities at 299.35 nm vs 297.2 nm with presumably higher laser intensities relative to the spectrum of Figure 1. There exist tabulated lines of Os at 299.21 and 299.36 nm in the optical wavelength tables for Os. However, these lines have lower levels at 13,365 and 12,774 cm$^{-1}$, respectively. These levels would have to become populated by some unknown optical process to explain their relatively high intensities. There are occasions when sensitivity and selectivity can be traded off in an isotope ratio measurement program. For instance, in the calibration of spike materials, where one deals with pure and large amounts of elemental material, selectivity is not required and one can use the increased sensitivity to make the highest precision measurement. In low-level, sample-limited measurements, however, where isobaric interferences, elemental and molecular, are a limiting source of error, it is vital to maintain both selectivity and sensitivity.

4. Analytical Applications

The RIMS procedures described here have been applied to the determination of blanks, loading and chemical, and to the calibration of spike and standard materials. The 192/190 and 187/186 ratios have been determined on sub-nanogram amounts of Os with precisions of 1.3 and 3.2 %, respectively (N=7) (Walker and Fassett, 1986). The chemical procedures for separating both Os and Re from rocks with high efficiencies is now being optimized and initial analyses of geologic materials begun.

REFERENCES

Allegre C J and Luck J-M 1980 Earth Planet. Sci. Lett. 48 148
Englert P and Herpers U 1980 Inorg. Nucl. Chem. Lett. 16 37
Fassett J D, Travis J C, Moore L J and Lytle F E 1983a Anal. Chem. 55 765
Fassett J D, Moore L J, Travis J C and Lytle F E 1983b Int. J. Mass Spectrom. Ion Proc. 54 201
Fassett J D, Moore L J, Shideler R W and Travis J C 1984a Anal. Chem. 56 203
Fassett J D, Travis J C and Moore L J 1984b Proc. SPIE 482 ed A B Harvey (Bellingham, Washington: SPIE) pp 36-43
Fassett J D, Moore L J and Travis J C 1984c Analytical Spectroscopy ed W S Lyon (Amsterdam: Elsevier) pp 137-42
Fassett J D, Moore L J, Travis J C and DeVoe J R 1985 Science 230 262
Fassett J D, Walker R J, Travis J C and Ruegg F C 1986 in preparation
Fehn U, Teng R, Elmore D and Kubik P W 1986 submitted for publication
Lindner M, Leich D A, Borg R J, Russ G P, Bazan J M, Simons D S and Date A R 1986 Nature 320 246
Luck J-M and Allegre C J 1982 Earth Planet. Sci. Lett. 61 291
Luck J-M and Allegre C J 1984 Earth Planet. Sci. Lett. 68 205
Moore L J, Fassett J D, Travis J C, Lucatorto T B and Clark C W 1985 J. Opt. Soc. Am. B, 2 1561
Simons D S 1983 Int. J. Mass Spectrom. Ion Proc. 55 15
Travis J C, Fassett J D and Moore L J 1984 Resonance Ionization Spectroscopy 1984 ed G S Hurst and M G Payne (Boston: Institute of Physics) pp 99-106
Walker R J and Fassett J D 1986 submitted for publication

*Inst. Phys. Conf. Ser. No. 84: Section 3*
*Paper presented at RIS 86, Swansea, Wales, 7–12 Sept. 1986*

121

# Resonance-enhanced multi-photon ionization mass spectrometry and unimolecular ion decay kinetics

H. Kühlewind, A. Kiermeier, H.J. Neusser, E.W. Schlag

Institut für Physikalische und Theoretische Chemie
der Technischen Universität München
Lichtenbergstr. 4, D-8046 Garching, Germany

## 1. Introduction

During the last decade it has been demonstrated that multi-photon ionization (MPI) can be used as the basis of a novel type of ion sources for mass
spectrometry (Boesl et al. 1978, Boesl et al. 1980, Zandee et al. 1978,
Antonov et al. 1978, Schlag and Neusser 1983). This technique has particular features not present in conventional ion sources like electron impact
etc. For example, at low light intensities soft ionization is possible
(Boesl et al. 1978), whereas at high light intensity a rich and highly
structure-specific fragmentation pattern is obtained (Zandee et al. 1978,
Boesl et al. 1980, Kühlewind et al. 1985a). In previous work (Boesl et al.
1980) we have shown that the mechanism responsible for soft ionization as
well as hard fragmentation is "ladder switching" from the neutral molecules
to the parent ions and then to fragment ions of decreasing size. Ladder
switching causes highly structure specific fragmentation even for those
isomeric cations which cannot be distinguished by electron impact and other
conventional ion sources (Kühlewind et al. 1985a). Another virtue of resonance-enhanced MPI is its high efficiency which leads to nearly complete
ionization of all molecules within the laser focus (Boesl et al. 1981a), a
precondition for sensitive trace analysis.

Since the intermediate state spectrum is sharp even in many polyatomic molecules resonance-enhanced MPI is extremely selective. It does not only
allow the selection of particular molecules from a mixture (Boesl et al.
1981b) – a feature of great practical interest for mixture and trace analysis – but it also permits the selection of molecules in a particular vibrational state, even though many states might be thermally populated. In this
paper it will be shown that resonance-enhanced two-photon ionization leads
to the production of vibrational state-selected polyatomic molecular ions.
State- and energy-selected ions are of particular interest for the precise
investigation of dissociation kinetics.

## 2. Photoelectron energy analysis

Even though the excitation of the resonant intermediate state guarantees
state-selection, a photoelectron energy analysis shows that the state
selection is not automatically maintained in the second absorption step
to the ionization continuum (Long et al. 1983). This is due to the excess
energy above the ionization potential present in a one-laser experiment
when the laser frequency is determined by the first resonant absorption

step to the intermediate state. Excess energy can distribute either to kinetic energy of the electrons or internal energy of the ions. In order to check, whether state-selected ions are produced, an analysis of photoelectron energy was performed with a home-built time-of-flight photoelectron analyzer (Kiermeier et al., to be published). In Fig. 1a the resulting photoelectron spectra of three benzene isotopes $C_6H_6^+$, $C_6D_5H^+$ and $C_6D_6^+$ are shown for ionization via the $6^0_1$ transition at $37481.6$ $cm^{-1}$, $37676.1$ $cm^{-1}$ and $37712.0$ $cm^{-1}$, respectively, which leads to the vibrationless $S_1$ state as an intermediate state. The energy scale represents the ion internal energy. It is seen that for the chosen intermediate state more than 90 % of the benzene ions are produced in a single vibrational state, i.e., the vibrationless ground state. Alternatively, when the laser frequency is tuned to the $6^0_1 16^1_1$ sequence band the $16^1$ vibrational state acts as an intermediate state in $S_1$. The photoelectron spectra in Fig. 1b show that in this case again state selected ions are produced, however, now in the $16^1$ vibrational state.

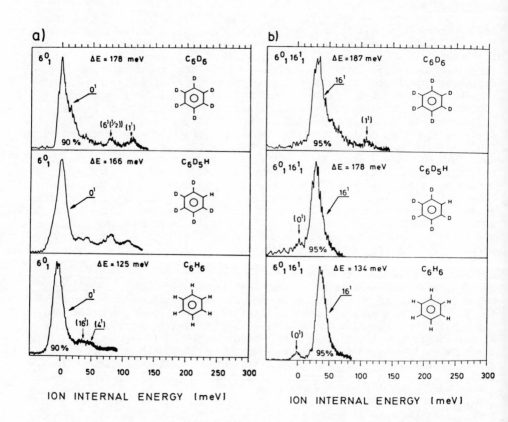

Fig.1 Photoelectron spectra of benzene $C_6D_6$, $C_6D_5H$, and $C_6H_6$ after resonance-enhanced two-photon ionization
a) via the $6^0_1$ band of the $S_1 \leftarrow S_0$ transition of neutral benzene.
b) via the $6^0_1 16^1_1$ band of the $S_1 \leftarrow S_0$ transition of neutral benzene.

## 3. Unimolecular Ion Decay in a Reflectron Time-of-Flight Mass Spectrometer

In our kinetic studies the state-selected benzene ions are produced in the molecular beam which is located within the acceleration field of a reflectron time-of-flight mass spectrometer (Boesl et al. 1982) (see Fig. 2).

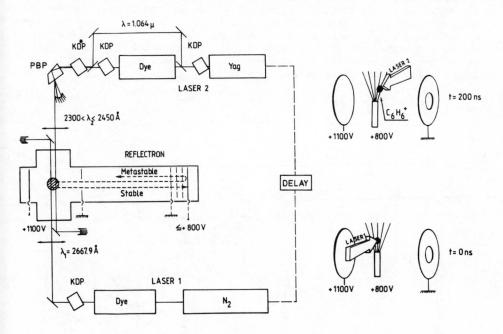

<u>Fig.2</u> Scheme of a two-laser pump-pump experiment for the production of internal energy-selected molecular ions in a reflectron time-of-flight mass spectrometer (from Kühlewind et al. 1984). Laser 1 produces state-selected molecular ions and 200 ns later laser 2 excites these ions to a well defined internal energy level above dissociation threshold. The dissociation rate constants of the energy-selected ions are measured by the technique of detection and energy analysis of metastable ions

200 ns later, when the ion cloud has left the molecular beam region, a second laser pulse of variable frequency (5.07 eV - 5.52 eV) further excites the ions to an energy level slightly above the thresholds for different dissociation channels of low threshold energy $E_0$. This leads to the production of $C_6H_5^+$, $C_6H_4^+$, $C_4H_4^+$ and $C_3H_3^+$ ionic fragments (Rosenstock et al. 1973) and their deuterated analogues, respectively. The energy of the ions is sufficient to induce a metastable decay on a μsec timescale, i.e., a decay of the ions on the way from the ion source to the reflecting field. Since the reflecting field acts as an energy analyzer,

daughter ions resulting from a metastable decay are separated from stable ions (Kühlewind et al. 1985b). From their kinetic energy the place of decay is determined. From this procedure the unimolecular decay rate constants are found according to the technique described in our previous work (Kühlewind et al. 1984). Since all decay channels of low energy, independent of their H-loss or C-loss character, are competing and originate from one electronic state (Kühlewind et al. 1984 and 1986), the directly measured rate constant is given by the sum of the individual rate constants of all competing decay channels (total decay rate constant). In separate experiments ions in the vibrationless state and in the $16^1$ state are excited to the metastable energy range with varying energy and their total decay rate constant of some $10^6$ s$^{-1}$ is measured. The experimental result is shown in Fig. 3 for three benzene isotopes.

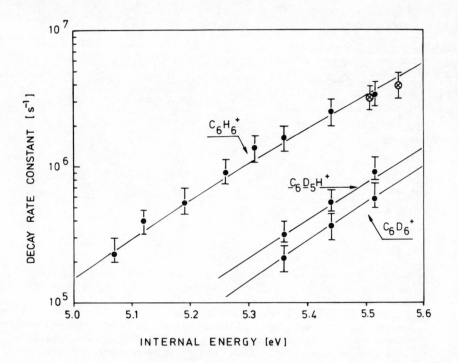

Fig.3 Total decay rate constants of internal energy selected benzene cations $C_6H_6^+$, $C_6D_5H^+$, and $C_6D_6^+$ as a function of their internal energy. The solid lines represent the result of RRKM calculations fitted to the experimental data. Vibrationless benzene cations are produced via the $6^0_1$ transition (●). Benzene cations in the $16^1$ vibrational state are produced via the $6^0_1 16^1_1$ transition (⊗). Subsequently both ion species are excited with the second laser pulse to the internal energy indicated on the abscissa.

Several points of interest are found:

i)   there is a smooth tenfold increase of the total decay rate constant in the observed internal energy range
ii)  no difference in the total decay rate constant is found for ions of the same internal energy resultig from different pathways of ion production. Ions produced via the vibrationless $S_1$ state ($\bullet$) and those ions produced via the $16^1$ state ($\oslash$) display the same decay rate constant when they are excited to the same internal energy.
iii) a pronounced kinetic isotope effect is found in the observed energy range

The results point to a dissociation which proceeds according to a statistical model of unimolecular dissociation. In particular no vibrational specificity of the decay rate constant is found when excitation proceeds via the two pathways described in ii).

For an explanation of our results we performed statistical RRKM calculations of the individual rate constants of all competing decay channels of low energy. The latter have been experimentally obtained from the directly measured total rate constants (see Fig. 3) and the branching ratios of the fragment ions under consideration (Kühlewind et al. 1986). For differently labelled isotopic species ($C_6H_6^+$, $C_6D_5H^+$, $C_6D_6^+$) a good simulation of experimental results is obtained with one set of parameters for the determination of the vibrational frequencies of the activated complexes (solid lines in Fig.3). (Isotope shifts of the vibrational frequencies are taken into account by use of the Redlich-Teller product rule).
This points to a high reliability of the set of parameters and yields detailed information on the structure of the activated complexes for the four decay channels under investigation (Kühlewind et al. 1986). In particular, the looseness of the activated complexes with respect to the complete set of frequencies as well as the threshold energies $E_0$ and activation entropies $\Delta S^{\#}_0$ of the decay channels under consideration are deduced from the experimental data. Furthermore the measured intramolecular and intermolecular isotope effects yield a microscopic probe of the looseness or tightness of C-H bonds in the activated complexes and give detailed insight into the character of the reaction coordinate (Kühlewind et al., to be published).

## 4. Conclusion

In conclusion, we have shown that resonantly enhanced two-photon ionization is a versatile method for the production of state-selected polyatomic molecular ions. In a reflectron time-of-flight mass spectrometer the total decay rate constants and individual rate constants of competing decay channels of internal energy selected ions have been measured for various defined energies. From our experimental results detailed information about the statistical character of the dissociation mechanisms and the structure of the activated complexes is obtained.
In the future it seems to be possible to study the ergodic character of the energy redistribution prior to dissociation for even larger polyatomic molecular ions.

References

Antonov V.S., Knyazev I.N., Letokhov V.S., Matiuk V.M., Morshev V.G. and
   Potapov V.K. 1978  Opt. Lett. 3 37
Boesl U., Neusser H.J. and Schlag E.W. 1978 Z. Naturforsch. 33A 1546
Boesl U., Neusser H.J. and Schlag E.W. 1980 J. Chem. Phys. 72 4327
Boesl U., Neusser H.J. and Schlag E.W. 1981a Chem. Phys. 55 193
Boesl U., Neusser H.J. and Schlag E.W. 1981b  J. Amer. Chem. Soc.
   103 5058
Boesl U., H.J. Neusser, R. Weinkauf and Schlag E.W. 1982 J. Phys. Chem.
   86 4857
Kiermeier A., Kühlewind H., Neusser H.J. and Schlag E.W., to be published
Kühlewind H., Neusser H.J. and Schlag E.W. 1984 J. Phys. Chem. 88 6104
Kühlewind H., Neusser H.J. and Schlag E.W. 1985a J. Phys. Chem. 89 5600
Kühlewind H., Neusser H.J. and Schlag E.W. 1985b J. Chem. Phys. 82 5482
Kühlewind H., Kiermeier A., Neusser H.J. and Schlag E.W. 1986 J. Chem.
   Phys. in press
Kühlewind H., Kiermeier A., Neuuser H.J. and Schlag E.W. to be published
Kiermeier A., Kühlewind H., Neusser H.J. and Schlag E.W. to be published
Long S.R., Meek J.T. and Reilly J.P. 1983 J. Chem. Phys. 79 3206
Rosenstock H.M., Larkins J.T. and Walker J.A. 1973 Int. J. Mass Spectrom.
   Ion Phys. 11 309
Schlag E.W. and Neusser H.J. 1983 Acc. Chem. Res. 16 355
Zandee L., Bernstein R.B. and Lichtin D.A. 1978 J. Chem. Phys. 69 3427

*Inst. Phys. Conf. Ser. No. 84: Section 3*
*Paper presented at RIS 86, Swansea, Wales, 7–12 Sept. 1986*

127

# High selectivity isotope analysis

Rolf Engleman, Jr., <u>Richard A. Keller</u>, Charles M. Miller and Denise C. Parent
Los Alamos National Laboratory, Los Alamos, NM 87545

William M. Fairbank, Jr., Robert D. LaBelle, Siu-Au Lee and Erling Riis
Department of Physics, Colorado State University, Ft Collins, CO 80521

## INTRODUCTION

There is continuing interest in increasing isotopic resolution and reducing isobaric interferences in mass spectrometric analysis. Progress in this area is needed for research in basic physics and practical applications. For example, the detection of super heavy elements, neutrinos, quarks, etc. often involves the accurate measurement of a very small amount of one isotope in the presence of a large excess of other isotopes of the same element and interfering isobaric species from other elements. Some of these applications are discussed elsewhere in this volume.

High resolution mass spectrometers at Los Alamos and elsewhere have been used to make accurate isotope ratio measurements at isotope ratios $\sim 1{:}10^7$ when chemistry can remove isobaric interferences(Rokop, 1982). It has been demonstrated that *element* selective photoionization in the source chamber of a mass spectrometer (RIMS) is an excellent technique for eliminating isobaric interferences (Nogar, 1984; Donohue, 1984; Moore, 1984;). In this case, isotope ratio measurements utilize the mass dispersion in the mass spectrometer. For example, isotope ratios of lutetium were accurately determined in 60-ng samples containing trace amounts of $^{173}Lu$ and $^{174}Lu$ and a thousandfold excess of the isobaric interfering isotopes $^{173}Yb$ and $^{174}Yb$. The ratio $^{173}Lu/^{175}Lu$ was measured to be $(0.44 \pm 0.07) \times 10^{-6}$ on a sample containing only $10^8$ atoms of $^{173}Lu$ (Nogar, 1984).

The dynamic range and accuracy of low abundance isotope measurements can be increased by: 1) isotope selective photoionization and 2) laser induced fluorescence (LIF) to interrogate ions in overlapped isotope peaks as they exit the mass spectrometer. In this paper we demonstrate the use of RIMS to acquire the high resolution spectra of rare isotopes necessary to affect isotope selective photoionization. We also present preliminary results on LIF of accelerated $Mg^+$ ions as a step towards probing ions as they exit a mass spectrometer. Isotope selective photoionization in combination with RIMS has been demonstrated in an atomic beam of barium (Cannon, 1985). LIF of accelerated ions has also been reported (Silverans, 1985).

## HIGH-RESOLUTION SPECTROSCOPY OF LUTETIUM ISOTOPES

In order to accomplish isotope selective photoionization it is necessary to know the isotope shifts and hyperfine structure of the optical transition(s) used for the isotopes of interest. We are particularly interested in rare and sometimes highly radioactive elements for which this information is not readily available. We demonstrate that the sensitivity and mass selectivity of the RIMS technique itself can be used to obtain the needed spectra.

We have chosen the lutetium system to demonstrate the feasibility of RIMS for determining the spectroscopy of rare isotopes. Analytically, this element is significant because of practical problems at Los Alamos associated with the mechanisms of nuclear reactions. Spectroscopically, the lutetium isotopes are of interest primarily because of the previously uncertain nuclear spin of $^{174}$Lu. The photoionization scheme for Lu is attractive for spectroscopic studies: a single photon at approximately 452-nm will promote a Lu atom to a resonant intermediate state that is more than halfway to the ionization limit. A second photon of the same wavelength ionizes the atom, in an energy region that previous studies have shown to be free of autoionization structure. This means that spectra observed in resonance ionization will be characteristic of the resonant transition alone $(^{2}D^{o}_{3/2} \leftarrow {}^{2}D_{3/2})$, considerably simplifying the analysis of results. Details of this work have been published recently (Miller, 1985).

Resonance ionization spectra of the lutetium isotopes were obtained using a scannable, single frequency, ring dye laser and a magnetic sector mass spectrometer. Output power of the laser was typically 60-mW in a 0.5-MHz bandwidth. The laser beam was focused to a diameter $\sim$ 0.050-mm with a confocal parameter of $\sim$ 4-mm. The laser beam propagated parallel to the sample filament about 2-mm above the filament. One microgram samples of mixed lutetium isotopes were evaporated from solution onto the rhenium filaments. Spectra of the individual isotopes were obtained by setting the mass spectrometer to pass only the mass of interest to the detector. Although $^{174}$Lu was present to only 30 parts per million in the sample (total amount $\sim$ 30-pg), it's spectrum could be observed without interference from the 97% abundant $^{175}$Lu or the 0.3% abundant $^{174}$Yb. The single color, laser spectra of $^{173}$Lu through $^{176}$Lu are shown in Fig. 1.

The data were analyzed by using a least squares procedure to match the observed spectra to theoretical predictions. The resulting fit yielded the hyperfine splitting constants and isotope shifts. The analysis was complicated by the presence of Doppler and power broadening of the observed lines, as well as the effects of differing optical saturation for each hyperfine component. In the case of $^{174}$Lu, only a nuclear spin of unity would reproduce the observed hyperfine splitting pattern.

Nuclear parameters for the lutetium isotopes are listed in Table I. The values underlined were determined in this work. Even though isotope shifts

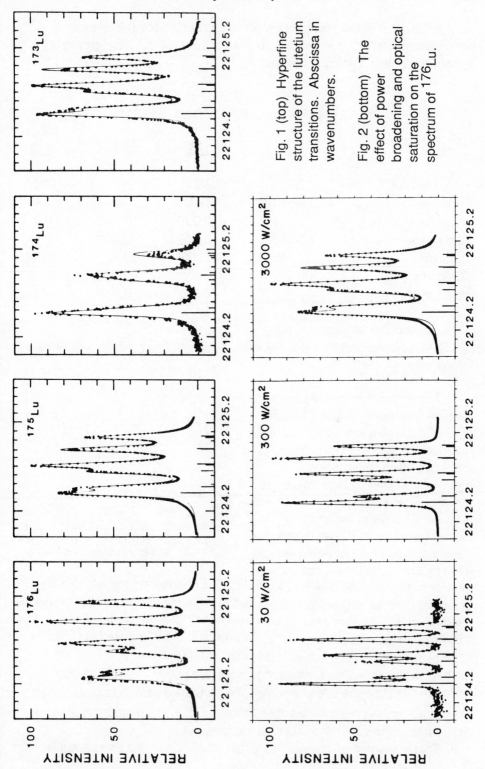

Fig. 1 (top) Hyperfine structure of the lutetium transitions. Abscissa in wavenumbers.

Fig. 2 (bottom) The effect of power broadening and optical saturation on the spectrum of $^{176}$Lu.

are very small, differences in the hyperfine pattern among the isotopes permit isotope selective ionization. This selectivity will increase greatly when Doppler broadening is reduced.

TABLE I

NUCLEAR PARAMETERS FOR LUTETIUM ISOTOPES

| Isotope | 176 | 175 | 174 | 173 |
|---|---|---|---|---|
| Nuclear spin | 7 | 7/2 | 1 | 7/2 |
| Half life (y) | $3.6 \times 10^{10}$ | stable | 3.3 | 1.4 |
| Doppler width $(cm^{-1})$ | 0.036[a] | 0.036 | 0.036[a] | 0.036[a] |
| Lorentz width $(cm^{-1})$ | 0.050[a] | 0.050 | 0.050[a] | 0.050[a] |
| $\mu$ (nm)[b] | 3.139 | 2.380 | 1.94 | 2.34 |
| Q (barns) | 8.0[b] | 5.68[b] | 0.1±0.5 | 5.7±0.6 |
| Isotope shift $(cm^{-1})$ | 0.01 | 0.00 | -0.009 | -0.037 |

[a] Assumed the same as $^{175}$Lu.

[b] C. M. Lederer and V. S. Shirley, eds., Tables of Isotopes. 7th ed.,(Wiley, New York, 1978).

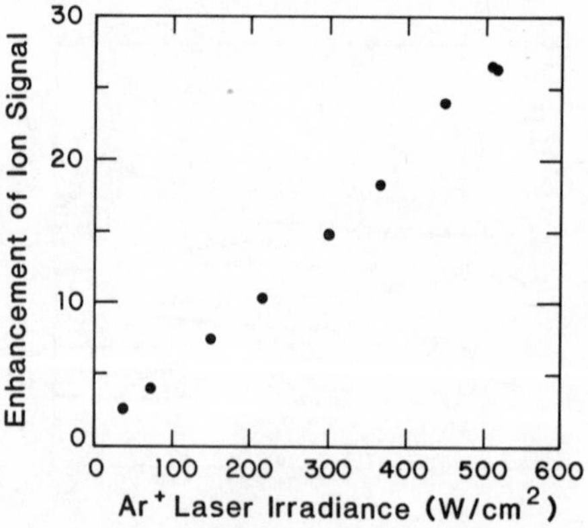

Transitions from the resonant intermediate state into the ionization continuum have a low cross section. In order to achieve reasonable ionization efficiencies, it was necessary to work at high laser powers. These high laser powers resulted in power broadening as displayed in Fig. 2. Power broadening can be reduced if a low power laser is used for populating the resonant intermediate state and a high power laser of a different wavelength is used for ionization from this state. The effect of using an intense Ar$^+$ laser beam for the ionization step in

Fig. 3 Enhancement of lutetium ion signal when both dye and Ar$^+$ lasers are used compared to dye laser alone. Dye power is 8.6 mW focused to 0.28 mm diameter (14 W/cm$^2$).

lutetium is shown in Fig. 3. In this figure the dye power is similar to that shown in Fig. 2a. The use of the Ar$^+$ laser enhances the ion signal ~ 30X. Note that there is no saturation observable in Fig. 3 which means that larger Ar$^+$ irradiance would further increase the ion yield.

These experiments demonstrate that RIMS is a useful and sensitive technique for high-resolution spectroscopy of rare isotopes. For lutetium, the presence of Doppler and power broadening precluded isotopically selective RIMS with an analytically useful selectivity. Doppler broadening may be reduced by the use of counterpropagating lasers to cancel velocity effects. As demonstrated above, power broadening may be reduced through the use of low-power excitation schemes in the resonant steps, followed by a high power laser to ionize from the highest level. We are presently exploring these and other alternative RIMS schemes to alleviate the broadening problems.

## LASER-INDUCED-FLUORESCENCE OF ACCELERATED IONS[*]

When the desired isotopic analysis exceeds the isotopic resolution of the mass spectrometer (or the mass spectrometer plus isotopically selective photoionization), unresolved tails of major isotopes will obscure ion signals from minor isotopes. It was suggested that LIF could be used to interrogate the exiting ions to provide further mass resolution (Keller, 1981). The technique of using LIF to probe accelerated ions exiting a mass separator is called collinear spectroscopy and has been developed to study the spectra of rare isotopes. The use of collinear spectroscopy in nuclear physics has been summarized in recent articles (Silverans, 1985; Neugart, 1985). The emphasis of collinear spectroscopy to this date has been on spectroscopy, not on improving mass and species resolution in mass spectrometric analysis.

A Colutron system was used to produce, extract, and focus $Mg^+$ ions. Typical ion currents were 200-300 nA. One tenth to one milliwatt of laser power at $\sim$ 280 nm was obtained from a Coherent Model 699-03 dye laser with a 2 cm long KDP doubling crystal placed inside the laser cavity. The crystal was mounted near the folding mirror and the uv beam was extracted from the cavity using a dichroic beam splitter mounted near the doubling crystal. The standard output coupler of the dye laser was replaced with a total reflector to increase the intracavity power. The laser beam traveled collinear to the $Mg^+$ ion beam and fluorescence was detected with a photomultiplier tube mounted $\sim$ 20 cm above the ion beam. No collection lens was used and the detection efficiency (photoelectrons/photons emitted) was estimated to be $\sim 4 \times 10^{-5}$.

The LIF excitation spectrum of $Mg^+$ accelerated to 600 eV is shown in Fig. 4. Note the large isotope shifts and excellent signal-to-noise. The observed half-width results mostly from the energy spread in the ion beam. The ion velocity and the beam current can be used to calculate that the probability of an ion being in the probe volume is less than unity; thus, we are processing ions individually. Our results indicate that with improved photon collection efficiencies, individual ions can be counted by the photon burst method.

The isotope splitting displayed in Fig. 4 is composed of the natural isotope shift ($^{24}$Mg - $^{26}$Mg ~3-Ghz) and an artificial isotope shift (~10-Ghz) caused by the mass dependent velocities of the ions which have all been accelerated to the same energy. The presence of this artificial isotope shift is very important for increased isotope resolution because it means that isotope selective analysis can be accomplished even in the absence of natural isotope shifts or different hyperfine patterns.

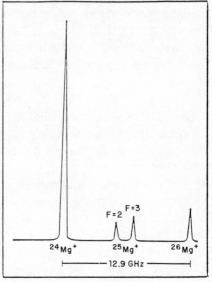

Applications envisioned for laser detection of accelerated ions in addition to large dynamic range isotope analysis include: 1) high resolution spectroscopy of rare isotopes in unseparated samples, 2) a new technique for isotope ratio measurements composed of a low resolution mass spectrometer (or velocity filter) combined with laser based detection schemes to provide good isotope resolution and no problem with isobaric interferences, and 3) super heavy element detection.

Fig. 4 LIF excitation spectrum of 600-eV magnesium ions. Instrumental time constant was 12-ms. Laser power ~ 100 μW.

REFERENCES

* Partially supported by the National Science Foundation.

Cannon B D, Bushaw B A and Whitaker T J 1985 J. Opt. Soc. Am. B 2,1542.

Donohue D L, Smith D H, Young J P, McKown H S and Pritchard C A 1984 Anal. Chem. 56, 379.

Keller R A, Bomse D S and Cremers D A 1981 Laser Focus, October, p 75.

Miller C M, Engleman R Jr. and Keller R A 1985 J. Opt. Soc. Am. B 2, 1503.

Moore L J, Fassett J D, and Travis J C 1984 Anal. Chem. 56, 2770.

Neugart R 1985 Hyperfine Interactions, 24, 159.

Nogar N S, Downey S W and Miller 1984 *Resonance Ionization Spectroscopy 1984*, Hurst G S and Payne M G eds. (Institute of Physics, Bristol, England), p. 91.

Rokop D J, Perrin R E, Knobeloch G W, Armijo V M and Shields W R 1982 Anal. Chem. 54, 957.

Silverans R E, Borghs G, de Bisschop P and Van Hove M 1985 Hyperfine Interactions, 24, 181.

*Inst. Phys. Conf. Ser. No. 84: Section 3*
*Paper presented at RIS 86, Swansea, Wales, 7–12 Sept. 1986*

133

# Studies of multiphoton ionization schemes for RIMS of La and Ba

L.W. Green, R.G. Macdonald, F.C. Sopchyshyn and L.J. Bonnell

Atomic Energy of Canada Limited, Chalk River Nuclear Laboratories
Chalk River, Ont., Canada, K0J 1J0

## 1. Introduction

Lanthanum 139 is a high yield fission product that is useful as a fission monitor in nuclear fuels (Meneghetti et al 1974). However, its determination by isotope dilution mass spectrometry with a thermal ionization source is made difficult by interference from contaminant Ba, since $^{138}$Ba interferes at the only mass of La available for spiking (Elliot et al 1986). Barium is a readily ionized ubiquitous contaminant, especially in the rhenium filaments used in thermal ionization.

Multiphoton ionization (MPI) with tunable dye lasers has the potential to selectively ionize La in the presence of Ba and attain very high sensitivity. The objective of this work was to study MPI processes of Ba and La and select suitable schemes for selective ionization of La in the presence of Ba. The spectral region 520 to 610 nm was chosen because of reported strong radiative transitions in this region (Bulos et al 1978; Miles and Wiese 1969) and its applicability to high repitition rate Cu vapour lasers.

## 2. Experimental

Two tunable dye lasers (.5 to 3 mJ/pulse) were pumped at 308 nm (10Hz, 50 mJ/pulse) by a Lumonics 860-2 XeCl excimer laser. One dye laser was a Lumonics EPD-330 laser whereas the other was constructed in this laboratory. The latter had the unique feature that the wavelength was resetable to within one laser bandwidth (FWHM $\approx 0.3$ cm$^{-1}$). The dye beams were overlapped in the source region of a small (20 cm drift tube) time-of-flight mass spectrometer (TOFMS) for most of this work. For a small portion of the work a recently commissioned TOFMS with a 1 m drift tube was used. The ion packets produced by the pulse lasers were detected by a 19 stage Vacumetrics ETP AEM 1000 electron multiplier. The ion signal was amplified by a series of Comlinear Corp. CLC 102 video amplifiers and connected to either a Biomation 6500 transient recorder and Nicolet 1170 signal averager combination for mass spectra or an Evans Associate Model 4130 gated integrator for optical spectra. For wavelength calibration about 10% of the scanning dye laser intensity was deflected into a neon filled hollow cathode lamp and the optogalvanic signals for Ne were monitored. This wavelength calibration technique yielded an absolute accurracy of about ±.02 nm for the wavelength scale.

Several different atom sources were used, the most frequent was a Ta ribbon folded into a canoe shape into which was placed a piece of either pure Ba or La metal or both.  In other cases a Ba or La salt was dried onto a Re filament and covered with a starch deposit.

3.  Results and Discussion

The types of multiphoton ionization (MPI) schemes investigated in this work for detection of La and Ba are shown in Fig. 1.  The dynamics of the inter- actions between the atomic energy levels and laser radiation for all 3 MPI schemes shown in Fig. 1 have been extensively described in the literature (Hurst et al 1979; Chin and Lambropoulos 1984) and will only be briefly considered here.  For the first two MPI schemes in Fig. 1 the rate deter- mining step is photoionization from the last excited bound energy level of the atom.  The cross section, $\sigma_I$, for such processes are not easily measured and hence little information is available about them.  Lorents et al (1973) and Cowan (1981) have given a simple estimate of $\sigma_I$:

$$\sigma_I \approx \frac{8}{Z\,(IP*/13.6)^{1/2}} \left( \frac{IP*}{E} \right)^3 x \; 10^{-18} \; cm^2 \tag{1}$$

where Z is the charge of the ion, IP* is the ionization potential (eV) from the excited state of the atom, and E (eV) is the energy of the ionizing photon.  Although Equation (1) is based solely on hydrogenic wave functions and neglects a great deal about multielectron continuum and bound state radial wave functions, Lorents et al (1973) have shown that it can provide useful estimates of photoionization cross sections.  Using the ionization potential of 5.577 eV for La (Meggers et al 1975), $\sigma_I$ for MPI scheme A is $\approx 3 \times 10^{-18}$ $cm^2$ and for scheme B is $\approx 9 \times 10^{-18}$ $cm^2$.  For a laser pulse of 1 mJ at 550 nm collimated to a beam diameter of 0.1 cm and with a pulse width of 10 ns, the flux of laser photons is $3.5 \times 10^{25}$ photons $cm^{-2}$ $s^{-1}$.  At such laser photon fluxes and at oscillator strengths greater than 0.001 the intermediate levels in MPI schemes A and B of Fig. 1 are completely equili- brated according to their degeneracies and the concentration of ions is represented by a simple exponential growth dependent only on $\sigma_I$ and the fluence (Hurst et al 1979).  Accordingly, if the u.v. and visible photon fluxes are comparable, scheme B is expected to be about 3 times more sensi- tive than scheme A; this factor of 3 is only approximate since the validity of Equation (1) decreases as the bound excited level lies further below the ionization potential.  For MPI scheme C (Fig. 1), the limiting cross section is the 2 photon excitation (Hurst et al 1979) and the sensitivity of this process is expected to be lower than for the other two.  The pre- ferred MPI process would appear to be scheme B; however, as will be discus- sed later, other considerations such as background ionization are important.

Fig. 2 shows an optical spectrum of Ba evaporated from a starch matrix supported on a Re filament.  For this particular RIMS experiment MPI scheme B (Fig. 1) was chosen.  The ionizing radiation ($\lambda_2$) was a small fraction of the XeCl pump laser at 308 nm and had an intensity of $3 \times 10^6$ watts/$cm^2$ for a beam diameter 0.1 cm in the source region of the TOFMS.  This u.v. laser beam was overlapped, by a dichroic mirror, with a Coumarin 540 dye laser beam which had an average intensity of $3 \times 10^7$ watts/$cm^2$.  Three interest- ing features in this spectrum are as follows:  first, many peaks other than that expected for scheme B are evident, second, only the transition corres- ponding to scheme B (553.548 nm) showed any dependence on the ionizing

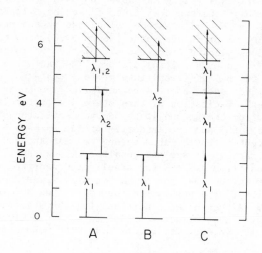

Fig. 1  Multiphoton ionization schemes for La.  (Similar for Ba).

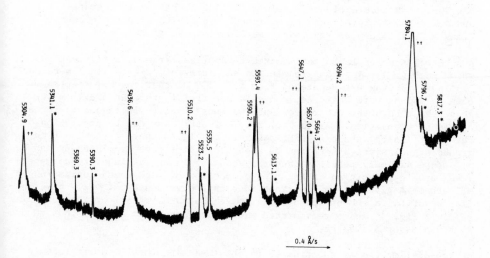

Fig. 2  Scan of Ba over range of Coumarin 540 dye ($\lambda_1$).  $\lambda_2$=308 nm
(excimer laser).

laser light intensity, and third, there was always a background ionization
signal which increased at the ends of the tuning range of the dye. Also,
the mass spectra at various wavelengths showed no mass peaks other than
those of the Ba isotopes. The spectral lines marked by * in Fig. 2 were
identified as electric dipole allowed one colour-two photon transitions to
upper states of $^1S$, $^1D$, $^3S$ or $^3D$ character (Moore 1978), and those marked
by tt were identified as allowed transitions from the $^1P°$ resonance state
of Ba, at 18060.3 $cm^{-1}$, to higher lying states. The radiation that
caused population of the Ba $^1P°$ (6s6p) intermediate state was discovered,
by dispersion of the dye laser beam, to be amplified spontaneous emission
(ASE) of the dye laser. Excitation of this state was particularly easy
since the transition from the ground state has an oscillator strength of
1.59 (Miles and Wiese 1968) and the intensity of the ASE was a few percent
of the total laser light intensity at wavelengths near the end of the
Coumarin 540 tuning range. It was not possible to accurately spatially
separate the laser beam from the ASE for wavelengths near the centre of the
tuning curve. The background $Ba^+$ ion signal in Fig. 2 was due to scheme B
ionization from the $^1P°$ (6s6p) population induced by the ASE independent of
the dye laser tuning. Although the conditions for the spectrum in Fig. 2
were chosen to emphasize the influence of ASE on RIMS experiments, the
results do show that ion selectivity may be poor with scheme B.

For La many possible MPI transitions corresponding to all 3 schemes shown
in Fig. 1 were found in the spectral region 530 nm to 600 nm. Some of the
stronger transitions of scheme A with $\lambda_1$ = 557.036 nm are shown in Fig. 3;
spectral assignments are from tabulated energy levels (Martin et al 1978).
Similar transitions were observed when $\lambda_1$ was tuned to populate other
intermediate states of La (eg. 18172.39 $cm^{-1}$) from either the ground state
or the $^2D_{5/2}$ metastable state at 1053.2 $cm^{-1}$; the latter transitions
showed evidence of thermal population of this metastable state. Because
the electronic energy level density of La is much higher than that of Ba,
configuration interaction severely mixes the highly excited states of La
and few of these levels can be assigned to definite electronic configurations. Several new energy levels were discovered in this work.

A spectral scan of La/Ba metal mixture vaporized from a Ta filament is
shown in Fig. 4. The mass resolution and integration gate width were such
that mass 138 was not separated from mass 139 so that the MPI spectra of
both La and Ba were detected, and a Ba spectral feature at 578.4 nm was
seen even though the other dye laser was not tuned to be resonant with an
atomic Ba transition. As in Fig. 2 this feature was due to population of
the Ba $^1P°$ (6s6p) state by ASE; however, in Fig. 4 there is no ion background signal since absorption of a photon from either dye laser by the Ba
$^1P°$ state was only possible when the scanning laser was tuned to an upper
level atomic transition in Ba. The supression of background ionization in
MPI scheme A gives it an important advantage over scheme B. Selective ionization of La in the presence of Ba is demonstrated by all of the remaining
peaks in Fig. 4, and a recommended excitation scheme is $\lambda_1$ = 557.036 nm and
$\lambda_2$ = 575.84 or 555.88 nm.

A time-of-flight mass spectrum of Ba obtained with the 1 m drift tube is
shown in Fig. 5, and corresponding isotope ratio data are in Table 2. Most
of the isotope ratios agree with literature values within the quoted precision but further work is required to eliminate peak tailing and background
caused by poorly focussed ions. Preliminary tests for a wavelength dependence bias of the isotope ratios showed no evidence of such a bias, and

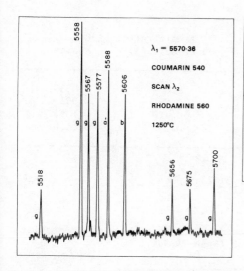

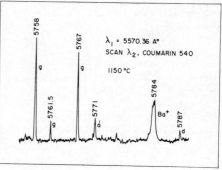

Fig. 4

Fig. 3    Scan of La with Rhodamine 560 dye ($\lambda_2$).    g, $\lambda_1$=557.036 nm
($^4G_{5/2}$, 17947.13 cm$^{-1}$); a' possibly $\lambda_1$=ASE 550.13 nm ($^2D_{3/2}$,
18172.39 cm$^{-1}$); b, upper state unknown.

Fig. 4    Scan of combined Ba and La with Coumarin 540 dye ($\lambda_2$).    g and a'
as in Fig. 3; d, $\lambda_1$=578.79 nm (scheme C).

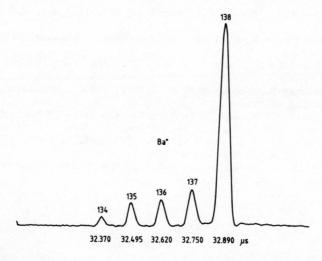

Fig. 5    RIMS of Ba with 1 m drift tube.    $\lambda_1$=553.548 nm; $\lambda_2$=578.41 nm.
Scheme A.

Table 1.  Ba Isotope Ratios

| Mass | Atomic ratio / 138 observed (±4%) | known[a] |
|------|-----------------------------------|----------|
| 134  | 0.031                             | 0.0337   |
| 135  | 0.091                             | 0.09194  |
| 136  | 0.110                             | 0.1095   |
| 137  | 0.156                             | 0.1566   |

[a] Holden N.E. 1979 J. Pure Appl. Chem. <u>52</u> 2349

this is consistent with the laser bandwidth of 9 GHz and an isotope shift of 150 MHz between isotopes 132 and 138 (Kuhn 1969).

## 4.  References

Bulos B.R., Glassman A.J., Gupta R. and Moe G.W. 1978 J. Opt. Soc. Am. <u>68</u> 842.

Chin S.L. and Lambropoulos 1984 "Multiphoton Ionization of Atoms" (Toronto: Academic Press) p.65

Cowan R.D. 1981 "The Theory of Atomic Structure and Spectra" (Berkeley: University of California Press) p.523

Elliot N.L., Green L.W., Recoskie B.M. and Cassidy R.M. 1986 Anal. Chem. <u>58</u> 1178

Hurst G.S., Payne M.G., Kramer S.D. and Young J.P. 1979 Rev. Mod. Phys. <u>51</u> 767

Kuhn H.G., 1969 "Atomic Spectra" 2nd ed., (London:  Longmans, Green and Co.) p.369

Lorents D.C., Eckstrom D.J. and Huestis D. 1973 SRI "Tech. Report No MP 73-2"

Martin W.C., Zalubas R. and Hagan L. 1978 NSRDS-NBS-60 (Washington, D.C.: US Govt. Printing Office)

Meggers W.F., Coaliss C.H. and Scribner B.F. 1975 N.B.S. Monograph 145 (Washington D.C.:  US Gov't Printing Office)

Meneghetti D., Ebersole E.R. and Walker P. 1975 Nucl. Technol. <u>25</u> 406

Miles B.M. and Wiese W.L. 1969 Atomic Data <u>1</u> 1

Moore C.E. 1978 Atomic Energy Levels vol. 3.  NSR DS-NBS 35 (Washington D.C.:  US Govt Printing Office)

*Inst. Phys. Conf. Ser. No. 84: Section 3*
*Paper presented at RIS 86, Swansea, Wales, 7–12 Sept. 1986*

139

# Pulsed ion beam produced via RIMS

J.K.P. Lee, V. Raut, G. Savard and G. Thekkadath
Foster Radiation Laboratory, McGill University,
Montreal, Que., Canada. H3A 2B2

T.H. Duong and J. Pinard
Laboratoire Aimé Cotton, 91405  Orsay, France.

## 1 - Introduction

During the past decade, important progress has been made in various
applications of lasers in nuclear science (for a general review, see
Bemis and Carter, 1982). The resonance ionization spectroscopic (RIS)
process, with its high sensivity for elemental and isotopic selection
is of particular interest for the development of an on-line laser-based
ion source to produce an isotopically pure beam, and for the study of
isotope shift (IS) and hyperfine structure (HFS) of long chains of radioactive isotopes, extending  away from the valley of beta stability.
In these applications, one is dealing with a limited quantity of radioactive atoms.  The general difficulty has been to devise an approach that
has suitable power for efficient RIS process and at the same time a good
temporal duty cycle. With the advent of high power, high repetition rate
copper vapour lasers, reasonably high overall efficiency could be expected if a multistep resonant excitation scheme can be used (Kluge 1985).
This was recently demonstrated for the case of Sr atoms with a 17% overall
efficiency (Andreev etal 1986). For IS and HFS studies, the absolute efficiency is not critical, and RIS methods have been applied to long isotopic
chains of Eu (Fedoseyev etal 1984) and Au (Wallmeroth etal 1986). In these
cases, the detection effficiencies were about $10^{-4}$ and $10^{-8}$ respectively.
In the present work, a different approach was  attempted. The atoms to
be analysed were first adsorbed onto a surface. A heating laser pulse
would desorb them and a synchronized dye laser beam would "catch" them
and selectively ionize atoms of the element (or isotope) of interest.
In this way, the atoms exist in a bunched form, and the temporal duty
cycle of the lasers is no longer a factor in the overall ionization
efficiency of the method. This approach is similar to that used in the
Kr ion counting work by Chen etal (1984), where a 1% detection efficiency
was reported. In the present work, gold atoms were used as a test case
and a single resonance two-photon ionization scheme was used.

2 - Experimental setup

The experimental setup is shown schematically in figure 1. Two pulsed
lasers were used: a Nd-YAG laser with 1064, 532, 355 or 266 nm
output,and an excimer-laser-pumped dye laser. Both lasers are
externally triggered, and the timing between them can be varied
continuously. In a typical measurement, the Nd-YAG laser pulse would
heat up the surface of a certain substrate (such as the rotor shown in
the figure), and desorb the atoms previously adsorbed there. A few
microseconds later, the dye laser beam would pass just in front of the
heated spot and selectively ionize the desired atoms. The ions would be
 accelerated by the electrode potentials, pass through a 1.2m flight
path, and be detected by a channelplate detector. The time-of-flight
(TOF) waveform output from the detector would be captured by the
transient analyser, yielding the information about the mass and
intensity distribution of the ions created by the laser pulses. A mass
resolution  (M/ΔM) of 300 can be readily achieved, adequate for resolving
the individual isotopes of the heavy elements.

For the present work, the dye laser output was frequency-doubled to
give 37358 cm$^{-1}$   (268nm) UV output, corresponding to the electronic
transition from $5d^{10}6s$ to $5d^{10}6p$ levels in the Au atom. Absorption of a
second photon of the same frequency will then be just adequate to
ionize the excited Au atoms (ionization limit $74410 cm^{-1}$).

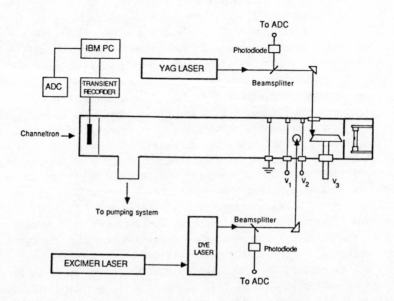

Fig. 1  Schematic layout of the experimental setup

## 3 - Saturation of ionization and power broadening effects

The dye laser output at 268nm has a maximum energy of 5mJ/pulse. At the corresponding power level, the ionization process may not be saturated, even though the broadening of the resonant transition will be severe. To investigate these effects separately, a continuous Au atomic beam was used, and the 266nm output of the Nd-YAG laser was adjusted to overlap the dye laser beam. In this way, the dye laser beam will resonantly excite the Au atoms while the Nd-YAG beam will supply the second ionization photon. The two laser power levels were varied independently and the results are shown in figure 2. As expected, the resonant excitation step was easily saturated while the saturation of the ionization step required a much higher but still moderate laser power density. Using a 5mJ dye laser pulse without the Nd-YAG 266nm output, an ionization efficiency of about 20% has been reached.

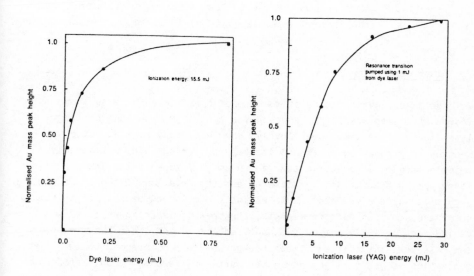

Fig. 2   Saturation effect of the resonant excitation step (left diagram) and the photoionization step (right diagram).

With both laser powers set at an appropriate level, the dye laser frequency was scanned across the resonant transition in Au atoms and the experimental linewidth was measured. Using a dye laser power level of 4KW/cm², we obtained a system resolution of 3.5GHz allowing us to resolve the HF splitting of the ground state (6.1GHz) of the stable Au isotope (Figure 3). The line width was broadened to about 60GHz with 4MW/cm² power density. Since the IS of Au is about 1.5GHz per mass unit (Streib etal 1985) for such studies, the dye laser power must be kept at a low level.

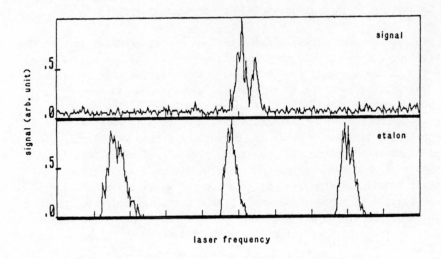

Fig. 3  HF splitting of the 5d106s to 5d106p transition in Au,
the etalon output (FST=50GHz) is given in the lower trace.

4 - Laser desorption

We have investigated the relative desorption efficiency of Au atoms
from a stainless steel (SS) surface for 1064, 532 and 355nm heating
laser beams at various power levels. For this study, the SS rotor was
mounted on a rotation shaft and its tapered surface was continuously
being coated by a Au atomic beam as shown in figure 1. The heating
laser spot was about 5mm in diameter and the rotation of the shaft was
such that successive laser shots slightly overlapped each other on the
rotor surface. For each wavelength, the heating laser power level was
varied and the TOF spectrum was analysed for Au and other background
ion mass peak intensities. These background ions correspond to the
ionization of residual gas and other particles desorbed from the SS
surface by the laser pulse. Results for the variation of the Au peak
intensities are shown in figure 4. The trend of variation is similar in
all cases. The 355 and 532nm heating beams were effective at desorbing
material from the surface, but more spurious ions were created by the
355nm than with the 532nm heating beam, increasing the intensity of the
background peaks. For the 1064nm heating beam, the desorption efficiency
was lower. We could only obtain a clean Au mass peak close to the
threshold power level. At higher energies, too much background material
was desorbed. For desorbing the material in the form of neutral atoms,
the 532nm beam proved to be better and we were able to obtain a clean
Au mass peak with little background ions at a power density of 5 to
10MW/cm$^2$.

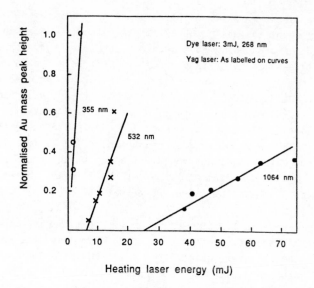

Fig. 4 Relative effectiveness for desorption of
Au atoms from a SS surface.

## 5 - Overall efficiency

To measure the overall desorption-ionization efficiency, the rotor was
replaced by a Ta foil, coated with a known amout of radioactive Au isotopes
(194-196,198 Au). These isotopes were produced via (p,n) reactions on a
natural Pt target using the internal beam of the McGill synchrocyclotron
and later evaporated onto the Ta foil in a separate oven. Repeated synchronized laser pulses were used to desorb and ionize the Au atoms. A
collection foil was placed at the detector position, and the collected
radioactivity was compared to that deposited on the original Ta foil. The
overall efficiency thus obtained was $(1 \pm 0.5) \times 10^{-4}$, and about 70% of the
activity on the Ta foil was removed. Considering the spatial distribution
of the desorbed atoms and the configuration of the dye laser beam, about
$3 \times 10^{-3}$ of the desorbed atoms would be irradiated by the ionization laser
beam. The average dye laser pulse energy was 3mJ, corresponding to an
ionization efficiency of about 12%. The ramaining 0.4 loss factor could
be due to incomplete ion collection, ion-atom collisions and/or that
the Au did not come out in the form of neutral atoms.

## 6 - Summary and conclusions

We have investigated the feasibility of using a simple single-resonance
two-photon RIS scheme to selectively ionize Au atoms in an environment
created by the desorption process induced by a heating laser pulse. We
have studied the power broadening effects and the laser power required for
saturation of the ionization process. This had led us to define suitable

operating conditions for various applications. At a low power level, the
laser frequency stability was adequate for the study of IS and HFS of
heavy elements. The broadening at moderate power level suggests that the
ionization efficiency of this RIS process will not be affected by slight
instability in the dye laser, nor by doppler shifts of the desorbed atoms
moving in different directions. From the studies of the desorption process of the Au atoms from a SS surface, the 532nm beam proved to be most
suitable to produce efficiently a pure Au ion beam. With these operating
conditions optimized, a combined desorption-ionization efficiency of $10^{-4}$
was demonstrated. This efficiency could certainly be increased considerably by improving the geometrical layout of the interaction region,
the laser beam profile and the laser  power output.

For practical application as an on-line ion source, an additional
loss factor -- the adsorption efficiency of Au atoms onto the substrate
surface-- must also be investigated. This will require a continuous Au
atomic beam of known intensity, or preferably, a radioactive beam with
adequate intensity. An alternative approach is to retard an existing
radioactive beam from an on-line isotope separator (ISOL) and implant
these ions onto the substrate. The desorption-ionization efficiency can
then be measured using the current method. If an adequate efficiency
can be reached, this approach can be readily adapted to any ISOL: while
the ISOL will perform the usual function of mass separation, the
proposed scheme of implanting the ions followed by desorption-ionization
method  will provide the function of elemental selection. This will
provide a convenient way to produce an isotopically pure beam for
further experimental applications.

## 7 - References

Andreev S.V., Mishin V.I. and Letokhov V.S. 1986 Op.Comm. <u>57</u>, 317
Bemis C.F. and Carter H.K. ed. 1982 Proceedings of Conference on Lasers
   in Nuclear Physics. Harwood Acad.Publ.
Chen C.H., Kramer S.D., Allman S.L. and Hurst G.S. 1984
   Appl.Phys.Lett. <u>44</u>, 640.
Kluge H.J. Proceedings of Accelerated Radioactive Beams Workshop,
   Parksville, B.C., Canada. 1985 TRIUMF publication (TRI86-1)
Streib J., Kluge H.J., Kremmling H., Moore R.B., Schaat H.W. and
   Wallmeroth K. 1985 Z.Phys.A. <u>321</u>, 53
Wallmeroth K. and ISOLDE collaboration, 1986 Private comm. and
   contribution to this conference.

*Inst. Phys. Conf. Ser. No. 84: Section 4*
*Paper presented at RIS 86, Swansea, Wales, 7–12 Sept. 1986*

# Surface studies using particle beam induced desorption and multiphoton resonance ionization

Nicholas Winograd

Department of Chemistry, 152 Davey Laboratory, The Pennsylvania State University, University Park, PA 16802

Abstract. Energetic particle beams are now important tools in a variety of areas ranging from chemical modification of electronic materials through reactive ion etching and ion implantation to probes of surface chemistry and high molecular weight mass spectrometry. This paper will focus on the fundamental aspects of the particle/solid interaction process and on how these aspects may be utilized to understand the chemistry of reactions resulting from high energy collisions. Our experimental strategy is to perform energy and angle-resolved measurements on atoms and molecules desorbed from single crystal surfaces. These measurements are made possible by a novel detector which utilizes the single-atom sensitivity of multiphoton resonance ionization (MPRI), the position-sensitivity of a microchannel plate and the timing capabilities of a time-of-flight analysis. Our measurements are then compared directly to results of molecular dynamics calculations of the particle-impact event. These comparisons help us to formulate mechanisms of molecular cluster formation and to elucidate the origin of angular anisotropies in the yields of desorbed species.

## 1. Introduction

Elucidation of the fundamental aspects of interactions of keV particles with solids has been of considerable interest during the past few decades. (Winograd 1981) Molecular dynamics calculations of the ion impact event are now available which help to formulate the mechanisms of cluster emission as well as to unravel the origin of angular anisotropies for particles ejected from single crystal surfaces. (Winograd 1981) Experimental measurements of particle trajectories, however, have not been possible at the level necessary to make direct comparisons to theoretical predictions. Furthermore, there exists a strong indication that a great deal more information is contained in data which are <u>concurrently</u> energy- and angle-resolved.

In this work, we report the design and application of a detector which can be used to collect, *simultaneously*, energy- and angle-resolved neutral-particle (EARN) distributions for material desorbing from ion-bombarded solid surfaces. The detector utilizes multiphoton resonance ionization (MPRI) as a postionization method which couples pulsed-laser and pulsed-ion-beam techniques to yield time-of-flight

(TOF) energy measurement. A microchannel-plate (MCP) element allows position-sensitive detection of the laser-ionized particles for determination of ejection angle. (Kobrin 1986) These components give the EARN detector the capability of continuous operation within an ultrahigh-vacuum (UHV) environment. Energy and angular distributions reported by this method therefore constitute the first reported complete neutral-particle ejection distributions for material desorbing from clean, well-defined surfaces.

## 2. Experimental Procedure

The experimental apparatus for obtaining the desired information has been described in detail (Kobrin 1986). Briefly, the sample is placed in a two-level UHV chamber equipped with a Physical Electronics 15-120 LEED/Auger system, a differentially pumped pulsed ion gun, quartz laser ports and an imaging detector for conducting the energy and angle-resolved neutral (EARN) experiments. The tunable photon radiation is provided by a Quanta-Ray PDL-1 dye laser pumped by a DCR-2A Q-switched Nd:Yag laser. Frequency mixing and doubling are accomplished using a WEX-1 wavelength extender. The resulting radiation takes the form of a 6-ns pulse which is repeated at a rate of 30 Hz. The 0.5-cm diameter laser beam is focused to the shape of a 0.1-cm thick ribbon using a 225-cm focal length cylindrical lense and is positioned 1.5 cm above and parallel to the target.

The general timing scheme used for the EARN experiments is controlled by an LSI 11/23 computer system through a CAMAC interface and two LeCroy model 2323 dual gate generators. In general terms, the experiment proceeds as follows: (1) a 200-ns pulse of approximately $2.5 \times 10^6$ ions is focused to a 2-mm spot on the sample; (2) upon impact of the ion pulse, an ion extraction field is activated for the duration of the measurement; (3) a laser pulse (~1 mJ by 6 ns) ionizes a small volume of neutral particles at a time $\tau_E$ after the ion-pulsed impact, thus defining the time-of-flight (TOF) of the probed particles; (4) the ionized particles are then accelerated by the ion-extraction field and arrive at the front of a microchannel plate (MCP) assembly at a time $\tau_m$ later, $\tau_m$ being governed primarily by the mass-to-charge ratio of the ionized particle. The impacting particles are detected on a phosphor screen by a RCA model TC2911 CCD series video camera during a time when the MCP is gated to an active state. The gate is used to discriminate against signals from scattered laser light and from ions striking the MCP at other times due to differences in either mass or ionization scheme. This image can then be accessed by an LSI 11/23 computer system for digital image data processing.

### 3.1 Results from polycrystalline targets

We have examined the EARN distributions from polycrystalline In and Rh foils. In general, as shown in Fig 1, these distributions may be

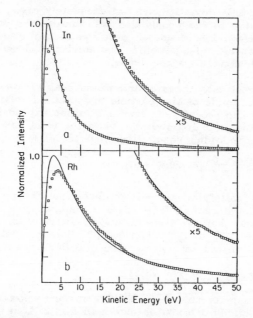

Figure 1. Angle–integrated energy distributions of atoms ejected from polycrystalline (a) In and (b) Rh bombarded by normally incident 5–keV Ar$^+$ ions. The experimental values ($\square$) as well as results from the predictions of eq. (1) (solid curve) are shown. The best fit for n=2 and peak energy $E_p=U/2$ was determined at 50 eV and extrapolated to 0 eV.

obtained with extremely high quality using primary ion dosages many times lower than previously reported. It should even be possible to record these distributions for sub–monolayer concentrations of overlayer species.

The shape of the distributions provide insight into the mechanism of energy dissipation by the lattice. According to the theory of this process, the distribution is expected to be of the form (Thompson 1968).

$$Y(E,\theta) = \frac{C\ E\ \cos\theta}{(E + U)^{n+1}}$$

(1)

where C is a normalization constant, E is the kinetic energy of the ejected particle, $\theta$ is the take–off angle measured from the surface normal, and n is typically close to 2.0. This expression predicts that, regardless of angle, $Y(E,\theta)$ maximizes at $E = U/n$ and decays as $E^{-n}$ when $E\gg U$. It also predicts that the angular distribution of particles is independent of E. This equation, as seen in Fig. 1, works well in the high KE regime, but deviates significantly at lower KE values. In addition, the peak position occurs at values greater than

U/2. The reasons for this deviation are not completely clear, although it appears to be related to the poorly defined concept of a surface binding energy (Garrison 1986b) and to problems in the assumptions regarding isotropic energy dissipation in the solid and those regarding binary collisions (Garrison 1986a).

It is of further interest that these distributions peak at lower KE values as the take-off angle of the particle is increased relative to the normal. Garrison has recently derived a modified expression which arbitrarily assumes that the distribution inside the solid has a $\cos^2\theta$ form to arrive at the following result (Garrison 1986a):

$$Y(E,\theta) = \frac{C \, E \cos \theta}{(E + U)^4} \, (E \cos^2\theta + U).\tag{2}$$

This formula is quite satisfying since it predicts that the polar angle distribution should change from $\cos \theta$ in nature when $E < U$ to $\cos^3\theta$ when $E \gg U$ and that the peak in the KE distribution should shift from U/2 when $\theta = 0°$ to U/3 when $\theta = 90°$. Both predictions are semi-quantitatively born out by our experimental EARN results.

### 3.2 EARN distributions from single Rh{111}

The EARN intensity map for the clean Rh{111} surface along the open crystallographic directions $\phi = \pm 30°$ is shown in Fig. 2. It is possible

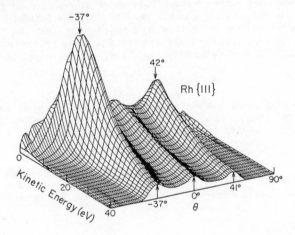

Figure 2. The EARN intensity map for clean Rh{111}. The positive values of $\theta$ are recorded along $\phi = 30°$ and the negative values of $\theta$ are recorded along $\phi = -30°$.

to understand the important features of this map using the molecular dynamics simulations to extract dominant collision sequences that lead to particle ejection. (Winograd 1986) These sequences begin with the alignment of atomic motions inside the solid. As these motions cause ejection of first-layer atoms, further focusing is caused by channeling or blocking by other first-layer atoms. For example, the highest intensity is observed along the open crystallographic directions ($\phi$ = ±30° in our case) and the minimum intensity is observed along the close-packed crystallographic direction ($\phi$ = 0°). If only surface processes were important, the peaks at $\phi$ = -37°, $\phi$ = -30° and $\theta$ = 42°, $\phi$ = 30° should be equal in intensity and not unequal as shown in Fig. 2. The additional intensity at $\theta$ = -37°, $\phi$ = -30° arises mainly from the ejection of atom 2 as denoted in Figure 3 by atom 1 with

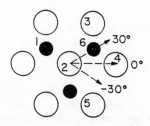

Figure 3. The open circles represent first layer atoms and the solid circles represent second layer atoms for the {111} surface. The atoms are labeled with arbitrary numbers for reference purposes. The atop adsorption site is considered to be directly above any Rh surface atom. The b-site refers to adsorption in the 3-fold hollow position above a second-layer atom, for example 1 or 6. The c-site is the 3-fold hollow position above the third layer atom (not shown).

first-layer focusing by atoms 4 and 5. The peak at $\theta$ = 42° and $\phi$ = 30° is lower in intensity by a factor of 0.5 at low kinetic energy (KE) since no such mechanism is available along this azimuth. The peak at $\theta$ = 0° arises mainly from ejection of the second-layer atom 6 which is focused upward by atoms 2, 3 and 4.

The important features of the EARN distributions are systematically altered by the presence of a p(2x2) oxygen overlayer as seen in Fig. 4. At low KE the intensity of the peak along $\phi$ = -30° is

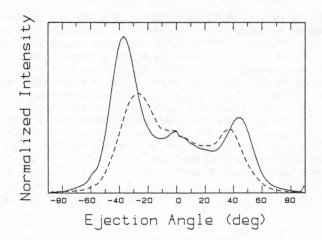

Figure 4. Comparison of the angular distributions for clean Rh{111} (——) and p(2x2)O/Rh{111} (---). The particles are selected to have KE's between 20 and 50 eV. The ejection angle is the same as that shown in Figure 2.

preferentially reduced relative to the peak along $\phi = 30°$ and its peak angle is shifted 10° closer to the normal relative to the clean surface ($\theta = -27°$ versus $-37°$). This simple observation suggests that the oxygen atom bonds preferentially in the c-site since ejection of atom 2 is blocked by oxygen atoms only along $\phi = -30°$. It is gratifying that extensive LEED calculations also suggest that c-site adsorption is slightly favored over other possible bonding geometries.(Wong 1986)

## Acknowledgement

The author is grateful to the National Science Foundation, the Air Force Office of Scientific Research, and The Office of Naval Research for partial financial support of this work. The contribution of many talented collaborators including Paul Kobrin, Alan Schick, Jyothi Singh, Jim Baxter, Fred Kimock, Dave Pappas, Curt Reimann and Barbara Garrison is also gratefully acknowledged.

## References

Garrison B. J., *Nuclear Instrum and Methods* (1986a).
Garrison B. J., private communication (1986b).
Kobrin P. H., G.A. Schick, J.P. Baxter and N. Winograd,
   *Rev. Sci. Instrum* 57, 1354 (1986).
Thompson M.W. *Philos. Mag.* 18, 377 (1968).
Winograd N. *Solid State Chem.* 13, 285 (1981).
Winograd N., P.H. Kobrin, G.A. Schick, J. Singh, J.P. Baxter and
   B.J. Garrison, *Surf. Sci. Lett.*, (1986).
Wong P.C., K.C. Hui, M.Y. Zhou and K.A.R. Mitchell, *Surf. Sci.* 165,
   L21 (1986).

*Inst. Phys. Conf. Ser. No. 84: Section 4*
*Paper presented at RIS 86, Swansea, Wales, 7–12 Sept. 1986*

151

# On the use of *nonresonant* multiphoton ionization for surface analysis

C. H. Becker, Chemical Physics Laboratory, SRI International, Menlo Park, California 94025 USA

Abstract. The utility of nonresonant multiphoton ionization for surface chemical analysis by mass spectroscopy is considered. Examples are presented where a nonresonant approach provides simultaneous detection of many species with high sensitivity and quantitative capability.

## 1. Introduction

The papers in these proceedings clearly show many important applications of resonantly-enhanced multiphoton ionization (MPI). In particular, resonance ionization has led to important advances in surface and material analysis; in addition to these proceedings, for references see e.g. Winograd et al. (1982), Mayo et al. (1982), Parks et al. (1983), Kimock et al. (1984), Fassett et al. (1984), Pellin et al. (1984), and Donohue et al. (1985). Nevertheless, the role that a nonresonant approach can play should not be overlooked.

The method developed at SRI, which we call surface analysis by laser ionization or SALI for shorthand, features nonresonant MPI of desorbed neutral atoms and molecules by an intense and untuned focused UV laser beam passing above (~ 1 mm) and parallel to the surface, followed by reflecting time-of-flight mass spectrometry (Becker and Gillen 1984). (The UV wavelength simply serves to minimize the number of photons needed for ionization, thus yielding more efficient ionization for moderate power lasers.) The stimulated desorption is caused by ion, (a second) laser, or electron beam, as suits the sample under investigation (Becker and Gillen 1985). Thermal desorption is also detected. In this arrangement it is the mass spectrometer, rather than laser, that serves to discriminate against background and separate the various chemical moieties. Such an approach to surface analysis is especially appropriate where many species need to be examined simultaneously, where one is ignorant of the possible species present, and where molecules with complicated or unknown spectroscopy or high density of occupied states are probed.

Even with the nonresonant character of the typical ionization, SALI analyses are generally very sensitive due to: (a) the large solid angle subtended by the laser ionization volume by working close to the surface, (b) efficient ionization (often saturating the ionization in the focal region) by using high power density UV sources, and (c) the high transmission of a time-of-flight mass spectrometer. A detection sensitivity of $10^{-8}$ monolayer has been demonstrated with a useful yield (transformation of surface molecule to detected ion) of $10^{-4}$ (Becker and Gillen 1984); a useful yield of $10^{-3}$ is anticipated for a new apparatus currently under construction.

Because the ionization is frequently saturated and desorbed neutrals typically represent the vast majority of desorbing species, the mass spectra are usually at least semi-quantitative in nature. Remaining uncertainties arise from varying velocity distributions of desorbing neutrals, effective laser beam ionization volumes for different moieties, and, for molecules, fragmentation patterns. These uncertainties are however subject to measurement or calibration.

Presented below are two examples where numerous species are detected simultaneously and with calibration providing quantitation of the results. One example is a sputter depth profile though the near surface region of a non-uniform thin film bimetallic alloy, and the other is the surface analysis of a plasma etch residue on a Si wafer.

## 2. Chemical Analyses

The experimental arrangement has been described in some detail previously (Becker and Gillen 1984). Sputtering was performed using a rastered and gated 3 keV $Ar^+$ beam of approximately 1 µA incident at 70 degrees from normal with a beam spot size of ~ 0.01 $cm^2$. The ion beam source was differentially pumped and the analysis chamber pressure was about $1 \times 10^{-8}$ torr with the ion beam on. Samples were mounted on a direct probe specimen-holder, to allow rapid sample exchange.

The laser ionization was performed with an excimer laser beam, pulsed at 8 Hz, with the output at 248 nm, 5.0 eV (KrF) focused by a 40 cm focal length suprasil lens. A stable resonator (giving rather high beam divergence and poor focusability) was used in these studies though better divergence is recommended in general.

Positive photoions are detected by a dual microchannel plate assembly, with the signal variably attenuated between 0 dB and 40 dB to maintain a linear response by the detection electronics. The data are recorded in analog form from the microchannel plate electron multiplier as the voltage drop across a 50 Ω resistor using a 100 MHz transient digitizer. Secondary ions sputtered directly from the sample are discriminated against by using the TOF reflector as a retarding-field energy analyzer (Becker and Gillen 1984). Calibration for elemental analyses was done with external standards as specified below. With all calibrants, care was taken to reach steady-state sputtering of the bulk.

An Al-Ti thin film alloy was prepared by co-sputtering Al and Ti metal at a ratio of 3:1, respectively, followed by an anealing cycle. The bulk composition was independently determined (Davies 1986) to be $Al_3Ti$ stoichiometry by Rutherford backscattering spectroscopy (RBS) as well as crystalline $Al_3Ti$ with tetragonal structure by transmission electron microscopy (TEM) microdiffraction studies (no inclusions were found). The SALI depth profile data were quantified by comparison with a Ti-Al-Nb fine powder 60:30:10 by weight (200 mesh powder, manufacturered by Alfa Products) pressed into an In foil. The depth scale was fixed by comparison to an earlier Auger electron spectroscopy (AES) study (Davies 1986); a sputter time of 12 minutes corresponds to a depth of approximately 500 Å.

The depth profiles for the Al-Ti thin film alloy are shown in Figure 1. The Al has significantly segregated to the surface region. Comparison with the Ti-Al-Nb powder standard gives a Al:Ti ratio in the bulk 2.92 ± 0.1:1; which is in good agreement with the accepted 3:1 ratio. Thus the absolute error in this case is estimated to be less than 5%. It is clear that Al has segregated to the surface. The $AlO^+$ photoion signal in Figure 1 shows oxidation of the Al surface. This diatomic is the strongest oxide-bearing

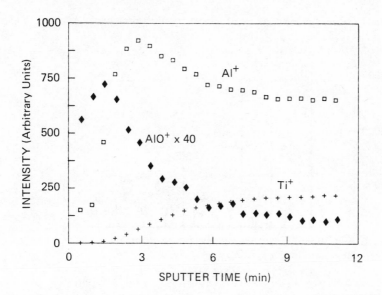

Figure 1. SALI depth profile of a co-sputtered Al-Ti thin film using 248 nm pulsed radiation at approximately $5 \times 10^8$ W/cm$^2$, accumulating signal over 200 laser pulses per depth point with 40 dB signal attenuation. The relative Al and Ti concentrations have been calibrated. The total sputtered depth shown is about 500 Å.

mass peak, though there are heavier $Al_xO_y^+$ peaks. Significant amounts of titanium oxide were not observed. An aluminum oxide standard was not examined to check stoichiometry though it is anticipated that the thermodynamically stable $Al_2O_3$ is present on the surface. The initial signals begin near zero due to the initial removal of atmospheric and any handling impurities on the surface.

An interesting submonolayer analysis studied by SALI is that of an etch residue on a silicon wafer from a plasma etcher using $C_2F_6$. Care was taken to detect impurities deposited by the etching process from various adjacent surfaces in the particular version of the plasma etcher used for this run; such impurities include Al, Cr, Fe, Ni, and Cu. Two standards were used for calibration: an Al-Si powder 88:12 by weight (325 mesh powder from Alpha Products) pressed on an In foil, and a NBS stainless steel standard C1154 to calibrate Si, Cr, Fe, Ni, and Cu. Metals can be a particularly troublesome impurity for Si processing because of their energy levels at mid-band gap.

Figure 2 shows a spectrum with static sputtering of the etch residue on a Si wafer, and Table 1 gives the data reduction of this figure. The relative calibrated amounts shown in Table 1 for the various elements can be considered only semi-quantitative because of the prevalent fluoride and oxide bonding, whereas the calibrations were performed for the reduced

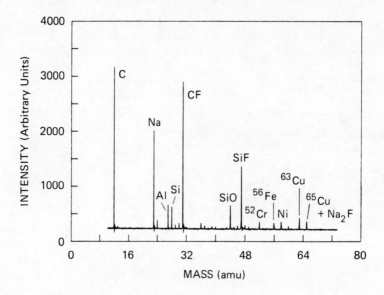

Figure 2. SALI time-of-flight mass spectrum taken from static (roughly $3 \times 10^{-6}$ monolayers) $Ar^+$ sputtering of an etch residue on a silicon wafer after a $C_2F_6$ plasma etch. The spectrum is a result of accumulating signal from 100 laser pulses with a 40 dB attenuated signal, and 248 nm radiation at approximately $5 \times 10^8$ W/cm$^2$.

Table 1. ANALYSIS OF A FLUOROCARBON PLASMA ETCH RESIDUE ON SI[a]

| m/e | Assignment | Relative Raw Signal | Relative Calibrated Atomic Signal[b] |
|-----|-----------|---------------------|--------------------------------------|
| 23 | Na | 1820 | |
| 27 | Al | 440 | 60 |
| 28 | Si | 400 | 880 |
| 31 | CF | 2670 | |
| 44 | SiO | 420 | |
| 47 | SiF | 1130 | |
| 52 | Cr | 130 | 100 |
| 56 | Fe | 100 | 7.6 |
| 58 | Ni | 120 | 170 |
| 63 | Cu | 190 | 160 |
| 65 | Cu + Na$_2$F | 130 | |

[a]Derived from the data of Figure 2.
[b]Using NBS stainless steel C1154 and Si-Al alloy powder as calibrants, see text.

elements. Nevertheless the relative calibrations provide a valuable approximate estimate of much of the residue composition. The carbon signal probably originates mostly from sputtering of $C_xF_y$ species. The amount of material sputtered to record this spectrum is estimated to be $\approx 3 \times 10^{-6}$ monolayer, though comparable data could probably have been recorded using only 1/20 to 1/100 of this amount, given the signal-to-noise ratios.

Calibrations with oxidized elements can be performed, including the relative sensitivity for both atomic and molecular photoions such as $Si^+$, $SiF^+$, $SiO^+$, $C^+$, and $CF^+$, including photodissociation/ionization channels; however, this is a more complex task and was not performed for this etch residue because this information was not considered necessary. Due to the "weak" laser power and relatively long wavelength UV light used (248 nm, 5.0 eV), negligible atomic O and F signals were observed. This can be dramatically overcome with about two orders of magnitude increase in power density (a practicality with a state-of-the-art laser), especially at 193 nm. The particularly high sensitivity to some species observed in Table 1 (e.g., for Fe) is due to an accidental resonance and the fact that the laser output using the stable resonator in this instance is not very tightly focused leading to a significantly larger effective ionization volume for easily ionized species. In the limit of very high UV power densities (say, $> 10^{11}$ W/cm$^2$ for pulse lengths $\approx 10^{-8}$ s) with good spatial beam quality, the raw data should accurately approach (because of uniform saturation) the density of the sputtered flux.

In conclusion, the SALI technique, which is a nonresonant ionization procedure coupled with stimulated desorption and time-of-flight mass spectroscopy, is seen to be of significant value for the analysis of many chemical species simultaneously. Very high sensitvity and substantial quantitative capability are also characteristics of the surface analysis. Examples have been presented of the sputter depth profile through the near surface region of an Al-Ti thin film and the submonolayer analysis of a plasma etch residue on a Si wafer.

## Acknowledgments

Support from NSF Division of Materials Research and Intel Corporation is gratefully acknowldged.

## References

Becker C H and Gillen K T 1984 Anal. Chem. 56 1671
Becker C H and Gillen K T 1985 J. Vac. Sci. Technol. A3 1347
Davies P W 1986, private communication
Donohue D L, Christie W H, Goeringer D E, and McKown H S 1985 Anal. Chem. 57 1193
Fassett J D, Moore L J, Shideler R W, and Travis J C 1984 Anal. Chem. 56 203
Kimock F M, Baxter J P, Pappas D L, Kobrin P H, and Winograd N 1982 Anal. Chem. 56 2782
Mayo S, Lucatorto T B, and Luther G G 1982 Anal. Chem. 54 553
Parks J E, Schmitt H W, Hurst G S, and Fairbank, Jr. W M 1983 Thin Solid Films 108 69
Pellin M J, Young C E, Callaway W F, and Gruen D M 1984 Surf. Sci. 144 619
Winograd N, Baxter J P, and Kimock F M 1982 Chem. Phys. Lett. 88 581

*Inst. Phys. Conf. Ser. No. 84: Section 4*
*Paper presented at RIS 86, Swansea, Wales, 7–12 Sept. 1986*

# Progress in analysis by sputter initiated resonance ionization spectroscopy

J E  Parks, D W  Beekman, L J  Moore, H W  Schmitt, M T  Spaar, and E H Taylor

Atom Sciences, Inc., 114 Ridgeway Center, Oak Ridge, Tennessee  37830

J M R  Hutchinson

National Bureau of Standards, Gaithersburg, Maryland 20899

W M  Fairbank, Jr

Department of Physics, Colorado State University, Fort Collins, Colorado 80523

Abstract.  Sputter initiated resonance ionization spectroscopy (SIRIS) has been found to be useful for the analysis of materials in varied fields, including semiconductor, medical, bioassay, health physics, and basic science communities. Progress in the development of SIRIS measurements for these uses are discussed and results are presented.

## 1.  Introduction

The development of sputter initiated resonance ionization spectroscopy (SIRIS) has progressed since the technique and initial results were presented at the last symposium on RIS (Parks 1984). The technique has continued to be applied to materials analysis for the semiconductor industry. In addition to measurements of impurities and dopants in silicon, measurements have been made in gallium arsenide, and recently the capability for depth profile analysis was added to the apparatus. Depth profiles of silicon implanted in gallium arsenide have been determined, and the performance of the instrument is now being optimized. At Atom Sciences, other applications for RIS and SIRIS are being developed and these include measurements in geophysical, medical, biossay, and health physics applications, as well as in basic science. The results of some of these measurements are presented and discussed.

The details of SIRIS have been well described in other works by Atom Sciences (Parks 1983a,b) as well as by other workers using the same technique, but calling it MPRI (Winograd 1982) and SARISA (Pellin 1984). Briefly, in the SIRIS process, a sample to be analyzed is placed in a target chamber under vacuum. An energetic, pulsed ion beam is then directed to the sample where it sputters a small portion of the sample, causing a cloud of neutral atoms and ions to form and drift away from the sample. The atoms of the vapor cloud are representative of the material being analyzed and these atoms are then probed with a synchronously pulsed beam of laser light, properly tuned to the precise wavelength(s) to excite and ionize atoms of the selected element via the RIS process. These ions are extracted and detected using a charged particle detector. A mass spectrometer can be incorporated when isotope identification is of interest, and at Atom Sciences both a magnetic sector and a time-of-flight mass spectrometer have been used. A schematic diagram of the apparatus using these two different types of

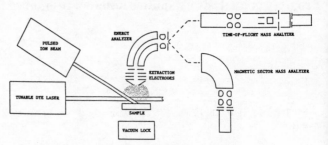

Figure 1    Schematic diagram of SIRIS apparatus showing two types of mass spectrometers.

mass spectrometers is shown in Figure 1. Secondary ions generated in the sputtering process constitute a backgound noise in SIRIS and are eliminated with the use of time and energy discrimination.

Atom Sciences has continued to use Standard Reference Materials (SRMs) from the National Bureau of Standards to test the measurement procedures in SIRIS. A number of elements, including aluminum, vanadium, boron, copper, molybdenum, and silicon have been measured in steel SRMs and in each case the SIRIS measurements have correlated very well with the certified concentrations. Recently, measurements of silicon in gallium arsenide and of copper in blood have been of interest, and the SRMs of steel were analyzed for silicon and copper. Figures 2 and 3 show the results of these measurements. Two other samples of lower silicon concentration are included in the plot with the steel SRMs, but the analyses from the supplier are of unknown accuracy. The SIRIS results agree very well with the reference values and with those from the niobium and tungsten samples although the latter are quite different matrices from steel. This indicates that the matrix dependence in the SIRIS measurement method for this subset of metal-based matrices is minimal; the generality of any apparent matrix independence in SIRIS measurements clearly requires further study.

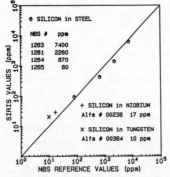

Figure 2  Comparison of SIRIS measurement with certified reference values of silicon in steel.

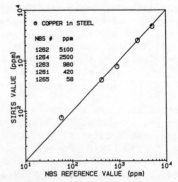

Figure 3  Comparison of SIRIS measurements with certified reference values of copper in steel.

Results were presented at the last RIS symposium which demonstrated a sensitivity for the detection of gallium in silicon of 2 ppb. In those measurements, a high purity silicon substrate was used to determine the background limitation of the measurements. Following those measurements in order to demonstrate an impurity measurement at the ppb level, a set of silicon samples containing boron in bulk concentrations ranging from 15.4 ppm down to 5.3 ppb were obtained and measured. These samples, grown by Wacker and supplied to Atom Sciences by Tektronix, Inc., had been characterized first by resistivity measurements, and later Atom Sciences had them characterized by FTIR measurements. The FTIR measurements were found to agree with the resistivity measurements and were compared with the SIRIS measurements of the same samples. This comparison is shown in Figure 4. The agreement between the SIRIS values and the reference values is good down to the 50 ppb level, but below that level the agreement suffers. Below 50 ppb, higher concentrations of boron in silicon are measured in the SIRIS measurements than are indicated by the reference values. The measurements were checked several times and sources of contamination within the system were investigated and eliminated as the cause. A background contribution was found inside system, most probably from a small amount of secondary sputtering, but this effect was small and the discrepency still existed.

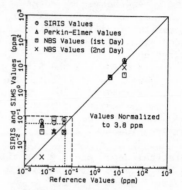

Figure 4 Comparisons of SIRIS and SIMS measurements with reference values of boron in silicon.

The samples were analyzed by Perkin-Elmer at their Physical Electronics Laboratory using their PHI 6000 Series SIMS instrument. Their results are shown in Figure 4 as triangles and were found to agree quite well with the SIRIS results. As a double check the samples were sent to the National Bureau of Standards and were run on their Cameca IMS 3F SIMS instrument. The results obtained by NBS during the first day of analysis, shown as squares, agree very well with the SIRIS results and the Perkin-Elmer SIMS results. However, NBS ran the samples a second day and milled the samples containing the smaller concentrations down to a depth of 8 microns. These data are shown as crosses in Figure 4 and are found to correlate very well with the reference values obtained by resistivity measurements. A check of the Perkin-Elmer measurement determined that their measurements were made to a depth of 2.7 microns and the results at that depth agreed with the NBS values of taken at that depth. These results indicate that a high concentration of boron had accumulated at the surface of the samples and the true bulk concentration could only be measured when the sample could be probed to a sufficient depth, free of the surface concentration and its effects when sputtered with an energetic ion beam.

The depth profiling capability enhances the usefulness of SIRIS to the semiconductor industry. This capability has recently been incorporated into the SIRIS instrument at Atom Sciences. Devices are made by implanting dopants into the substrate materials and the behavior of the device depends on the concentration as a function of depth. Also, devices are constructed of different layers of materials and it is sometimes possible for one of these materials to diffuse into the other and affect the electrical response of the other material and hence the device. Because SIRIS is a pulsed technique, SIRIS only analyzes the surface concentration of the material being probed. Typically, about 1000 ion pulses are required to sputter a monolayer of material from the surface. As a result, SIRIS has the advantage of being capable of measuring very small samples and samples deposited as a thin layer.

To measure the concentration of an analyte in the bulk of a material, the material can first be sputtered away with a dc ion beam and then analyzed with the pulsed beam. By repeating this procedure, depth profiles of implanted samples can be measured. Figure 5 shows early results of a depth profile of silicon implanted in gallium arsenide. A sample was implanted with silicon-29 atoms at an energy of 200 keV, such that a peak concentration of $5 \times 10^{18}/cm^3$ was established. The sample had been depth profiled with SIMS, and the SIRIS and SIMS results are compared in Figure 5. The agreement is excellent. The background in the SIRIS measurement appears to be about $3 \times 10^{16}/cm^3$, arising most probably from natural silicon in the sample and/or the apparatus. Neither the SIRIS measurement nor the SIMS measurement had been optimized for detecting silicon in gallium arsenide.

A theoretical depth profile for silicon implanted into gallium arsenide at 200 keV is also shown in Figure 5. While the SIRIS and SIMS measurements agree with each other, the SIRIS and SIMS measurements also agree with the the theoretical distribution on the small depth side of the distribution, but the measurements deviate from the theory at the higher depths. This could be due to background and/or resolution of the instruments. SIRIS has the potential for measuring depth profiles with high resolution. Since less than a monolayer of material is sputtered during a typical measurement time, the range of depths over which sputtered material is analyzed is a minimum, and hence the resolution is optimum. It is possible that an ion beam that is too energetic can cause

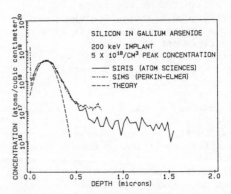

Figure 5    SIRIS depth profile of silicon in gallium arsenide.

mixing between the layers during the measurement and ion milling times. In the present case the energy of the ion beam was 18.5 keV. Experiments are planned at lower energies, down to 0.5 kev.

The RIS technique is well-suited for some applications in physiology and medicine, one being the measurement of trace elements in small blood samples. Because SIRIS is a pulsed technique, very small and thin samples can be analyzed without sputtering away the entire sample. Recently, Atom Sciences demonstrated that trace amounts of copper in a 0.25 ml serum sample could be measured in SIRIS. A bovine serum, Reference Material 8419, was obtained from the National Bureau of Standards, and analyzed using the inherent accuracy of isotope dilution coupled with the RIS technique. After spiking the serum with isotopically enriched copper, the sample was chemically prepared and deposited onto a high purity gold substrate. Bombardment with argon ions in the SIRIS apparatus allowed the sputtered copper atoms to be ionized and detected by means of resonance ionization. From the isotopic ratio determined by the double focusing magnetic spectrometer, the concentration of the copper in the bovine serum was determined.

Using larger samples, typically 1–3 ml, this bovine serum had been extensively characterized for trace element content by numerous analytical techniques and had been determined to have a copper concentration of $0.73\pm0.1$ µg/g (Veillon 1985). The SIRIS measurement gave a value of $0.76\pm0.1$ µg/g, which agrees well with the reference value within the experimental uncertainties of $\pm15\%$. Other blood serum samples were measured and were found to agree with values determined by the technique of atomic absorption on larger, 2 ml samples. The experiments were repeated for molybdenum as the analyte, and a detection limit of only 20 pg was demonstrated.

In addition to semiconductor and medical applications, Atom Sciences has been involved in the use of SIRIS for bioassay, health physics, materials composition, and basic science applications. SIRIS in some cases is an attractive alternative to radiological counting methods, especially where the radioactive species has a long half-life. The detection of uranium in urine with SIRIS has been investigated to determine whether SIRIS could provide a rapid, sensitive, and cost effective means of analysis. In this study (Parks 1986), it was found that 1.0 µg/l of uranium in urine could be detected, with some advance chemical preparation. The minimum detection limit in this study was limited by reagent impurities and the fact that the RIS detection efficiency for uranium from oxides is much lower than that for other elements.

The determination of trace amounts of uranium in soils is of interest. As a result, the characteristics of SIRIS were investigated as a means of analysis. Solid, conducting samples compatible with high vacuum were made by compacting soil with 20 percent or greater graphite binder. The conducting graphite alleviated any adverse effects due to charging of the sample by the ion beam. For this case, it was found that matrix effects on the absolute magnitude of uranium SIRIS signals, can be as large as two orders of magnitude. This precludes direct comparisons of measured uranium concentration through uranium SIRIS signal levels. In order to quantify the measurements, the method of isotope dilution was explored. Systematic errors caused by different molecular forms or different microscopic physical locations of the two isotopes were found to be less than 20 percent in this method. It was found that for samples in which these effects are minimized, the capablility of the instrument is such that uranium concentrations can be determined successfully to 5 to 10 percent accuracy.

The physical properties of uranium alloys are highly dependent on the concentrations of the alloying metals. Atom Sciences has used SIRIS to measure carbon in samples which had been characterized with photoabsorption spectroscopy analyses. The samples were found to be very heterogeneous and showed different concentrations at different depths and positions within the samples. However, the SIRIS analyses correlated very

well with the photoabsorption
values, and the results are
shown in Figure 6. The one
renegade data point was later
found to be correlated with
a carbon inclusion at a
grain boundary.

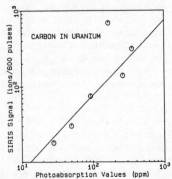

Figure 6   Comparison of SIRIS
measurements of carbon in uranium
with photoabsorption measurements.

In other work, SIRIS is being used at Atom Sciences to study basic
science phenomena. The SIRIS apparatus has been used to search for
superheavy atoms and a detection limit of $1 \times 10^{-11}$ has been demonstrated
in searching for superheavy lithium atoms. Also, work is underway to
investigate the occurrence of isotope fractionation in the sputtering of
neutral atoms as is known to occur in thermal vaporization and in SIMS
with the sputtering of ions. Atom Sciences uses SIRIS to study the RIS
process itself and to date, the SIRIS process and apparatus is a most
convenient and effective way to atomize the samples.

Acknowledgements:

This work was supported in part by the National Bureau of Standards, the
National Science Foundation, the Department of Energy, and the Avionics
Laboratory, Air Force Wright Aeronautical Laboratories, Aeronautical
Systems Division (AFSC), United States Air Force, Wright-Patterson AFB,
Ohio 45433.

References:

Parks J E, Schmitt H W, Hurst G S and Fairbank W M 1983a Thin Solid Films
    108 69
Parks J E, Schmitt H W, Hurst G S and Fairbank W M 1983b Laser-based
    Ultrasensitive Spectroscopy and Detection V, Richard A. Keller, Editor,
    Proc. SPIE 426 32
Parks J E, Schmitt H W, Hurst G S and Fairbank W M 1984 in Resonance
    Ionization Spectroscopy 1984, Invited Papers from the Second
    International Symposium on Resonance Ionization Spectroscopy and Its
    Applications, (Conference Series Number 71, The Institute of Physics,
    Bristol and Boston) pp 167-174
Parks J E, Taylor E H, Beekman, D W and Spaar M T 1986 U.S. Nuclear
    Regulatory Commission Report NUREG/CR-4419
Pellin M J, Young C E, Calaway W F and Gruen D 1984 Surf. Sci. 144 619
Veillon C, Lewis S A, Patterson K Y, Wolfe W R, Harnly J M, Versieck J,
    Vanballenberghe L, Cornelis R and O'Haver T C 1985 Anal. Chem. 57 2106
Winograd N, Baxter J P and Kimock F M 1982 Chem. Phys. Lett. 88 581

*Inst. Phys. Conf. Ser. No. 84: Section 4*
*Paper presented at RIS 86, Swansea, Wales, 7–12 Sept. 1986*

163

# Trace surface analysis via RIS/TOF mass spectrometry

C.E. Young, M.J. Pellin, W.F. Calaway,
B. Jørgensen,** E.L. Schweitzer, and D.M. Gruen
Materials Science and Chemistry Divisions
Argonne National Laboratory, Argonne IL 60439

Abstract.   An energy- and angle- refocusing time-of-flight (TOF)
instrument for Surface Analysis by Resonance Ionization of
Sputtered Atoms (SARISA) has been tested by analysis of an Fe-
implanted silicon sample ($10^{11}$ atoms/cm$^2$, 60 keV). Data obtained
were: 5% collection efficiency and <2 ppb sensitivity (for S/N=1,
1000 s experiment; effective repetition rate = 10 Hz, 0.9 monolayer
total removed in an individual measurement). Ion pulse: 2μA for 2
μs, area 0.05 mm$^2$ for RIS data, 2x2 mm$^2$ raster for depth profiling.

## 1. Introduction

This paper reports construction and performance data for an
isochronous, energy- and angle-refocusing TOF mass spectrometer in
which special attention was devoted to reduction of noise reaching
the detector. Resonance ionization techniques are used to enhance
selectivity in the ionization of atoms sputtered from a sample of
interest. The early realization of the sensitivity of RIS processes
(Ambartsumyan and Letokhov 1972, Hurst et al. 1975) was followed by
detailed analysis of the optical physics involved (Letokhov
et.al.1977; Hurst et.al.1979; Payne et.al.1981; Letokhov 1983; Payne
1984).The use of resonance ionization for analytical purposes has
become a major activity, recently reviewed (Alkemade 1981, Zare 1984,
Fassett 1985).The application of RIS to atoms sputtered from a solid
target has considerable importance for surface analysis. The first
report of this variant was quite recent (Winograd 1982).

* Work supported by the U.S. Department of Energy, BES-Materials
Science, under contract W-31-109-Eng-38.

** Fysisk Institut, Odense Universitet, Denmark

Initial work has emphasized sensitive surface analysis and low damage (Parks et.al.1983,1984, Kimock 1983,1984,1985, Pellin et.al.1984,1986). To the extent that sputtering yields are well established (Behrisch 1981,1983) and the ionization process can be saturated, absolute measurements on the elemental composition of the first layer of solids becomes possible, even for relatively minor constituents.

In the present paper, a practical design for a time-of-flight instrument with isochronous energy- and angular-refocusing properties is described. The physical configuration and operating parameters have been selected to optimize the overall collected fraction of impurity atoms removed from the sample. High collection efficiency is achieved through the use of an extended laser volume, near the sample surface, intercepting a substantial range of the possible directions of ejection, together with a strong extraction field. The geometry and component-design for the detection flight path was optimized for suppression of noise, based on experience obtained with previous versions of this apparatus (Pellin et.al.1984,1986).

## 2. Apparatus Design Features

The essential features of the present apparatus design are depicted schematically in Fig.1. A key feature is the einzel lens operating at high negative potential in the vicinity of the target with the consequence that: (1) the primary ion beam is focused onto the sample with minimum aberration, (2) laser-produced photoions are extracted with high efficiency while their energy-spread due to the draw-out potential in the laser volume is held to an acceptably small value. A second important design improvement has been the introduction of sector-electric-field deflection elements into the flight path of the product ions, allowing isochronous TOF operation, i.e. compensation of the overall flight times of photoions for the variation induced by distributions in initial energy and direction of the sputtered atoms. The improvement in mass resolution results directly in additional suppression of noise,

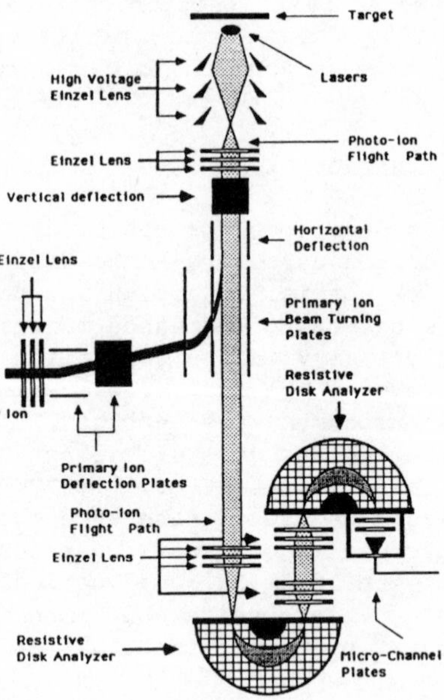

Fig.1.Schematic diagram of SARISA apparatus.

since detection gates can be narrowed without attendant loss of signal. Basic principles of energy/angle isochronous TOF have been

discussed previously (Poschenrieder 1972) and recently a
comprehensive analysis of the design principles of such devices has
been given and tested (Sakurai et.al. 1985a,b). As will be indicated
in the detailed discussion below, the present apparatus takes
advantage of the isochronous refocusing principle, while retaining a
simplicity of fabrication for the individual components.

Other aspects of operation follow a previous description (Pellin
et.al.1986). A mass-selected, pulsed ion source capable of delivering
several μA of ion current to the target in a continuous mode is now
introduced into the sample chamber by means of the primary ion beam
turning plates of Fig.1. Base pressure in the ion-pumped sample
chamber is $5x10^{-11}$ mbar. Sample condition can be monitored by AES and
LEED instruments on the chamber, and an airlock is provided for rapid
sample introduction.

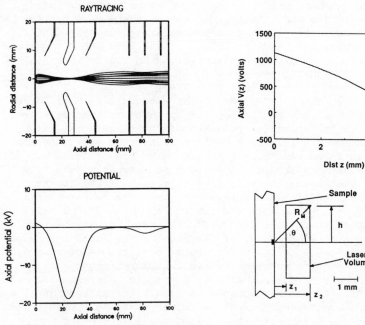

Fig.2. Sample, extraction and
collimation region.Upper panel:
electrodes and typical photoion trajectories;lower panel:
axial potential.

Fig.3. Detail of laser ionization
region. Above: near-linear potential in laser volume; below:
typical dimensions. Ion-bombarded
region is 0.25 mm dia. (dark
area).

The critical region adjacent to the sample is depicted in Fig.2. In
the upper panel the extraction electrode configuration is shown, with
a typical axial electrostatic potential indicated below. The three-
element group adjacent to the target is operated at high negative
potential (-21kV) in order to focus the primary ion beam onto the
sample. The electrode directly adjacent to the target (Fig.2) was
given the sharp-edged conical shape in order to minimize redeposition

problems. As Liebl has pointed out in the context of SIMS (Liebl 1981), it proves impossible to collimate returning low-energy ions (of the same charge sign as the primaries) without the aid of an additional lens. The same situation obtains in the case of photoions, produced in a volume near the surface. A triple aperture lens at lower potential (-2.8 kV) was employed as a photoion collimator, while having little effect on the high energy primary beam.

Particle optics properties of the extraction lens system were determined via ray-tracing with the aid of the program EGUN (Herrmannsfeld 1981). The EGUN numerical code generates electrostatic potentials, given the electrode boundary  conditions. Subsequently, ion trajectories are calculated for a variety of starting conditions.In regions of cylindrical symmetry, the code gives both spatial orbit and flight time for each ray. Flight paths through the 180°. spherical sectors are in a region of lower symmetry than cylindrical but there exact analytical formulas are possible for both the orbits and times of flight (Landau and Lifschitz 1960). By combining analytical expressions and numerical code data, a value for the total transit time from the laser volume to the detector is obtained.

Since the positive time dispersion in the spherical sectors (dt/dT > 0, t=time, T=initial kinetic energy) is opposite to that on near-linear flight paths, a stationary value of the flight time can be achieved, relative to variations in initial kinetic energy of ions in the laser volume. The condition is achieved by proper selection of flight path dimensions and is optimized experimentally by a trim-adjustment of the nominal potential in a portion of the linear flight path. The angular refocusing property of the spherical sectors is also necessary to achieve high resolution. With two 180° units, coupled by afocal imaging, a full 360° path results and the same flight time accrues independent of the sign of any initial angular deviation. Large-gap spherical  deflector  devices are constructed to utilize  exact boundary-matching in the region between spherical conductors by means of current flow in resistive materials (Siegel and Vasile 1981). An outer electrode of highly transparent metal mesh has been found advantageous for allowing the escape of potential noise sources (e.g. scattered ions, electronically excited metastables).

The large extraction fields required for efficient photoion collection result in substantial energy variation across the laser-volume (152 volts for the conditions of Fig.3). The spread in photoion energies thus induced dominates that due to the energy width characteristic of sputtering. Potential energy variation near the target is quite linear, however, and optimization of machine parameters for isochronous operation relative to initial ion position in the laser volume results in an experimental mass resolution of 200.

## 3. Results and Discussion

Silicon (111) crystal samples implanted with low doses of $^{56}Fe$ ($10^{11}$ atoms/cm$^2$ at 60 keV) were used as a test case. The Fe/Si$_2$ isobaric interference (to 1 part in 3000) is avoided by resonant excitation of the Fe atoms at 302.065 nm ($a^5D_4$ to $y^5D_4$ transition). Additional 302 nm photons can ionize the intermediate state,but excimer radiation at 308 nm is employed to bring the ion signal near-saturation (at ~$10^7$ W/cm$^2$). The depth profile between data points of Fig.4 is performed with continuous bombardment by 4 keV argon ion current of 2 µA in a 2x2 mm$^2$ raster area. Measurements

at a given depth, however are performed in a "static" mode with 2 µs ion pulses synchronized to the 10 Hz repetition rate of the Nd:YAG-pumped resonant laser system. The implant profile of the Fe occurs at the expected depth (~50 nm) at a calculated peak concentration of 400 ppb, which was used to scale the data of Fig.4. Noteworthy is the substantial surface impurity peak, despite considerable efforts to avoid contamination. No matrix effect is seen in the transition from oxide

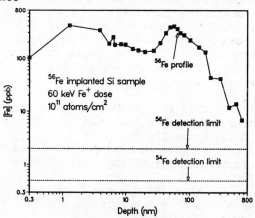

Fig.4. SARISA depth-profile of Fe-implanted Si (111).

to bulk at ~ 6 nm. The signal and statistics at the greatest depth sampled (700 nm) imply a 2 ppb sensitivity limit,at a signal/noise ratio of unity, for $^{56}Fe$ in a 1000 second experiment (20,000 laser shots, including background measurements with the resonant laser off). With a sputtering yield of 1.2 for 4 keV argon ions on silicon, 0.9 monolayers are removed in the 0.25 mm dia. spot. Reduced interference with Si$_2$ dimers ionized non-resonantly by 308 nm light leads to a lower detection limit for $^{54}Fe$. Overall collection efficiency (counts detected/atoms sputtered) is 5%, based on the 400 ppb peak calibration of Fig.4. An analytical model for sputtered atomic density in the cylindrical detection volume depicted in Fig.3 (Young et.al.1986) gives an upper limit of 9.4%, which must be multiplied by the fraction of Fe atoms in the ground state (~0.67) for comparison with experiment. The EGUN calculations indicate very low additional losses in particle transit.

## References

Alkemade C Th J 1981 Appl. Spectrosc. <u>35</u> 1
Ambartsumyan R V and Letokhov V S 1972 Appl. Opt. <u>11</u> 354
Behrisch R ed <u>Sputtering by Particle Bombardment I</u> (Topics in Appl. Phys. <u>47</u>, Springer-Verlag, Berlin, 1981)

Behrisch R ed <u>Sputtering by Particle Bombardment II</u> (Topics in Appl.
　Phys. <u>52</u>, Springer-Verlag, Berlin,1983)
Fassett J D, Moore L J, Travis J C and DeVoe J R 1985 Science <u>230</u> 262
Herrmannsfeldt W B 1981 Report SLAC-226, UC-28 (A), Stanford Linear
　Accelerator Center, Stanford University, Stanford CA 94305
Hurst G S, Payne M G, Nayfeh M H, Judish J P and Wagner E B 1975
　Phys. Rev. Lett. <u>35</u> 82
Hurst G S, Payne M G, Kramer S D and Young J P 1979 Rev. Mod Phys <u>51</u>
　767
Kimock F M, Baxter J P and Winograd N 1983 Surface. Sci. <u>124</u> L41
Kimock F M, Baxter J P, Pappas D L, Kobrin P H and Winograd N 1984
　Anal. Chem. <u>56</u> 2782
Kimock F M, Pappas D L and Winograd N 1985 Anal. Chem. <u>57</u> 2669
Landau L D and Lifshitz E M <u>Mechanics</u> (Pergamon, Oxford, 1960),Ch.3
Letokhov V S <u>Nonlinear Laser Chemistry</u> (Springer Ser. Chem.
　Phys. <u>22</u>, Springer-Verlag, Berlin, 1983)
Letokhov V S, Mishin V I and Puretzky A A 1977 Prog. Quant. Electron.
　<u>5</u> 139
Liebl H 1983 Int. J. Mass Spectrom. Ion Phys.<u>46</u> 511
Parks J E,Schmitt H W, Hurst G S and Fairbank W M, Jr. 1983 Thin
　Solid Films <u>108</u> 69
Parks J E, Schmitt H W, Hurst G S and Fairbank W M, Jr., <u>Resonance
　Ionization Spectroscopy</u> (Inst. Phys. Conf. Ser. No.71, Bristol,
　1984)
　p.167, ed Hurst G S and Payne M G
Payne M G, Chen C H, Hurst G S and Foltz G W 1981 Adv. At. Mol.
　Phys. <u>17</u> 229
Payne M G <u>Resonance Ionization Spectroscopy</u> (Inst. Phys. Conf.
　Ser. No. 71, Bristol, 1984), p 19, ed Hurst G S and Payne M G
Pellin, M.J., Young, C.E., Calaway,W.F., and Gruen, D.M. 1984 Surf.
　Sci. <u>144</u> 619
Pellin, M.J., Young, C.E., Calaway, W.F. and Gruen, D.M. 1986 Nucl.
　Instr. Methods <u>B13</u> 653
Poschenrieder W P 1972 Int. J. Mass Spectrom. Ion Phys. <u>9</u> 357
Sakurai T, Matsuo T and Matsuda H 1985a Int. J. Mass Spectrom. Ion
　Proc. <u>63</u> 273
Sakurai T, Fujita Y, Matsuo T, Matsuda H, Katakuse I and Miseki K
　1985b
　Int. J. Mass Spectrom. Ion　Proc. <u>66</u> 283
Siegel M W and Vasile M J 1981 Rev. Sci. Instrum. <u>52</u> 1603
Winograd N 1982 Chem. Phys. Lett. <u>88</u> 581
Young C E, Pellin M J, Calaway W F, Jørgensen B, Schweitzer E L and
　Gruen D M 1986 Proceedings of the Sixth International Workshop on
　Inelastic Ion-Surface Collisions, NIM-B to be published.
Zare R N 1984 Science <u>226</u> 298

<u>Acknowledgement</u>

The authors wish to thank Dr. J.M. Anthony, Materials Science
Laboratory, Texas Instruments, Dallas TX for supplying implanted
silicon samples and related data.

*Inst. Phys. Conf. Ser. No. 84: Section 4*
*Paper presented at RIS 86, Swansea, Wales, 7–12 Sept. 1986*

169

# Small sample analysis using sputter atomization/resonance ionization mass spectrometry

W. H. Christie and D. E. Goeringer

Analytical Chemistry Division, Oak Ridge National Laboratory, Oak Ridge, TN 37831, USA

Abstract. We have used secondary ion mass spectrometry (SIMS) to investigate the emission of ions via argon sputtering from U metal, $UO_2$, and $U_3O_8$ samples. We have also used laser resonance ionization techniques to study argon-sputtered neutral atoms and molecules emitted from these same samples. For the case of U metal, a significant enhancement in detection sensitivity for U is obtained via SA/RIMS. For U in the fully oxidized form ($U_3O_8$), SA/RIMS offers no improvement in U detection sensitivity over conventional SIMS when sputtering with argon.

## 1. Introduction

Environmental monitoring for trace elements in small samples such as airborne particles requires ultrahigh-sensitivity detection techniques. Depending on particle size and composition, they may contain only picogram to femtogram amounts of any given element. SIMS has been shown to be a useful method for locating and analyzing such small samples in certain instances. Although the sputter-ionization process can be highly controlled by varying the energy, current density, and composition of the sputtering ion beam, the ionization efficiency is generally quite low and is greatly affected by the chemistry of the sputtering region. The sputtered neutral atom and molecule density typically ranges from 2-5 orders of magnitude greater than that for sputtered secondary ions.

The technique of resonance ionization mass spectrometry (RIMS) has recently been developed for isotope ratio measurements. Studies performed in our laboratory by Donohue et al (1982) and Young et al (1984) and elsewhere by Miller et al (1982), Downey et al (1984), and Fassett et al (1984), have been reported for a number of elements using thermal vaporization sources to produce neutral atoms for laser-induced resonance ionization. Other methods of atomization such as glow discharge (Savickas et al 1984), laser ablation (Williams et al 1984), and ion beam sputtering (Winograd et al 1982, Donohue et al 1985), have been reported.

Research sponsored by the U. S. Department of Energy, Office of Basic Energy Sciences, under Contract DE-AC05-84OR21400 with Martin Marietta Energy Systems, Inc.

A commercial ion microprobe mass analyzer (IMMA) has been interfaced with a tunable, pulsed dye laser for carrying out resonance ionization mass spectrometry of sputtered atoms. The IMMA instrument has many advantages for this work, including a micro-focused primary ion beam (2 μm diameter) of selectable mass, complete sample manipulation and viewing capability, and a double-focusing mass spectrometer for mass separation and detection of secondary or laser-generated ions.

## 2.  Experimental

The secondary ion mass spectrometer used in this work was manufactured by Applied Research Laboratories (Sunland, CA) and is based on the design of Liebl (Liebl 1967). In essence, the instrument is two mass spectrometers. In the primary spectrometer, a beam of ions is generated, mass analyzed, and focused onto the surface of interest. Ions sputtered from the sample surface are directed into a double-focusing mass spectrometer where mass analysis is accomplished.  Modifications to the ion extraction lens system, which allow discrimination against sputtered ions while allowing efficient collection of resonance ions have been described in another publication (Donohue et al 1985).

### 2.1  Laser Source and Light Optics

The laser system used in this study was a tunable dye laser (Quanta Ray PDL-2) pumped by a pulsed Nd:YAG laser (Quanta Ray DCR-2A). The dye laser was operated in fifth order resulting in a bandwidth of approximately 0.3 $cm^{-1}$ as determined by a 1 $cm^{-1}$ Fabry-Perot etalon.   Rhodamine 610 dye used in our experiments gave a useful tuning range from about 582-598 nm.  Dye laser wavelength scans were performed under computer control via a programmable motor controller (Quanta Ray MCI-1) with parallel line interface option.  Wavelength calibration was confirmed by recording the optogalvanic spectrum of a Ne-filled uranium hollow cathode lamp.

### 2.2  Signal Processing/Control Electronics

The signal from the Daly detector's photomultiplier was simultaneously processed by both analog and digital electronic systems.   The analog system consisted of a fast current amplifier and gated integrator (Stanford Research Systems SR250).  The signal produced by a packet of ions generated from a single laser shot was integrated by the current amplifier into a single pulse approximately 15 μs wide.  This pulse was sampled by the gated integrator which was triggered by the laser and delayed for the flight time of the ion packet through the mass spectrometer.  The resultant dc signal was further processed by the data system (see below).  The analog signal from the current amplifier was also used to drive video amplifiers in the standard ARL video display system for real-time observation of mass spectra. The digital system was composed of the standard ARL preamplifier/discriminator, pre-scaler, and fast pulse counting (30 ns pulse-pair resolution) system.   The digital system was mainly used to determine relative signal strengths for SIMS signals.

The primary ion beam was pulsed by applying a +450V pulse to opposing alignment plates located at the top of the primary lens column.  The ion beam gate width and delay were set by gate/delay generators triggered by the laser flashlamp sync pulse.

## 2.3 Data System

Data acquisition and control for the experiments were performed by a microcomputer system (MDB Micro-11) based on the LSI-11/23-Plus (Digital Equipment Corp.) central processing unit and containing a 10 MB Winchester disk and dual floppy disk drives. The system housed an analog-to-digital converter (ADC)/digital-to-analog converter (DAC) interface board; the ADC was used to convert the dc output of the gated integrator, and the DAC was used to control the secondary magnet. Another direct memory access DAC interface was used to drive an x-y scope for real-time display of spectra. A parallel interface was used to control the dye laser motor controller (see above). Spectra could also be sent to a digital plotter (Hewlett-Packard 7475A) via serial line interface for hard copy output.

## 3. Results and Discussion

Earlier studies in our laboratory demonstrated that argon sputtering of uranium metal in the presence of residual oxygen produced both SIMS and RIMS spectra rich in $UO^+$ and $UO_2^+$ as compared to the atomic ion (Donohue et al 1985). The present study shows that for U metal samples the molecular species are essentially eliminated if the partial pressure of oxygen and oxygen-containing species (e.g. $H_2O$) is sufficiently reduced. To accomplish this in our instrument, the existing 400 L/s ion pump was replaced with a 1200 L/s cryosorption unit. As expected, for any given instrument pressure, primary ion beam current density also affected the $UO^+$ and $UO_2^+$ signals seen from U metal targets.

## 3.1 RIMS Sensitivity Improvement for U Metal Samples

For maximum sensitivity in the SIMS analysis of electropositive metals, oxygen bombardment or oxygen flooding of the sputtering region is commonly used to maximize secondary ion yields. Table 1 presents a comparison of laser-generated $U^+$ signals using $Ar^+$ and $O_2^+$ sputter-atomization of U metal with a time normalized SIMS $U^+$ signal generated by an $O_2^+$ primary beam on the same sample. The time normalization for the SIMS signal was done by using the gated integrator/ADC electronics and a gate width (6 µs) identical to that for the RIMS signals rather than the pulse counting system. The primary beams were also run in cw mode for these experiments. As expected, a given flux of $Ar^+$ primary ions was significantly more effective at generating sputtered neutral U atoms from a U metal target than $O_2^+$ primary ions were.

Table 1

| | SA/RIMS | | SIMS |
|---|---|---|---|
| Primary Beam (10 nA) | $Ar^+$ | $O_2^+$ | $O_2^+$ |
| Net Signal | 6.44 | 0.03 | 0.02 |
| Relative Strength | 320 | 1.5 | 1.0 |

### 3.2  Pulsed Primary Ion beam Experiments

For small sample analysis via SA/RIMS it is necessary to operate the primary sputtering beam in pulsed mode for efficient utilization of the sample. We have found that for highly oxidizable metal samples such as U, the duration of the sputter pulse must be long enough to remove adsorbed impurities that accumulate on the sample surface when the primary beam is not impinging on the sample. Figure 1 shows a low level $U^+$ resonant ion signal detected when the $Ar^+$ sputtering beam is 5 μs long and pulsed 10 times per second in sychrony with the laser pulse. The region of the U metal surface being pulse-sputtered had previously been sputter-cleaned with a cw $Ar^+$ beam. The primary beam was then switched to pulsed operation and the detected resonance $U^+$ signal allowed to stabilize as equilibrium was achieved between arrival rate of adsorbable species from the vacuum system and $Ar^+$ from the pulsed sputtering beam. When the $Ar^+$ sputtering beam was switched to cw mode, an immediate increase in resonance ion formation was detected. When the sputtering beam was again switched to pulsed mode, the resonance ion signal was seen to decay as the surface was again recovered with chemisorbed impurities.

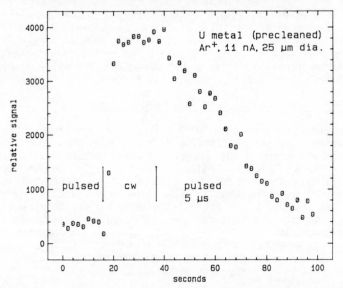

Fig. 1.  Effect of adsorbed species on the sputtering of neutral U atoms from U metal target

For the particular $Ar^+$ current density being used, we show that it takes a pulse width of 120 μs or longer to provide enough presputtering to restore the maximum neutral uranium population that can be sputtered from the surface under these conditions. Samarium is considerably more resistant to oxidation than U. In another experiment, the same $Ar^+$ beam current was used to sputter a Sm target, but the beam diameter was increased to 60 μm so that the current density was reduced to about 20% of that used in the U experiment. Even with the lowered current density which should enhance surface coverage effects and with only a 5 μs sputter pulse, the fall off in $Sm^+$ resonant ion production was much less severe than for U. The results of these experiments reflect the fact that oxide formation on U is considerably faster than on Sm.

## 3.3 SIMS and RIMS Comparisons

Table 2 compares the SIMS signals determined under similar conditions for U metal, $UO_2$, and $U_3O_8$ samples and also compares the resonance ion signals measured under similar conditions for U and $UO_2$ samples. No data for resonance ion formation on the $U_3O_8$ sample is shown because no resonant ion signal was obtained under the conditions used for these comparisons. Where $UO^+$ and $UO_2^+$ signals were seen in the RIMS experiments, they were found to be non-resonant photoionization effects. The $U^+$ RIMS signal on the other hand was always sharply tunable with wavelength. The case of $U_3O_8$ is interesting in that despite the highly oxidized state of the U, it is the $U^+$ ion that appears in greatest abundance in the SIMS spectra. Viewed in terms of the RIMS results for $U_3O_8$, it would appear that relatively little neutral U is sputtered from this sample, and for that matter, little $UO^0$ and $UO_2^0$. It would appear from these experiments that conventional SIMS offers more sensitivity for the detection of U in this material.

Table 2

| | SIMS counts/s x $10^3$ | | | RIMS volt-s x $10^{-6}$ | |
|---|---|---|---|---|---|
| Ion | U metal | $UO_2$ crystal | $U_3O_8$ | U metal | $UO_2$ crystal |
| $U^+$ | 93 | 102 | 750 | 17.0 | 0.39 |
| $UO^+$ | 6 | 575 | 90 | 0.21 | 6.3 |
| $UO_2^+$ | 2 | 895 | 52 | - | 0.57 |
| $UO_3^+$ | -- | --- | 0.5 | - | - |

## 4. Conclusions

Resonance ionization combined with pulsed $Ar^+$ sputtering provides increased detection sensitivity for U from metal samples. From oxidized samples of U it is seen that the population of sputtered neutrals is significantly reduced. At one time it was felt that the laser offered the promise of quantitative SIMS because one could conceivably ionize with 100% efficiency the atomic neutral species sputtered from a surface.

This work clearly indicates that the population of neutrals above a sputtered surface is not always representative in a direct way of the composition of the surface. Just as chemistry in the sputtering region dominates the relative ion yields in SIMS experiments, it also controls the relative neutral yields available for RIMS experiments.

## References

Donohue D L, Christie W H, Goeringer D E and McKown H S 1985 Anal. Chem. 57 pp 1193-1197

Donohue D L, Young J P and Smith D H 1982 Int. J. Mass Spectrom. Ion Phys. 43 pp 293-307

Downey S W, Nogar N S and Miller C M 1984 Anal. Chem. 56 pp 827-828

Fassett J D, Moore L J, Shideler R W and Travis J L 1984 Anal. Chem. 56 pp 203-206

Leibl H J 1967 J. Appl. Phys. 38 pp 5277-5283

Savickas P J, Hess K R, Marcus R K and Harrison W W 1984 Anal. Chem. 56 pp 817-819

Williams M W, Beekman D W, Swan J B and Arakawa E T 1984 Anal. Chem. 56 pp 1348-1350

Winograd N, Barter J P and Kimock F M 1982 Chem. Phys. Lett. 88 pp 581-584

Young J P, Donohue D L and Smith D H 1984 Int. J. Mass Spectrom Ion Processes 56 pp 307-319

*Inst. Phys. Conf. Ser. No. 84: Section 4*
*Paper presented at RIS 86, Swansea, Wales, 7–12 Sept. 1986*

# The development of a SIRIS system for exploration in China

D.Y. Chen, G.Y. Xiao, K.L. Wen

Physics Department, Qinghua University, Beijing, PROC

Abstract. The research project of RIS group of Qinghua University was described. A SIRIS apparatus is under installing.

## 1. Introduction

In recent years the experiments in several laboratories has confirmed the belief that RIS would be an ultrasensitive, highly selective analysis technique and applicable to nearly every elements. RIS has found applications to various fields. Geochemical exploration may become another field that RIS will find many fruitful applications to it. The RIS group of Qinghua University is installing a new apparatus and will try to use it to analyze geological samples for geochemical explorations.

## 2. Geochemical exploration and analysis technique

Humankind requires more and more minerals everyday, but the mineral deposits on developed areas and surface of earth have been almost exploited out. The exploration for mineral resource on undeveloped areas or in the depth of earth becomes an important task today. In these cases the difficulty and cost of exploration increase greatly. This is one of the motivation to develop new technique of prospecting, such as the geochemical exploration.

Geochemical technique has special advantages in search for

mineral deposit in depth, especially the precious, rare or nonferrous metal and rare earth minerals. Its development depends on the progress of analysis technique. There are two technical generations already. The analysis technique in the first generation had the detect limit of about 1 ppm, such as the X-ray fluorescence spectroscope. In that time, only certain elements with high abundance in the crust could be effectively measured, geochemical prospecting could be used to explore some superficial or shallow mineral deposits.

In the second generation the detect limit has been down to tens ppb. For example, the limits of ICP (Induced Coupling Plasma Spectroscope) for Ba, Sc, Sr and Y are 20, 30, 50 and 80 ppb. Many elements, including mineralizing and indicator elements can be effectively measured. Geochemical prospecting became a powerful method and has been widely used.

According to the viewpoint of certain geochemists a new period of geochemical prospecting will emerge if some requirements has been met:
(1). Increase the analysis sensitivity of some elements, such as Au, Pt, Ag, As, Te,···· to be better than 1 ppb.
(2). Increase the analysis sensitivity of some elements, such as Sn, W,···· It is difficult to measure them by conventional method for various reasons.
(3). Decrease the required amount of the sample. Simplify the chemical prepairing of samples.
(4). Develop a technique to measure the isotope ratio rapidly for some elements.
(5). Increase the microscope analysis sensitivity from 100 ppm to about 1 ppm. If possible, measure the microscope isotope ratio to a precision of about 1% .

Apparently, RIS is a hopeful candidate to satisfy these demands.

3. The apparatus

According to the former reasoning a SIRIS apparatus was de-

signed and manufactured. It was anticipated to realize a detect limit less than 1 ppb. Fig.1 is the schematic diagram.

Usually, a solid sample can be atomized by one of three means:
electric heating, laser ablation and ion sputtering. From
the consideration of following three factors:  matrix effect,
sample exchange and universal to most elements, ion sputtering was choosed. An argon ion source of 20 KeV energy with
50 $\mu$A intensity and 1mm$^2$ beem section is under adjustment.
It will meet the requirement for sensitivity.

The eximer laser, model 202 EMG from Lamda Physic Inc., was
choosed. It will produce the laser pulse of 400 mJ energy
and 3.08 nm wavelength with a repetition rate of 150 Hz.
This is higher than that of YAG laser in a factor of 5, and
increase the analysis sensitivity accordingly. After the dye
lasers and frequency doubling crystal two laser beams of 100
$\mu$J ~ 10mJ energy, linewidth 0.01Å, wavelength range from red
to ultraviolet can be obtained. So various elements (except
He and Ne) can be analyzed.

In order to raise the ratio of signal to background, mass
spectroscope is absolutely necessary. 20 ns laser pulse
supplied a natural zero time signal, so choosing the time of
flight mass spectroscope is an obvious thing. Since the
difference of initial time, energy, position and move direction of ions, the estimated mass resolution is only about 250.
This is probably good enough.

The schematic diagram of ion detector and record system is
shown in fig.2. The detector is a microchannel plate. The
current pulse signal is converted to digital by QDC (charge
digital converter) and then fed to the microcomputer through
a GPIB bus. The second QDC can be used to monitor the background or measure the isotope ratio.

4. <u>Future improvements to the apparatus</u>

We plan to run some preliminary experiments in next spring.

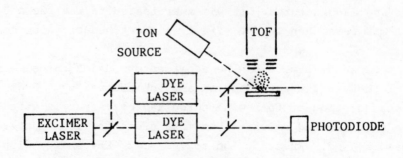

Fig. 1        Apparatus

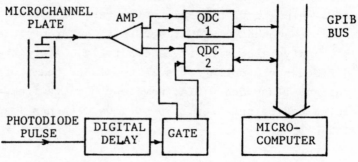

Fig. 2        Detector and record system

If the sensitivity is not enough, we shall try to increase the
ion beam intensity to 250μA ∼ 1mA, perhaps use a liquid metal
ion source.  If the background is too high, we shall add an
energy analyzer or use a reflection TOFMS.  We also plan to
decrease the ion beam section to about ten μm, so that we can
do micro analysis.

*Inst. Phys. Conf. Ser. No. 84: Section 4*
*Paper presented at RIS 86, Swansea, Wales, 7–12 Sept. 1986*

# RIMS diagnostics for laser desorption/laser ablation

E. C. Apel, N. S. Nogar, and C. M. Miller and R. C. Estler*

Los Alamos National Laboratory, Los Alamos, New Mexico 87545, *Ft. Lewis College, Durango, Colorado 81301

Abstract. Laser desorption mass spectrometry is a useful method for interrogating materials and events at or near surfaces. Laser desorption/ablation combined with Resonance Ionization Mass Spectrometry (RIMS) provides a powerful tool to obtain information on chemical composition and speciation and, in some cases, internal and translational energy distributions. The application of this technique to the interrogation of materials and interfaces is discussed for several systems, including the analysis of conventional analytical samples, and the study of optical damage events.

## 1. Introduction

RIMS is a powerful tool for elemental and isotopic analysis. The multistep laser photoionization process, when coupled with conventional mass analysis, can provide exceptional performance in detectivity, dynamic range, and discrimination against interfering species (Nogar et al 1985a, Fassett et al 1985). These properties can be used to great advantage both in the analysis of conventional materials (Bekov and Letokhov 1983) and in the interrogation of interfacial phenomena (Kimock et al 1983). In this report we will discuss the use of laser desorption/ablation in combination with RIMS analysis.

The motivations for this work are threefold. First, we desire to improve the effective duty cycle, relative to continuous sample evaporation, for the examination of routine analytical samples. Second, this method will allow the direct analysis of materials without extensive sample preparation. And third, we wish to study the mechanism of laser-material interactions, particularly optical damage, and characterize the desorption/ablation/damage process.

## 2. Experimental

The apparatus has been described previously (Nogar et al 1985b, Estler et al 1986), and will only be summarized here. Briefly, a Q-switched Nd:YAG laser equipped with beam-filling optics is used for desorption of a sample mounted on a manipulator in the source region of a time-of-flight mass spectrometer. Typical laser parameters include 10 nsec pulse length, 5 to 50 mJ pulses, 0.2 to 1 mm spot size at the sample, and 10 Hz repetition rate. After a variable time delay, the interrogation laser (XeCl excimer-pumped dye laser , typical parameters: ~12 nsec pulse, 1 mm diameter, 0.5 to 3 mJ, ~0.3 $cm^{-1}$ bandwidth) is passed through the spalled plume. At a fixed time delay following the excimer trigger, ions are detected with a channel electron multiplier, and the signal processed with a gated integrator and standard analog signal processing electronics.

## 3. Results and Discussion

Two-photon excitation. In order to expedite diagnostics for our laser desorption/ablation studies, we wished to minimize the complexity of the laser ionization process and associated hardware. We have therefore used single-color ionization processes at the fundamental output frequency of the interrogation dye laser. This, in turn, required n-photon excitation, $n \geq 2$, for moderate to high ionization potential elements. We have examined (Apel et al 1986) a number of atomic (and some molecular) systems ionizable via "2+1" processes. A list of observed transitions is shown in Table 1. For the atomic systems, the two-photon excitation typically saturated at ~2 mJ/pulse, while the ionization step required somewhat higher energies.

### TABLE 1
### Observed Two-Photon Transitions

| Element | Ground State Configuration | Term | Excited State Configuration | Term | Energy Difference[a] $(cm^{-1})$ |
|---------|---------------|------|---------------|------|--------------------|
| Ca  | $(4s^2)$       | $^1S_0$        | $(4s4d)$       | $^1D_2$        | 37298 |
| CaF |               | $X\ ^2\Sigma^+$ |               | $F\ ^2\pi$     | 37550 |
| Ta  | $(5d^36s^2)$   | $^4F_{3/2}$    | $(5d^36s7s)$   | $J=3/2$        | 44096 |
|     | $(5d^36s^2)$   | $^4F_{3/2}$    | $(5d^36s7s)$   | $^4F_{3/2}$    | 43964 |

[a]$T_o$ is given for CaF.

Analytical Applications. The use of pulsed lasers with continuous sample evaporation can result in substantial loss of analyte because of the low effective duty cycle. For a probe laser beam diameter of 0.5 cm, and an atomic velocity of $5 \times 10^4$ cm/sec, the rate of sample turnover in the beam volume is ~$10^5$ $sec^{-1}$. For a laser repetition rate of 10 $sec^{-1}$, the effective duty cycle is ~$10^{-4}$. The use of lasers (or particle beams, Kimock et al 1984) to pulse desorb the sample can substantially improve this value. We have recently (Nogar et al 1985b) demonstrated laser desorption/RIMS for tantalum samples. Tantalum atoms were detected via a "2+1" ionization through the $^2F_{3/2}$ state. The results were quite promising: the pulse of desorbed material was sufficiently narrow (<4 μsec) that the effective duty cycle was improved to ~$10^{-1}$, resulting in a substantial increase in sample utilization efficiency. The observed energy distributions were somewhat anomalous: for high intensities ($10^8$ W/cm$^2$) the hydrodynamic temperature was measured to be 8000 K, while the kinetic temperature was 400 K, and the internal (electronic) temperature was ~2000 K. At lower intensities (4 x 10$^7$ W/cm$^2$) the observed distributions were cooler, and more nearly thermal, as shown in Fig 1. These results are consistent with the existance of a thermal desorption barrier (Nogar et al 1985b).

<u>Laser Damage Studies</u>. RIMS has also been used as a monitor of the interaction between lasers and dielectric materials (Estler et al 1986). In initial experiments, optical damage on uncoated CaF$_2$ substrates was initiated with 1.06 $\mu$m pulses at fluences of 1 to 10 J/cm$^2$. Interrogation of the spalled plume revealed Ca atoms and CaF radicals only when damage events occured. Both species are monitored by "2+1" ionization processes through the $^1$D$_2$ and F $^2\pi$ states, respectively. This choice of intermediates states was made so that both species could be detected with a single laser dye. In subsequent experiments, CaF$_2$ damage was also induced at 355 nm and 266 nm. For 1.06 $\mu$m irradiation, we observed thermal (850 K) velocity distributions for both Ca and CaF. In addition, the CaF radical exhibited significant amounts of internal

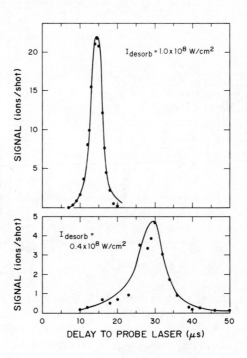

Fig. 1. Tantalum atom flight time distributions for high intensity (~10$^8$ W/cm$^2$) and low intensity (~4 x 10$^7$ W/cm$^2$ laser desorption.

(rotational and translational) excitation. For both 355 nm and 266 nm irradiation, the velocity distributions were bimodal, with a fraction ($\geq$50%) of the spalled material exhibiting very high (4000 K) kinetic temperatures (see Fig. 2), while the remainder exhibited a temperature similar (800 to 1000 K) to that observed for the 1.06 $\mu$m experiments. In addition, both the vibrational and rotational temperatures of the CaF radicals decreased with decreasing damage wavelength. Lastly, the threshold for damage decreased slightly with decreasing wavelength.

These results suggest that while the 1.06 $\mu$m experiments can be adequately modeled in terms of a single damage mechanism (likely avalanche breakdown), the short-wavelength results suggest the onset of second mechanism, perhaps multiphoton absorption. This is consistent with both the bimodal velocity distribution, and with the decrease in CaF vibrational and rotational excitation.

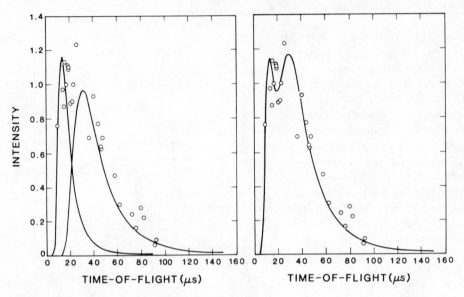

Fig. 2. (a) Calcium atom flight time distributions (dots) for damage induced at 266 nm, 25 J/cm$^2$ . The solid lines are Boltzmann fits for 4000 K and 850 K distributions. (b) Same as (a), where the solid line now represents a composite of the fast (33%) and slow (67%) contributions.

References

Apel E C, Anderson J E, Estler R C, Nogar N S, and Miller C M 1986 Appl. Opt. (submitted)

Bekov G I, and Letokhov V S 1983 Appl. Phys. B <u>30</u> 161

Estler R C, Apel E C, and Nogar N S 1986 J. Opt. Soc. Amer. (accepted)

Fassett J D, Moore L J, Travis J C, and DeVoe J R 1985 Science <u>230</u> 262

Kimock F M, Baxter J P, Pappas D L, Kobrin P H, and Winograd N 1984 Anal. Chem. <u>56</u> 2782

Kimock F M, Baxter J P, and Winograd N 1983 Surf. Sci. <u>124</u> L41

Nogar N S, Downey S W, and Miller C M 1985a Spectroscopy <u>1</u> 56

Nogar N S, Estler R C, and Miller C M 1985b Anal. Chem. <u>57</u> 2441

*Inst. Phys. Conf. Ser. No. 84: Section 5*
*Paper presented at RIS 86, Swansea, Wales, 7–12 Sept. 1986*

# Two-color, Doppler-free ionization of molecular hydrogen for isotopic analysis

Jan P. Hessler and Wallace L. Glab

Chemistry Division, Argonne National Laboratory
9700 South Cass Avenue, Argonne Illinois 60439

## 1. Introduction

Although resonant ionization techniques have been applied to many atomic systems for ultrasensitive isotopic analysis, very little work has been done on even the simplest molecular system, molecular hydrogen, and its isotopic variants: deuterium hydride, tritium hydride, molecular deuterium, tritium deuteride, and molecular tritium. We present a three-step ionization scheme which will selectively ionize any isotopic variant. The projected isotopic selectivity is sufficient to detect tritium hydride at ambient concentrations without the need of mass analysis. The efficiency of the ionization process is high enough to allow work on small samples. This technique can be applied to many areas of research, e.g., isotopic exchange reactions in the atmosphere and isotopic effects on surfaces. The physical mechanisms which govern the conflicting objectives of both high isotopic selectivity and detection efficiency, discussed below, can be applied to any molecular system.

## 2. Two-photon Excitation to the E-electronic State

Vasilenko et al. (1970) pointed out that two-photon excitation with counter-propagating beams not only produces sub-Doppler line widths, but also excites a species regardless of its velocity with respect to the propagation axis of the light. To obtain both isotopically selective and efficient excitation of molecular hydrogen in the first step of the ionization process, two-photon excitation with counter-propagating photons must be employed. To illustrate this point, we show in figure 1 the line profiles of the $Q_1$ two-photon transition of molecular hydrogen obtained with both co- and counter-propagating photons. The co-propagating photons were produced by the fourth anti-Stokes component of a hydrogen Raman shifter. The counter-propagating photons were taken from the third and fourth anti-Stokes components of the same shifter. The line profile for co-propagating excitation beams is Gaussian with a Doppler width of $1.30$ cm$^{-1}$. The line profile obtained with counter-propagating beams is nearly Lorentzian with a fwhm of $0.42$ cm$^{-1}$. This width and the slight asymmetry of this profile are due to the spectral profile of the excitation source which is broadened and shifted by the a.c. Stark effect in the hydrogen Raman shifter (Glab and Hessler, 1986a).

To estimate the isotopic selectivity of the two-photon excitation step we have measured the energies of the lowest rotational components of the $v = 0$ vibrational component of the E-electronic state of the isotopic species of molecular hydrogen. The results are summarized in Table I. The isotopic selectivity of the two-photon excitation step will be determined by the intensity (power broadening) and pulse duration (uncertainty principle) of the light beam used to induce the two-photon transition.  At ambient concentrations the species HH, HD, DD can all be selectively excited with a 5 nanosecond duration pulse without interference from other species.  To measure HT at ambient concentrations will require additional selectivity with respect to molecular hydrogen of between 3 and 4 orders of magnitude.

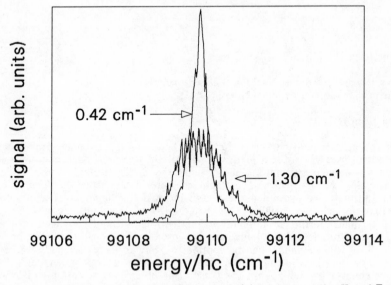

Fig. 1   Line profiles of the $Q_1$ two-photon transition between the X and E states of molecular hydrogen.   The broader profile is produced with co-propagating beams and the narrower profile with counter-propagating beams. The areas under each curve are equal.

Table I.  Excitation energies/hc $(cm^{-1})$ of the Q-branch two-photon transitions to the E(v = 0, J) electronic state of the isotopic species of molecular hydrogen.

| Species | J = 0 | J = 1 | J = 2 |
|---------|---------|---------|---------|
| $H_2$ | 49582.5 | 49554.9 | 49500.2 |
| HD | 49650.8 | 49630.1 | 49588.8 |
| HT | 49675.8 | 49657.4 | 49620.6 |
| $D_2$ | 49730.8 | 49717.0 | 49689.3 |
| DT | 49761.3 | 49749.8 | 49726.9 |

### 3.  Excitation to High Rydberg States

To obtain additional isotopic selectivity we use light from a second dye laser to excite species from the E-state to a high Rydberg state.   We have measured Doppler-free spectra of the Rydberg series of each isotopic variant by using co-propagating photons to excite molecules with specific velocity components within the Doppler profile, and then applying a second light pulse, which propagates in the opposite direction, to further excite molecules to high Rydberg states.   A short section of the np R(0) Rydberg series of HT is shown in figure 2.

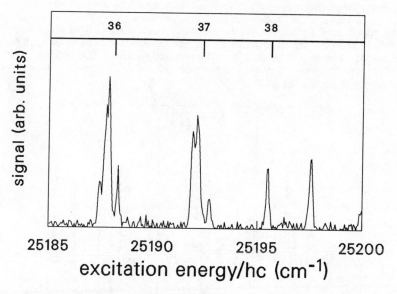

Fig. 2 Excitation spectrum of the Rydberg series of tritium hydride from the E(v = 0, J = 0) state. The calculated positions are from a two-channel quantum defect calculation.

By performing the spectral measurements of the Rydberg series in an electric-field free region and applying a pulsed electric field to ionize the species, we have measured the series out to approximately n = 90. This data has been analyzed with a two-channel quantum defect theory to deduce the adiabatic ionization potential (A.I.P.) of molecular hydrogen as 124417.61 ± 0.06 cm$^{-1}$ with quantum defects of -0.082 and 0.196 for the pi and sigma series respectively (Glab and Hessler, 1986b). For example, if the 37p Rydberg state of HT is used as the final state for the second excitation step, additional isotopic selectivity of up to 3 orders of magnitude with respect to molecular hydrogen may be obtained.

## 4. Two-step, Bi-polar Electric-field Ionization

To remove any ions created by the initial two-photon excitation step and to provide additional isotopic selectivity we use two-step, bi-polar electric field ionization as the final step in the process. The process is possible because the A.I.P. for HT is approximately 205 cm$^{-1}$ above the A.I.P. of molecular hydrogen. For example, an electric field of 104 V/cm may be used to ionize selectively the hydrogen molecules in n > 42 Rydberg states and to accelerate all ions produced in the initial two-photon excitation step away from the ion detector. If the amplitude of the field is increased to 175 V/cm in the second part of the field-ionization step and the polarity of the field is reversed, the HT molecules in the 37p Rydberg state will be ionized and accelerated toward the ion detector. Bekov et al. (1984) have, by applying this technique, obtained selectivities between 3 and 4 orders of magnitude for boron in the presence of germanium.

## 5. Predissociation of High Rydberg States

One mechanism not encountered in atomic systems is dissociation. This may turn out to be a very important loss mechanism in ultrasensitive detection schemes which utilize high Rydberg states of molecular systems. Figures 3 and 4 show the spectra for the production of both molecular hydrogen ions and atomic hydrogen atoms in the 2s state. The molecular ions were created by collisional ionization and the hydrogen atoms were ionized by adding a Balmer alpha light source to the system. There are significant changes in the predissociative rate as the principle quantum number changes and, within a single configuration, the predissociative rate for different branches may vary significantly. To our knowledge this effect has never been previously observed.

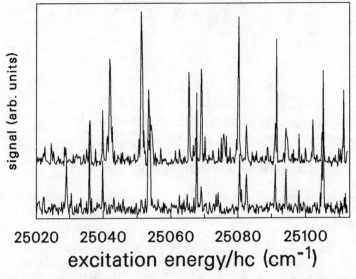

Fig. 3 Predissociation of the Rydberg states of molecular hydrogen excited from the E(v = 0, J = 1) state. The lower curve represents the production of $H_2$ ions by collisional ionization. The upper curve represents the production of H(2s) atoms which are ionized by adding Balmer-alpha light to the excitation region.

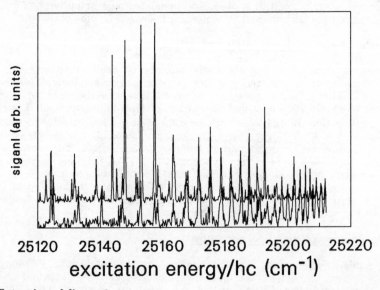

Fig. 4 Extension of figure 3.

## 6. Future Directions

Clearly, the linewidth of the two-photon transition and the dynamical properties of the high Rydberg states will determine both the isotopic selectivity and the detection efficiency of this ionization scheme. We are currently investigating both of these areas. We believe this scheme will find many applications in the area of ultrasensitive isotopic analysis.

## Acknowledgments

The expert technical help of Bert Ercoli and Al Svirmickas are gratefully acknowledged. James F. Kelly, University of Idaho, and David W. Braddock III, a graduate student at Cornell University, participated in the early phases of this work. This research was supported by the U.S. Department of Energy under contract GC-01-01-06-1.

## References

Bekov G I, Maksimov G A, Nikogosyan D N, and Radaev V N 1984 Sov. J. Quantum Electron. 14, 852.
Glab W L and Hessler J P 1986a and 1986b to be published.
Vasilenko L S, Chebotaev V P, and Shishaev A V 1970 Sov. JETP Lett. 12, 161.

*Inst. Phys. Conf. Ser. No. 84: Section 5*
*Paper presented at RIS 86, Swansea, Wales, 7–12 Sept. 1986*

# Rotationally resolved zero kinetic energy photoelectron spectroscopy of nitric oxide

M.Sander, L.A.Chewter[*] and K.Müller-Dethlefs
Institut für Physikalische und Theoretische Chemie
der Technischen Universität München
Lichtenbergstr.4
D-8046 Garching, Germany

Abstract. We use "Zero Kinetic Energy Photoelectron
Spectroscopy" to fully resolve rotational transitions
in the $(NO^+)$ X ← (NO) A photoionization process.

## 1.Introduction

Resonance enhanced multiphoton ionization (REMPI) combined
with zero kinetic energy photoelectron spectroscopy
(Müller-Dethlefs et al. 1984a) is a powerful tool to obtain
detailed information about rotationally resolved molecular
photoionization processes.
This field is of particular interest, because contrary to
the strict selection rules observed in bound-bound photoabsorption, in molecular photoionization processes the
emitted electron can carry away an angular momentum:
Of course, the transition from a particular rovibronic level of the molecule to the system ion + photoelectron has
to obey the strict selection rules given by overall symmetry considerations and conservation of angular momentum.
But as there are no a priori restrictions for the orbital
angular momentum of the outgoing electron, the problem of
finding propensity rules for the change of angular momentum
molecule → ion, depending on Hund's coupling cases for molecular and ionic state, seems not to be solved generally.

## 2. Experimental

The molecular beam apparatus contains a skimmed supersonic
jet-system, using a pulsed nozzle. The two colour experiment employs two synchronously Nd:YAG pumped pulsed dye lasers, both frequency doubled, and the second laser is
delayed by 10 ns relatively to the first laser.
What is not standard in the experimental set-up is the use
of our technique of "Zero Kinetic Energy-Photoelectron
Spectroscopy (ZKE-PES)". It does not only allow for an
electron energy resolution of about $1cm^{-1}$, but also for a
transmission of about 30 to 100%,which is important to
avoid space charge effects (Müller-Dethlefs et al. 1984a)-
(typical values for conventional PES techniques: 10meV=
$80cm^{-1}$; $10^{-4}$). This was first demonstrated for the
rotational state selective photoionization of nitric oxide
(Müller-Dethlefs et al. 1984b) with the C state as intermediate resonance:

$$(NO^+) \ X \ ^1\Sigma \ (v^+=0, J^+=N^+) \ \leftarrow \ (NO) \ C \ ^2\Pi \ (v=0, T_y(1/2))$$

In the present 1+1 photon experiment the NO A state is chosen as the intermediate resonance: this is a 3s-Rydberg state for which Hund's coupling case b) applies. The small spin-rotation interaction splits each N level into two levels: $F_1$: $J=N+1/2$ and $F_2$: $J=N-1/2$ of different J, but same parity $(-1)^N$. Though the spin-rotation doublet cannot be resolved by our laser system, it is possible, by choice of the $P_1$ A←X transition to populate only the $F_1$ rotational levels in the A state of definite N, $J=N+1/2$ and parity $(-1)^N$. The first laser is used for this purpose and $F_1$ levels of $N=0,1,2,3$ are selected in the experiments. In order to study the photoionization process

$$(NO^+) \; X \; {}^1\Sigma^+ \; (v^+=0,J^+=N^+) \; \leftarrow \; (NO) \; A \; {}^2\Sigma^+ \; (v=0,J,N)$$

the second laser is tuned around the ionization threshold and the transitions into single rotational ionic states are detected by measurement of the ZKE- photoelectrons.

## 3. Results

The ZKE-PE spectra taken for the various rotational levels in the NO A state ($N=0,1,2,3$) are shown in fig.1 to 4. The rotational transitions into the ionic ground state with $N^+=0,1,2,3..$(energy spacing $4,8,12cm^{-1}$; $B^+\approx 2cm^{-1}$) are clearly resolved.
For all spectra the $\Delta N=N^+-N=0$ contribution has the highest intensity . For $N=0$ (fig.1) we also find strong contributions with $\Delta N=2,1,3$. With increasing N, however, a drastic decrease in the intensity of the $\Delta N \neq 0$ transitions is observed (fig.2,3). For $N=3$ (fig.4) they nearly vanish completely.

## 4. Discussion

Assuming a quasi atomic picture with a molecular core + Rydberg electron with orbital angular momentum l in the NO A state one would expect an electric dipole transition to change the orbital angular momentum of the electron by one leaving the molecular rotation unchanged: $\Delta l=\pm1$, $\Delta N=0$. This obviously fails to explain the $\Delta N \neq 0$ contributions for low N.
The partial wave picture supposes a mixing of angular momenta: the outer electron in the NO A state does not have pure s-character, but also, to a smaller extend, p-, d- and even f-character. Therefore $\Delta N \neq 0$ transitions become allowed, their strengths depending on the magnitudes of the higher angular momentum contributions in the A state and their specific electric dipole transition moments. With such a picture the appearance of a variety of Rydberg-Rydberg transitions (Cheung et al. 1983) were interpreted. What cannot be explained in this picture is the strong N dependance of the angular momentum transfer that appears near ionization threshold and which is not observed in the Rydberg-Rydberg transitions.
We therefore favourize an interaction model in which the outgoing electron interacts with the long range potential

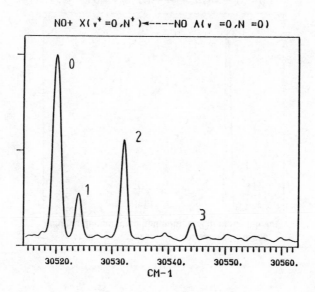

Fig.1

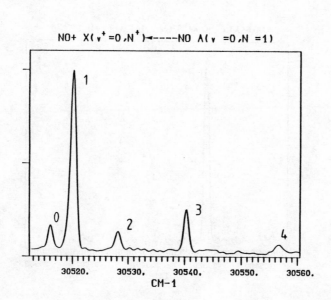

Fig.2

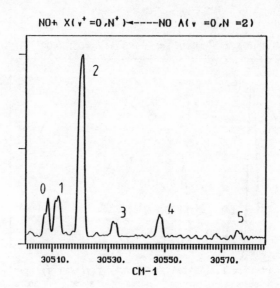

NO+ X( v⁺ =0 ,N⁺ )◄-----NO A( v =0 ,N =2 )

Fig.3

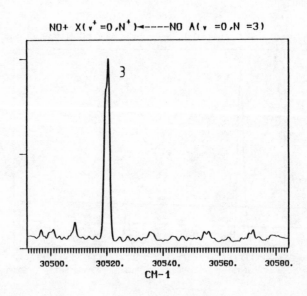

NO+ X( v⁺ =0 ,N⁺ )◄-----NO A( v =0 ,N =3 )

Fig.4

of the ion core. Such a long range potential, used to explain autoionization effects (Eyler), includes dipole and quadrupole moments of the ionic core as well as its polarizability. The interaction Hamiltonian can be written in the form:

$$V = H_{dipol} + H_{quadr} + H_{pol}$$

If, for convenience, one assumes a pure s-electron for the NO A state (in general 1), according to the quasi-atomic picture the outgoing electron should have pure p-character (in general 1±1). An interaction of the outgoing electron with the core would mix in contributions of s-, d- and f- character thereby changing the rotation of the molecule. Each neutral → ionic rotational transition would then correspond to a specific orbital angular momentum $1^+$ of the final electron state.

Ignoring interference effects the probability P for a $\Delta N \neq 0$ transition can be expressed as the product of the probability $P_1$ for exciting a $\Delta N = 0$ Rydberg- or continuum state and the probability $P_2$ that this state interacts with $\Delta N \neq 0$ states:

$$P_2 \propto |<N^+,1^+|V|N,1\pm1>|^2 \ / \ (B^+ \cdot N^+(N^++1)-B^+ \cdot N(N+1))^2$$

$$P_1 \propto |<N,1\pm1|\mu|N,1>|^2 \cdot \rho(E) \ ; \quad P = P_1 \cdot P_2$$

$B^+$ is the rotational constant for the ion, $\mu$ is the electric dipole operator in the matrix element between the NO A state and the NO+ ground state, and $\rho(E)$ corresponds to the density of $\Delta N = 0$ Rydberg- or continuum states.

According to the quantum defect theory the matrix elements go smoothly through the ionization potential and are therefore only weakly dependent on the energy near threshold (at least on our scale). The spacing between adjacent rotational levels, however, increases linearly with $N^+$. This does not only increase the denominator in the formula above; for $\Delta N < 0$ contributions it also decreases the density of Rydberg states, which can isoenergetically interact with ionic levels of lower rotational quantum number. Both effects together can therefore explain the drastic decrease of $\Delta N \neq 0$ contributions with increasing N.

## 4. Conclusions

ZKE-PES is an adequate tool for the investigation of rotationally resolved photoionization processes. Near the ionization threshold we observe relatively strong transitions with high angular momentum transfer.

The drastic decrease of the intensity of these transitions with increasing rotational quantum number can be understood by an interaction of the outgoing photoelectron with the long range potential of the core. Further work on this problem is in progress.

Acknowledgements: Financial support of this research from the Deutsche Forschungsgemeinschaft is gratefully acknowledged. L.A.C. thanks the Royal Society (London) and the Humboldt Stiftung for grants.

References:
Cheung W.Y.,Chupka W.A.,Colson S.D.,Gauyacq D.,Avouris P.,
    Wynne J.J. 1983 J.Chem.Phys.78 3625
Eyler E.E. submitted to Phys.Rev.A
Müller-Dethlefs K.,Sander M.,Schlag E.W. 1984a
    Z.Naturforsch. 39a 1089
Müller=Dethlefs K.,Sander M.,Schlag E.W. 1984b
    Chem.Phys.Lett. 112 291


* now at: University of Birmingham, UK

*Inst. Phys. Conf. Ser. No. 84: Section 5*
*Paper presented at RIS 86, Swansea, Wales, 7–12 Sept. 1986*

195

# Collisional and molecular related processes in laser-pumped magnesium vapour and argon mixture

Zhang Jing-yuan

Department of Physics, Graduate School
Academia Sinica, P.O.Box 3908, Beijing, P. R. C.

## 1. Introduction

The research on collisional and molecular related processes in laser-pumped dense metal vapours is a subject of great interests   not only because of itš intrinsic interest but also because of the profound application background. Extensive studies on one-electron alkali-metal vapours, such as K, Na, Rb and Cs, have been performed and studies on alkali-earth metal vapours have only recently received more attention. In this paper we will present our recent experimental results on collisional and molecular related processes in low-lying excited states and high Rydberg states of MgI and excited states of MgII, respectively, by using resonant multi-photon ionization and laser-induced fluorescence techniques.

## 2. Experiment and discussion

The experimental set-up has been described previously (Zhang 1986a). Briefly, the experiments were carried out in a laser-pumped four-way crossing heat-pipe oven, which contains Mg vapour and noble gas Ar mixture.  An YAG-pumped tunable pulsed dye laser and its frequency doubling component were focused into the centre of the heat-pipe, a typical thermionic diode was used to detect the ion-signal produced by the laser, the laser-induced fluorescence and the collision-induced fluorescence were collected and analysed from the side window of the oven in a direction pependicular to the laser beam through a monochromator and detected by a PMT at the exit slit of the monochromator.

It is well known that a resonant intermediate state will significantly enhance the multi-photon ionization.  But, in the case of Mg vapour, this is true only when the vapour density is very low and the laser intensity is quite high. At high vapour pressure (>0.1 Torr) and laser intensity of $I < 10^9 W/cm^2$, a resonant ionization dip with $3s3p\,^1P$ as the intermediate state could always be observed.  Fig. 1 shows the variation of ionization lineshape with Mg vapour. It is seen that the ionization was normally enhanced at $3s^2\,^1S$–$3s3p\,^1P$ resonance wavelength ($\lambda_L = 2852.1$ Å) when vapour pressure was very low (0.002 Torr).  When vapour pressure was increased up to 0.1 Torr or higher, a ionization dip centred at 2852.1 Å could be seen.  On the other hand, when the laser was scanned over a very wide range,  from 3243 Å to 2770 Å or shorter,  a green fluorescence consisting of three components (5183.7 Å, 5172.7 Å

and 5167.4 Å) corresponding to 3s4s $^3$S–3s3p$^3$P$_J$ transitions was seen inside the heat–pipe along the laser beam path. Among those  the component
at 5183.7Å could be significantly enhanced  and  become  a stimulated
emission in the forward directions (although a weaker emission  in  the
backward  direction  could also be observed) when laser was tunned  to
near 2851.6 Å, the 3s3p$^3$P$_o$–3s5d$^3$D resonance,  and the thresholds  for
vapour pressre (>0.25 Torr) and laser in intensity (>10$^7$W/cm$^2$)  were sat-
isfied.  The  experiment showed  that the intensity of this emission is
the  nonlinear functions of both laser intensity and vapour  pressure.
With increasing vapour pressure the ionization dip at 2852.1Å became
deeper, a wide and red–shifted ionization dip,  which is superposed on
the narrow dip at 2852.1Å,  could be observed and a ionization peak at
2851.6Å would appear  if  laser  intensity is high enough (see Fig. 1).

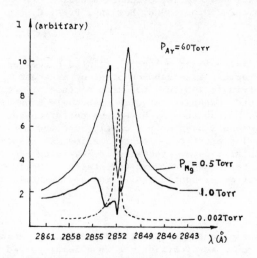

Fig. 1.  The lineshape of ionization as a function of Mg vapour pressure
showing the occurence of resonant dip.

The occurence of ionization dip could be explained as a result of com-
petetion between ionization and collisional transfer of excited 3s3p$^1$P
Mg atoms.  The A$^1\Sigma_u^+$ Mg$_2$ excimer could be formed via collision between
3s3p $^1$P  and  3s$^2$ $^1$S atoms, then the excimer can make a transition to the
triplet $^3\pi_u$  state or  be  collisionally transferred to $^3\Sigma_g^+$  state and
the repulsive $^3\pi_u$ or $^3\Sigma^+$ excimer could be dissociated into a 3s3p$^3$P atom
and a 3s$^2$ $^1$S atom (Scheingraber and Vidal, 1977).  Within the wavelength
range used,  ionization of 3s3p $^1$P atom or A$^1\Sigma_u^+$ excimer  needs  only
another photon, while  ionization of 3s3p$^3$P atom needs two photons.  If
the rate of such a predissociation (proportional to N$^2$,  N  is  vapour
density)  become  comparable with or even larger than that of ionization (proportional to N),  then  a ionization dip would appear.  On the
other hand, two–photon ionization of the predissociation–produced 3s3p
$^3$P atom should be resonantly enhanced at 2851.6 Å. Taking into account
of above processes, the intensity of ionization could be  expressed as
following

$$I(\Delta)=N\frac{R_{3^1S-3^1P}(\Delta)\cdot R_{3^1P-c}}{2\,R_{3^1S-3^1P}(\Delta)+A+R_{3^1P-c}}-N^2R_{3^1S-3^1P}(\Delta)\left[1-R_{3^1S-3^1P}(\Delta)\right]\sigma\,\bar{v}\left[1-R_{3^1P-5^3D}(\Delta+\delta)\right]R_{5^3D-c}]$$

where R$_{i-j}$ is the transition rates from |i> to |j>state, A is the spontaneous emission rate of 3s3p $^1P$ state, subscript C is ionization continue and $\delta$ is the frequency difference of $\omega_0(3^3P-5^3D)$ and $\omega_0(3^1S-3^1P)$, $\sigma$ is the collisional transfer cross-section from $3^1P$ atate to $3^3P$ states and $\bar{V}$ is the mean collison velocity of the atoms. In above expression, the first term is the two-photon ionization via 3s3p $^1P$, the second term is the collisional transfer to triplet state and the last term is the multi-photon ionization via 3s5d$^3$D. At high vapour pressure, excitation and ionization of ground state $X^1\Sigma_g^+$ Mg$_2$dimer could become considerable and the predissociation in molecule-excitation would lead to a wider and red-shifted ionization dip as described previously.

The green fluorescence is attributed to the population in 3s4s$^3$S due to plasma recombination, part of the forward stimulated emission could be interpreted as the four-wave parametric processes with 3s5d$^3$D as a resonant state and the backward stimulated emission is attributed to a Optical Pumped Stimulated Emission (OPSE) due to the population inversion between 4$^3$S and 3$^3$P state established through depleting the population in 3s3p$^3$P state by laser near 2851.6 Å when the threshold for the laser intensity is met. The related energy levels and potential curves showing the proposed processes are ploted in Fig. 2.

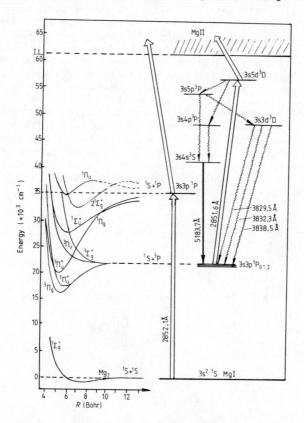

Fig. 2. Potential curves of some molecular states of Mg$_2$, energy levels of MgI and relevant transitions.

When laser was tuned to two-photon resonances from ground state $3s^2\,^1S$ to Rydberg states of MgI. The resonance enhanced ionization with $3sns$ $^1S$ and $3snd^1D$ Rydberg series as two-photon resonant states could be observed. At low vapour pressure and low perturbing gas pressure, the observable transitions are those from ground state to Rydberg states of even parity. Comparing with $3s^{2\,1}S-3snd^1D$ transitions the intensities of $3s^{2\,1}S-3sns\,^1S$ transitions are rapidly decreased with increasing principal quantum number n. This is, according to Lu (1974), due to the much weaker interaction between $3sns'S$ series and the doubly excited state $3p^2\ ^1S$ than that between $3snd'D$ series and $3p^{2\,'}D$ doubly excited state. In low partial pressure, the intensity ratio of $I(3s^{2\,1}S-3sns^1S)/I(3s^{2\,1}S-3snd\ ^1D)$ for n>11 is no more than 5% and no forbidden component could be observed. At high vapour pressure and high perturbing gas (Ar) pressure, two-photon forbidden transitions from $3s^2\ ^1S$ state to "odd" parity $3snp^1P$ states could be observed with considerable relative intensities, depending on n value and experimental parameters. Fig. 3 is a reproduction of ionization spectrum at high vapour pressure showing the collision-induced fobidden $3s^{2\,1}S-3snp\ ^1P$ transitions. The observed forbidden components have following properties: the intensity ratio of forbidden transition to the nearby allowed transition depends on the perturber density (Mg or Ar) for a given n value and increases with n for a constant perturber density; the minimum n value, at which the collision-induced transition could be observed, decreases with increasing perturber density. Similar ionization spectrum with forbidden components could be obtained at relative low vapour density ($\sim$0.5 Torr) and high gas pressure. On the other hand, with increasing vapour pressure or gas pressure, the intensity ratio of $I(3s^2\ ^1S-3sns^1S)$ to $I(3s^2\ ^1S-3snd^1D)$ is significantly increased. Fig. 4 shows the dependences of above intensity ratio on vapour pressure and perturbing gas pressure for a given n. From Fig. 4 one can see that the intensity ratio of $I(3s^{2\,1}S-3s12s^1S)/I(3s^{2\,1}S-3s11d^1D)$ is increased from a initial value of 5% at P(Ar)=25 Torr to a final value of 44% at P(Ar)=440 Torr and simliar behaviour could be seen when increase Mg vapour. It

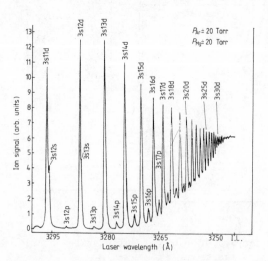

Fig. 3.  Two-photon resonant ionization spectrum at high vapour pressure showing the forbidden $3s^2\ ^1S-3snp\ ^1P$ transitions.

is important to note that the forbidden transitions can induced by
applying a D. C. field of E>10V/cm, this is due to L-mixing of high
Rydberg states caused by external electric field. But our observation
is carried out under nearly field-free enviroment (E<0.3 V/cm) and is
essentially caused by collision.

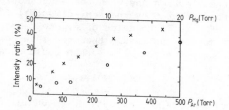

Fig. 4. The dependences of intensity ratio of I $(3s^2\ {}^1S-3s12s^1S)/$
$I(3s^2\ {}^1S-3s11d^1D)$ on $P_{Ar}$ (-x-) and of $I(3s^2\ {}^1S-3s9s\ {}^1S)/I$
$(3s^2\ {}^1S-3s8d^1D)$ on $P_{Mg}$ (-o-) showing the collisional enhancement
of $3s^2\ {}^1S-3sns\ {}^1S$ transitions.

Qualitatively, above phenomena could be explained in term of the
formation of $Mg_2$ excimer and $Mg_xAr_y$ Fermi-complex as following:
At high partial pressure, for example high vapour pressure, the ground
state $Mg_2$ dimer could be formed through collision of two ground state
Mg atoms and the only possible state is $x^1\Sigma_g^+$. In molecular sense, two-
photon transition from ground state $x^1\Sigma_g^+$ to Rydberg state $n^1\Sigma_g^+$ with $3s^2$
${}^1S$ and 3snp ${}^1P$ atoms as its asymptotic states is dipole-allowed though
two-photon transition from $3s^2{}^1S$ state to 3snp ${}^1P$ Rydberg state is pari-
ty-forbidden in atomic sense. On the other hand, the formation of $Mg_2$
molecule and its participating in two-photon transition from molecular
ground state to its Rydberg states can, to some degree, reduce the
significant difference in interactions between 3sns ${}^1S$ and 3snd ${}^1D$
series with the doubly excited states 3pnl in atomic sense thus result
in a collisional enhancement of $3s^2{}^1S-3sns\ {}^1S$ transition. Similar ex-
plaination could be made from the viewpoint of $Mg_xAr_y$ molecules for
collision-induced phenomena due to collision with Ar, in which the Mg
atom or $Mg_2$ excimer is in a high Rydberg state, while the Ar atoms are
staying in the ground state within the Rydberg electron orbit.
Quantitatively, the collision-induced forbidden transition and the
collisional enhancement of $3s^2{}^1S-3sns^1S$ transition could be interpreted
in term of collision-induced mixing of Rydberg states, whose orbit angular momentum L are different. For high Rydberg states the energy
defect of nearby states with different L are very small, a wavefunction mixing among the near-degenerated Rydberg states could be induced
through a thermal collision of Rydberg atom with other collisional
partner, such as the ground state Mg or Ar atoms. In this way, for
example, a pure $|3snp\rangle$ state wavefunction could be hybridized through
collision as following.

$$|3snp\rangle \rightarrow |3snp\rangle^* = |3snp\rangle + \sum_{n'} (\alpha_{n'}\ |3sn'\ s\rangle + \beta_{n'}\ |3sn'\ d\rangle) \quad (n'=n,\ n\pm1,\ ...)$$

where $\alpha_{n'}$ and $\beta_{n'}$ are the mixing coeffeciencies and could be estimated
by perturbation theory:

$$\alpha_{n'} = \frac{\langle 3sn'S|V(r,R)|3snp\rangle}{E(3sn's)-E(3snp)} \quad ; \qquad \beta_{n'} = \frac{\langle 3sn'd\ |V(r,R)|3snp\rangle}{E(3sn'd)-E(3snp)}$$

where V(r,R) is the collisional perturbation potential. According to
Omont (1977), all treatments of the interaction of a very excited atom
A* with a neutral perturber B are based on the fact that the dimentions
of A* are much larger than those of B. The probrem is then reduced to
the highly localized interaction of B with the excited electron. The
situation is particularly simple when B is located in the region of
classical motion of the electron and when the wavefunction $\psi(\vec{r})$ of
the electron is well reprensented by the semi-classical (JWKB) approximation. In the case of very large value of n, one needs consider only
the region of classical motion and assume that the interaction is zero
outside. The perturbation potential V(r,R) then could be expressed as

$$V(\vec{r},\vec{R}) = 2\pi L \delta(\vec{r}-\vec{R}) + \mathcal{M}_z^2/R^3 - \alpha/(2R^4) + \ldots$$

where the first term is so-called Fermi pseudo-potential describing e-
B scattering by the approximation of the scattering length L, the second
and the third term are the first order despersion interaction (if A=B)
and polarization of B by A*, respectively, describing the long-range
interaction between A* and B, which are usually smaller
than that of Fermi scattering. Since $3s^2\ ^1S\text{-}3sns\ ^1S$ transitions would
become very weak for large value of n, one should only consider the
collisional mixing of $|3snp\rangle$ with with nearby $|3sn'd\rangle$ (n'=n and n-1).
Considering only Fermi scattering the mixing coefficiency can then be
given by

$$\beta_{n'=n} \doteq \beta_{n'=n-1} \doteq L\sqrt{2/R - 1/n^2} / \left[\pi n^3(E_{n'd} - E_{nd})\right]$$

and the intensity ratio K of two-photon forbidden transition to the
nearby two-photon allowed transition can be written as

$$K(n,R) = I(3s^2 {}^1S\text{-}3snp^1P)/I(3s^2 {}^1S\text{-}3snd^1D) \doteq (\beta_n + \beta_{n-1})^2$$
$$= 4L^2(2/R - 1/n^2)/\left[n^6\pi^2(E_{nd} - E_{np})^2\right]$$

on the other hand, the collisional enhancement of $3s^2\ ^1S\text{-}3sns\ ^1S$ transition could be interpreted as collisional mixing with strong $3s^2\ ^1S\text{-}3snd$
$^1D$ transition and thus borrow from $3snd^1D$ series. The collisional mixing
coefficiency between $|3sns\rangle$ and $|3sn'd\rangle$ could also be estimated in a
similar way. The theoretical estimations by using above expression are
in good agreement with the experimental results.

When the laser is tuned to single- or two-photon resonances of MgII
from the ground state 3s $^2S$ to excited 4s $^2S$, $3p^2P$, and $3d^2D$ states, the
resonance enhanced double ionization could be observed. Fig. 5 is the
ionization spectrum and the identification of related lines. The
double ionization is subject of interests although its mechanism is
still obscure. As showed in Fig. 5, the obvious resonant character of
double ionization in MgII strongly suggests that the doble ionization
is essentially a two-step process, in which the first step excites the
singly charged ion in its ground state or in excited states and the
second step resonantly drives the ion up to the second ionization
level via one or two resonant intermediate states by the same laser
pulse or via collisional ionization, including associative ionization,
of excited MgII atoms. The possible collisional ionization processes
can be summarized as follows

1). for the $3s^2S-4s^2S$ resonance:

$MgII(4s^2S) + MgII(4s^2S) \rightarrow Mg^{2+} + MgII(3s^2S) + \triangle E$

2). for the $3s^2S-3d^2D$ resonance:

$MgII(3d^2D) + MgII(3d^2D) \rightarrow Mg^{2+} + MgII(3s^2S) + \triangle E$

3). for the $3s^2S-3p^2P_j$ resonances:

$MgII(3p^2P_j) + MgII(3p^2P_j) \rightarrow MgII(4s^2S) + MgII(3s^2S) + \triangle E$

$\quad\quad\quad or \rightarrow MgII(3d^2D) + MgII(3s^2S) + \triangle E$

these are then followed by the double ionization processes 1) and 2). The above-proposed processes were supported by the experimental dependence of different ionization peaks on vapour pressure as can be seen in Fig. 6. In above process 1) and 2) the intensity of double ionization is proportional to the number density of the excited states MgII, while in process 3) the intensity is proportional to the square of the number density. So that the intensity of double ionization with $3p^2P$ state as intermediate state should be a more nonlinear function of vapour density than those with $3d^2D$ and $3s^2S$ as the intermediate state. Fig. 6 showed such a dependence as expected. the above-proposed processes were also supported by the collision-induced fluorescence, which is described elsewhere (Zhang et al, 1986b).

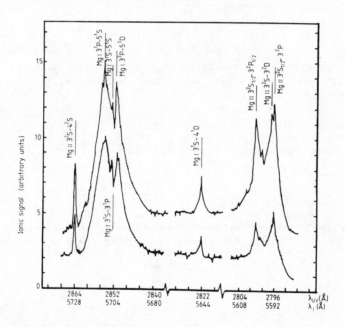

Fig. 5. The ionisation spectrum of Mg at a vapour pressure of 0.7 Torr (upper curve) and 0.5 (lower curve). The UV laser power density was about $10^9$ W cm$^{-2}$.

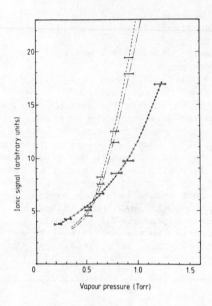

Fig. 6   The dependences of double ionic peaks on vapour pressure -o-o-, ion signal at 286.4 nm, -..-..-, ion signal at 280.3 nm; ------, ion signal at 279.6 nm.

The author is grateful to Prof. Lu K.   T. and Dr.   Sun Jun-qiang for helpful discussion and to Li Qiong-ru, Yang Jing and Zhou hai-tian for their assistances during the course of this work.

References

Lu K T 1974 J. Opt  Soc. Am. 64 706
Omont A 1977 J. Physique 38 1343
Scheingraber and Vidal 1977 J. Chem. Phys. 66 3694
Zhang Jingyuan, Li Qiongru, Yang Jian, Zhao Lizeng and Nie Yuxin 1986a
                    J. Phys. B: At. Mol. 19 L75-L80
---- 1986b (submitted to J. Phys. B: At. Mol.)

*Inst. Phys. Conf. Ser. No. 84: Section 5*
*Paper presented at RIS 86, Swansea, Wales, 7–12 Sept. 1986*

203

# Angular distributions and electron correlation from resonance ionization spectroscopy of barium atoms

R. Stephen Berry

Department of Chemistry, The University of Chicago, Chicago, Illinois 60637, U.S.A.

Abstract. Measurements of photoelectron energies and angular distributions give indications of the degree of electron correlation in excited states of alkaline earth atoms.

## 1. Introduction

This report is a summary of experiments on resonance ionization spectroscopy (RIS) that are part of a continuing effort to develop the technique into a quantitative tool for analyzing electron correlation in atoms. The stimulus for this effort is the evidence that in some states of atoms having two valence electrons, the correlation between these electrons is so strong that the quantization may be much more collective, like atoms vibrating and rotating in a molecule, than independent-particle-like, as the traditional planetary model of Bohr suggests. We must ask questions such as these: 1) Does either the independent-particle model or the collective, molecular model describe experimental data more accurately than the other? 2) Are some states of some atoms described equally well by both models? 3) Are some states of some atoms poorly described by both models?

These questions cannot yet be answered. For example to answer Question 1, we must look for systematic relationships among the properties of several states that would distinguish one model from the other. As yet, neither the experimental data nor the theoretical predictions display such relationships. Despite the large body of theoretical work on, there is not yet a set of predictions of the experimental results that a collective model would imply for two-electron or quasi-two-electron systems. There is, however, some theoretical information implying that the alkaline earth atoms and alkali negative ions in their ground and low-lying excited states may be better described by a collective model than by the independent-particle model (Krause and Berry, 1985, 1986; for a review of this topic, see Berry, 1986). Furthermore there is now experimental evidence of strong correlation of the valence electrons in low-lying and intermediate excited states of the barium atom, and some of extensions of theory that must be made in order to use resonance ionization spectroscopy to probe electron correlation. This experimental evidence is the substance of this report.

## 2.   Low-lying States of Ca, Sr and Especially Ba

The experiments described here have all involved laser-induced resonant two-photon or multiphoton ionization of alkaline earth atoms. Monitoring has been done by (time-of-flight) energy selection of the photoelectrons, analysis of their angular distributions and sometimes separate analyses of ion yields. Most experiments involved two or more frequencies of radiation, and variation of the angle between the polarization directions of the different-colored coaxial beams. The most straightforward of the experiments, resonant ionization of an alkaline earth from its ground state through an nsnp intermediate state to the ground state of the ion plus a free electron, yield no information concerning correlation in the intermediate state unless a microscopic model is introduced. Since this has not yet been done, the experiments concerning the $ns^2 \to nsnp \to ns + \varepsilon s,d$ process for Ca, Sr and Ba only illustrate how microscopic parameters may be extracted when a simple, independent-particle model is employed (Mullins et al., 1985a). However when these results are combined with those from experiments in which the ion is left in its first excited $^2D$ state e.g., the 5d $^2D$ state of $Ba^+$, qualitative inferences can be drawn concerning electron correlation without reference to a specific collective model, particularly from branching ratios, the simplest data to interpret.

Barium atoms were excited through each of three low-lying states to the continuum, at energies high enough to leave the $Ba^+$ ions in either the ground 6s $^2S$ or first excited 5d $^2D$ state (Mullins, 1985a,b). The intermediate states were the (6s6p) $^1P_1$, (5d6p) $^3P_1$ and (5d6p) $^3D_1$; for the latter two, the fine structure components corresponding to ions in $^2D_{3/2}$ and $^2D_{5/2}$ final states were resolved. Furthermore, only one-color, two-photon resonant two-photon ionization was used for those states, so the angle $\eta$ between polarization vectors for the two photons was fixed at 0 .

The first evidence for significant configuration mixing is the appearance of the so-called 5d6p triplet states in

| State | $\lambda$(nm) | to $^2S$ ion | to $^2D$ | j= 3/2:5/2 |
|---|---|---|---|---|
| (5d6p) $^3P_1$ | 389 | 7 | 93 | (49:51) |
| (5d6p) $^3D_1$ | 413 | 4 | 96 | (88:12) |
| (5s6p) $^1P_1$ | 308 | 83 | 17 | -- |

Table 1. Branching ratios to the ground and first excited states of $Ba^+$ for photoionization of low-lying states of Ba.

electric dipole excitation. These are forbidden in the independent-particle picture not only the basis of spin selection rules but on being nominally 2-electron excitations. Next are the branching ratios which are as shown in

Table 1.  These could be taken as quantitative measures of
configuration mixing in the intermediate states, if one were
willing to suppose that final-state interactions and configuration mixing in the ground state are negligible.

The simplest interpretation of these ratios is that the
(6s6p) $^1P_1$ state has a significant but not extreme mixing
with such configurations as 5d6p, and that the complementary
mixing of the 5d6p triplets with the 6s6p configuration is
notably less.  The amount of mixing with  $6p^2$ cannot be
inferred from those branching ratios because the $6p$ $^2P$
channel could not be reached in these experiments.  The
branching ratios to the fine structure components also cannot
be accounted for by a simple, independent-particle model.

More subtle but more telling implications of electron correlation can be found in the angular distributions of photoelectrons in these 2-photon experiments.  If the ion is left
in an s-state, the angle $\eta$ between photon polarization is
enough to determine the general form of the distribution, at
least if hyperfine and LS coupling are neglected.  If $\eta$ = 0,
the resonant ionization through a P state must produce an
angular distribution that is a sum of an isotropic (s-wave)
component and a d-wave of the "$z^2$" form, i.e., with large
maxima at $\theta$ = 0 and $\pi$, and a subsidiary maximum at $\pi/2$.  If $\eta$
= $\pi/2$, then the outgoing wave is an "xz" d-wave.  For $^3P_1$
intermediate states,  the distributions for $\eta$ = 0 and $\pi/2$ are
like those of the $^1P_1$ for $\eta$ = $\pi/2$ and 0, respectively.  This

feature is clear in Fig. 1,
for Sr.  However this is not
all that these distributions
tell us.  At least two other
kinds of information can be
extracted, even at a qualitative level.

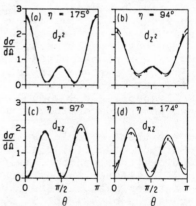

First, the anisotropy is
greater for the electrons
produced via the $^1P_1$ state
than for those produced via
the $^3P_1$, for both Sr and Ba.
The implication is that either
the s-wave branching is
greater for the triplets, as
is found if the data are
fitted to the parameters $|\sigma_s/\sigma_d|$
and $|\delta_s - \delta_d|$, or the triplets
show greater LS coupling, or

Fig. 1.  Photoelectron angular distributions from the photoionization of
the singlet [curves (a) and (c)] and
triplet [curves (b) and (d)] (5s5p)
$P_1^o$ curves in strontium.

both.  Second, if the angular distribution data for each are
fitted to its own spherical harmonic expansions and then all
the data for different values of $\eta$ are used to determine
best values of $|\sigma_s/\sigma_d|$ and $|\delta_s-\delta_d|$ for each state, the computed
curves and the experimental data are all in excellent
agreement for the singlets but the model-determined curves
show small but significant deviations from the experimental
data and the best-fit phenomenological curves.  Again, Figure
1 illustrates this for Sr; Ba is similar, although the data

for both xz waves for Ba, from the $^1P_1$ for' $\eta = \pi/2$ and the $^3P_1$ for $\eta = 0$, show greater isotropy than is implied by the curves from the model-dependent fit. The deviations imply greater spin-orbit coupling for the $^3P_1$ states than for the $^1P_1$'s, that is, greater breakdown of the one-electron model. Ionization of Ba to give Ba$^+$ in its $^2D_{3/2}$ or $^2D_{5/2}$ state shows similar deviations in the angular distributions. Figure 2, taken from Mullins et al., 1985b, shows the angular distributions for production of these states of the ion via both the (5d6p) $^3P_1$ and (5d6p) $^3P_1$. In all cases, the best-fit phenomenological curves (solid) fit the data within the latters' error bars but the curves based on the simple 2-channel (s+d, spin-preserving) model of ionization-- the dashed curves-- deviate by significant amounts from the data. The implication of these deviations is again some degree of breakdown of the independent-electron picture for the $^3P_1$ states. Inclusion of either L or J dependence in the photo-ionization step gives enough flexibility to account for the deviations, but such dependence could occur by several means, for example L or J dependent cross sections, or different phase shifts for final (total system) singlet and triplet channels, with

$$\frac{d\sigma}{d\Omega}$$

$$\frac{d\sigma}{d\Omega}$$

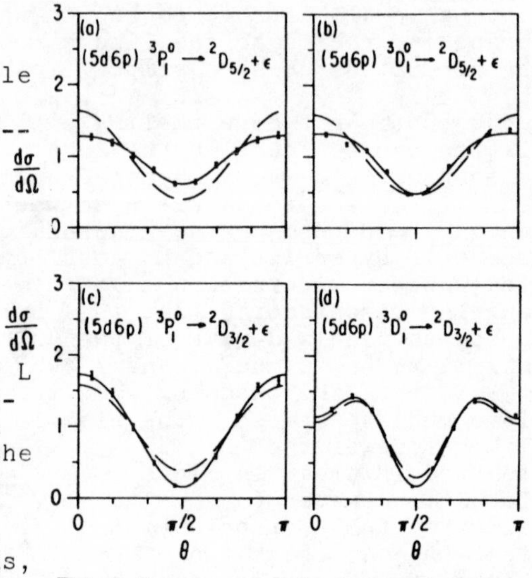

FIG. 2. Photoelectron angular distributions resulting from the process $(6s^2)^1S_0 \xrightarrow{\nu} (5d\,6p)^3L_1^o \xrightarrow{\nu} (5d)^2D_j + \epsilon s, d$.

some LS coupling. How much each of these effects contributes cannot yet be determined from the data.

A more extreme example is the so-called (5d7s) $^1D_2$ state of Ba (Mullins et al., 1985c). This state was prepared by both two-color, three-photon, twice-resonant ionization (via the (6s6p) $^1P_1$ state) and by one-color, three-photon, once-resonant ionization. The two-color experiments gave branching ratios of 43% 6s $^2S_{1/2}$, 15% 5d $^2D_{3/2}$ and 42% 5d $^2D_{5/2}$, the only channels open at the energy of one 553.7 nm photon plus two 635.5 nm photons. The distributions corresponding to the $^2S$ channel are $\eta$-dependent and yield reasonable values for $|\sigma_s/\sigma_d|$ and $|\delta_s - \delta_d|$, and an estimate of 9±5% triplet mixing in the $^1D_2$ state. These results alone imply significant mixing of the (6snd) $^1D_2$ channel.

However, the angular distributions for the processes leaving the 5d $^2D$ ions give much stronger evidence for mixing. The one-electron, independent-particle model implies that the 7s electron should leave as a p-wave, albeit with a mixture of m components dependent on how the (5d7s) $^1D_2$ state is produced. Pure p-waves should give distributions of the form $(1 + \beta P_2)$, with no higher harmonics. In fact the coefficients of the spherical harmonics fitted to these distributions are large for both the $P_4$ and $P_6$ harmonics. There must be significant mixing with 5dnd configurations to account for such extreme anisotropy when the ion is left in a 5d $^2$D state. The conclusion of the experiments with the "5d7s" $^1$D$_2$ state is that this is a very strongly mixed state, with extreme correlation, at least a large part of which is angular. This is in direct contradiction with an inference based simply on fitting the energy level into a multichannel quantum defect (MQDT) pattern (Aymar and Robaux, 1979), which suggested that the state is at least 90% 5d7s. For such an intermediate state, MQDT analyses can only be indicators of whether or not Rydberg assignments must be rejected for specific states; they cannot be used as indicators of the validity of a Rydberg assignment.

Another indication of the limitations of MQDT assignments and of strong configuration mixing comes from the photoionization of several states of Ba which were, for some time, all assigned as 6p$^2$ states (Moore, 1958) some of which were later assigned as Rydberg states by Aymar and Robaux (1979). Branching ratios and angular distributions were measured (Hunter et al., 1986) for the "6s8s" $^1$S$_0$, "6p$^2$" $^3$P$_0$, "6p$^2$" $^3$P$_1$, "6s7d" $^1$D$_2$ and "6p$^2$" $^3$P$_2$ states, all between 34370 and 35620 cm$^{-1}$ above the ground state. They were produced by three-photon, three-color, twice-resonant excitation via the (6s6p) $^3$P$_1$ state and ionized with the 337.1 nm light of a nitrogen laser, energetic enough to leave the Ba$^+$ as 6s $^2$S, 5d $^2$D or 6p $^2$P.

The branching ratios of all these states are strongly dominated by formation of 6p $^2$P ions. The largest contribution from the 6s and 5d final configurations occurs with the 6s8s $^1$ state, where it is only 25%. The minimum is 12%, for the "6s7d" $^1$D$_2$ state. Hence the branching ratios imply significant configuration mixing and dominance by--and hence assignment as--6p$^2$ levels for all five states.

All told, the evidence is strong but not conclusive that these states are all significantly mixed but dominated by the 6p$^2$ configuration. For example the "6s8s" ionization process could yield a branching ratio favoring the 6p $^2$P state because of a small cross section for the 6s8s → 6s + $\epsilon$p process at precisely the energy we used. Measurement of the dependence of the branching ratio on the ionizing wavelength would test this hypothesis. Another probe would be the measurement of the branching ratio for the states assigned as "6s7d" $^3$D and 6s8s $^3$S. There are no 6p$^2$ $^3$D or $^3$S states so none of these large fractions of ions in the 6p $^2$P state. If such were found, then one would have to suspect that final-state interactions were important, or that angular correlations are extremely important in the "6s7d" $^3$D and "6s8s" $^3$S states.

The last results to be cited here are the analyses of angular distributions from some autoionizing levels of Ba in the

region 44-45,000 cm$^{-1}$ above the ground state, first studied
by 2-photon photoionization spectroscopy (Wynne and Hermann,
1979).  Three of these states have been examined in detail
now (Keller et al., 1986) by preparing them through two different intermediate states, the 6s6p $^1P_1$ and 5d6p $^1P_1$.  Both
photoionization spectra and angular distributions were
measured.  The profiles and the angular distributions depend
markedly on intermediate state.  Most of these dependencies
could be interpreted adequately from a simple configurational
model, assigning a simple, spin-conserving final ionized
state for each process.  One result could not be so simply
explained:  the values of the phase shift differences $\delta_s-\delta_d$
implied by the data depended on the intermediate states.
This could not be, if the same final state were reached
whatever path was used.  The implication of this result is
that the final "state" is really a composite of singlet and
triplet compound states, that singlet and triplet have their
own phase shifts and that their pro portion in the final
states depends on the intermediate state through which the
final state is reached.

Aymar M and Robaux O 1979 J. Phys. B $\underline{12}$ 531

Berry R S 1986 in The Lesson of Quantum Mechanics (Amsterdam:
  North Holland)

Hunter J E III, Keller J S and Berry R S 1986 Phys. Rev. A $\underline{33}$
  3138

Keller J S, Hunter J E III and Berry R S 1986 (in
  preparation)

Krause J L and Berry R S 1985 J. Chem. Phys. $\underline{83}$ 5153; Phys.
  Rev. A $\underline{31}$ 3502

Krause J Land Berry R S 1986 Comments At. Mol. Phys. $\underline{18}$
  91

Moore C E 1958 Atomic Energy Levels (Washington, D.C.:  U.S.
  Gov't. Printing Office)

Mullins O C, Chien R-L, Hunter J E III, Keller J S and Berry
  R S 1985a Phys. Rev. A $\underline{31}$ 321

Mullins O C, Chien R-L, Hunter J E III, Jordan D K and Berry
  R S 1985b Phys. Rev. A $\underline{31}$ 3059

Mullins O C, Hunter J E III, Keller J S and Berry R S 1985c
  Phys. Rev. Lett. $\underline{54}$ 410

Wynne J Jand Hermann J P 1979 Opt. Lett. $\underline{4}$ 106

*Inst. Phys. Conf. Ser. No. 84: Section 5*
*Paper presented at RIS 86, Swansea, Wales, 7–12 Sept. 1986*

209

# Resonance enhanced ionization of clusters: spectroscopy and dynamics

A. W. Castleman, Jr., P. D. Dao, S. Morgan, and R. G. Keesee

Department of Chemistry, The Pennsylvania State University, University Park, PA 16802 USA

Abstract. Resonance enhanced ionization of clusters provides a detailed way of investigating the molecular aspects of condensation phenomena and the molecular properties of condensed matter at the microscopic level. Such studies also contribute to a further understanding of intermolecular energy flow and energy disposal following multiphoton ionization. Information on the spectroscopic shifts of phenylacetylene and paraxylene clustered by a number of solvent molecules are presented. While the spectral features resemble those in the condensed or matrix state, the ionization potentials are found to differ considerably from those of bulk condensed matter. Data is presented on the internal Penning ionization of a cluster following resonance enhanced absorption in a chromophore as a result of electron transfer within the aggregate; surprisingly long time constants ranging up to 200 ns were observed for ionization through excitations to low Rydberg states of the paraxylene chromophore.

## 1. Introduction

Cluster research is a rapidly growing field which offers the exciting prospect of bridging the gap between the gaseous and condensed phase by probing the details of condensation and nucleation phenomena at the molecular level. Studies of spectroscopic shifts upon successive clustering are especially interesting with regard to the onset of liquid or sold-like features in the spectra (1-6). Investigations of the processes of cluster ionization and dissociation are of particular interest since they contribute to a further understanding of the evolution of changes in ionization potentials as a system approaches the bulk work function. The details of intramolecular energy flow and energy disposal following ionization can also be revealed.

Studies of the molecular properties, reactions, and behavior of clusters generally require ionization in one of the steps as either a probe and/or a method of detecting clusters through mass spectrometry. Although ionization can be accomplished through electron impact as well as single photon techniques, resonance enhanced multiphoton ionization often enables selective ionization of clusters in particular states. More detailed and specific information can be obtained through resonance-enhanced ionization spectroscopy, and is the preferred method when such processes can be readily accomplished. Herein, examples are drawn from three studies.

In the first example, a detailed investigation of the spectroscopic shifts of two probe molecules, phenylacetylene and p-xylene, clustered by a series of rare gases, $CO_2$, $H_2O$, $N_2$, $O_2$, $NH_3$, and $CCl_4$ shows that clusters of specific composition can be selectively ionized. In the most part, the spectral shifts are toward the red of the main $S_1$ state of the unclustered parent molecule, although in a few cases, most notably with $H_2O$, the shifts are to the blue. Through use of various expansion and ionizing conditions, fragmentation can be readily assessed and investigated. Comparison of one- and two-color studies reveal the importance of excess energy in cluster fragmentation. In a second example involving p-xylene clustered with argon, the change in ionizaton potential as a function of the degree of aggregation is considered. The results provide insight into the extent of interaction of the ion of the chromophore with the "solvating matrix".

In the final example, cluster ionization is accomplished through high Ryberg states of the p-xylene chromophore whereby ionization is accomplished through an intracluster process having analogy to gas-phase Penning ionization processes. Contrasting results between systems comprised of p-xylene bound to trimethylamine and to ammonia detail the molecular processes involved. Evidence for a slow ionization process is presented.

## 2.  Spectral Shifts

Resonance enhanced multiphoton ionization through the specific excitation of an electronic state of a chromophore contained within a cluster is a powerful method of ascertaining the properties of clusters in relating these to their counterparts in the condensed and isolated gas phases. Generally, the clustering of atoms or molecules onto a chromophore result in a perturbation of the electronic states of that chromophore. The spectral shift of a given electronic transition from that of the isolated chromophore is a measure of the relative differences between the lower and upper states of the energetic perturbation induced by clustering. This is analogous to the spectral shift of electronic transitions observed for molecules in solutions or matrices from their gas-phase transitions. A red shift implies that complex formation has reduced the energy difference between the two states, whereas a blue shift indicates an increase in the difference. The magnitude and direction of the shift are due to a combination of effects, including dispersive and repulsive interactions, hydrogen bonding, and electrostatic forces involving such processes as dipole-induced-dipole or dipole-dipole interactions.

We have employed both one- and two-color resonance enhanced multiphoton ionization to investigate the $S_1 \leftarrow S_0$ $\pi$-electron transition in phenyl-acetylene and p-xylene as clusters with various solvents. Single color multiphoton ionization studies of the perturbed $L_b(^1B_2)$ states of phenylacetylene (PA) bound with Ne, Ar, Kr, and Xe were all found to induce a lowering of the $S_1$ resonance with respect to the ground state. These observed red shifts have been attributed to dispersive interactions with solvent molecules (7,8). In general, a spectral shift is governed by three factors; (1) short-range electronic repulsive interactions which result in a blue shift, (2) electronic dispersive interactions which result in a red shift, and (3) differences of zero point energies between the excited and ground states. Our results support the finding (7) that in aromatic molecule-rare gas atom systems the spectral shifts are dictated by atom polarizability, i.e., the important role of dis-

persive forces in the perturbation of the $S_1$ excited state.   Figure 1
displays the results of our study which shows a direct linear dependence
of the spectral shift on the electrostatic polarizability of the rare gas
atom.   The results conform to the Onsager model, but on a microscopic
level (9).

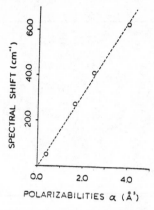

Fig. 1.   Spectral Shift of PA•R
(relative to the nascent PA)
versus the polarizability of the
rare-gas atom.   $\alpha$ = 0.40 (Ne),
1.63 (Ar), 2.48 (Kr) and 4.01 $\text{Å}^3$
(Xe).

Spectral shifts of phenylacetylene due to aggregation by rare gases are
given in Table 1.   Investigations with large-ringed systems have
generally shown an approximate additivity of spectral shifts based on
the number of rare gas atoms clustered on to the aromatic (7,10).   In our
own work, we find that this additivity is apparently additive only up to
the clustering of two atoms per aromatic ring as seen from the data in
Table 1.   Since the additivity is nearly exact for the two-atom case, the
spectral shifts shown in Figure 1 are identical on a spectral shift per
atom basis for the phenylacetylene system in the case of the two-atom
containing rare gas complex.

Particularly interesting are the trends seen for larger clusters.   Figure
2 shows a selected set of data for the spectral shifts relative to the $S_1$
electronic origin of phenylacetylene for clusters containing four to ten
argon atoms.   First it is interesting to note that the major feature
asymptotically approaches a shift of approximately 50 cm$^{-1}$.   Clearly the
additivity rule does not apply.   Secondly, the van der Waals modes to the
right side of the main resonance begin to fill in for large cluster
sizes.   The spectra are broadening in analogy to those seen in the
condensed phase and the features to the right resemble photon modes for
a system of infinite lattice.

Other interesting spectral shifts have been observed for the clustering
of $N_2$, $O_2$, $N_2O$, $NH_3$, $H_2O$, $CCl_4$, and $CH_4$ to phenylacetylene.   Some
representative spectral shifts are given in Table 2.   In most cases the
main resonance is also red-shifted, although in a few a substantial blue
shift is observed, most notably for $H_2O$.   The striking difference between
the isoelectronic molecules $H_2O$ and $NH_3$ can be rationalized in terms of
the excitation of the $\pi$ system leading to a repulsive interaction with
the two long-pair electrons of the $H_2O$ molecule (11).

## 3.   Shifts in Ionization Potential with Degree of Aggregation

The ionization potentials of p-xylene bound with argon (PX•Ar$_n$) were
determined through studies in which the energy of one photon was fixed at

Table 1.  Spectral shifts of the electronic origin of the $S_1$
excited state of the complex $PA \cdot R_n$ (R = rare gas atom).

| Species | Spectral Shifts[a] $\delta\nu$ (cm$^{-1}$) | Species | Spectral Shifts[a] $\delta\nu$ (cm$^{-1}$) |
|---|---|---|---|
| $PA \cdot Ar$ | $-27.8 \pm 0.5$ | $PA \cdot Ar_8$ | $-57.2 \pm 0.8$ |
| $PA \cdot Ar_2$ | $-53.3 \pm 0.5$ $-25.4 \pm 0.5$[b] | $PA \cdot Ar_9$ | $-56 \pm 1$ |
|  |  | $PA \cdot Ar_{10}$ | $-67 \pm 1$ |
| $PA \cdot Ar_3$ | $-49.1 \pm 0.5$ $-22.2 \pm 0.5$[b] | $PA \cdot Ar_{11-15}$ | $(-54$ to $-47)$ |
| $PA \cdot Ar_4$ | $-49.9 \pm 0.5$ | $PA \cdot Ne$ | $-5.0 \pm 0.8$ |
| $PA \cdot Ar_5$ | $-53.5 \pm 0.5$ | $PA \cdot Kr$ | $-42.2 \pm 0.5$ |
| $PA \cdot Ar_6$ | $-57.0 \pm 0.5$ | $PA \cdot Kr_2$ | $-79.0 \pm 0.8$ |
| $PA \cdot Ar_7$ | $-57.5 \pm 0.5$ | $PA \cdot Xe$ | $-63.5 \pm 0.8$ |

[a]Energy shift with respect to the unperturbed PA $S_1$
resonance.  A negative value corresponds to a red-
shift.
[b]From stagnation pressure studies, we tentatively
assigned these two features to different conformers.

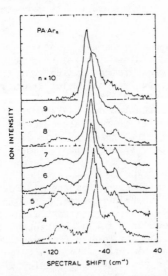

Fig. 2.  R2PI current versus
one-photon energy.  The ion
currents are recorded at the m/e
ratios corresponding to $PA \cdot Ar_n$
($4 \leqslant n \leqslant 10$).  The energy scale is
relative to the $S_1$ electronic
origin of PA.  The ion current
scale is relative and different
for each spectrum and $p_0 = 300$
Torr.

the $L_b$ resonance and the wavelength of the second laser was scanned.
Resonance-enhanced ionization with a single-color laser results in
significant fragmentation due to the fact that the absorbed energy is
substantially above the ionization threshold since the $S_1$ state lies more
than halfway to the ionization continuum.  Cluster ionization was found

Table 2.  Spectral shifts of the vdW complex PA·M.

| Species | Spectral Shift[a] $\delta\nu$ (cm$^{-1}$) | Assignment[b] |
|---------|---------------------------|---------------|
| PA·NH$_3$ | −81.9 ± 0.5 | (0)[c] |
|           | − 2.0 ± 0.5 | (0)[c] |
|           | +55.0 ± 0.5 | (0)[c] |
|           | +12.5 ± 0.5 | $\nu$ |
| PA·CCl$_4$ | −58.3 ± 0.5 | (0) |
| PA·CH$_4$ | −50.0 ± 0.8 | (0) |
| PA·N$_2$ | −11.2 ± 0.5 | (0) |
|          | + 3.7 ± 0.5 | ? |
| PA·(N$_2$)$_2$ | −22.3 ± 0.8 | (0) |
| PA·O$_2$ | − 3.0 ± 0.8 | (0) |
| PA·H$_2$O | +13.6 ± 0.5 | (0) |
| PA·CO$_2$ | + 2  ± 1 | (0) |
| PA·N$_2$O | + 4  ± 1 | (0) |

[a]Energy shift with respect to the unperturbed S$_1$ state
of PA.  A negative value corresponds to a red shift.
[b](0) denotes the electronic origin of the vdW
complex.
[c]We tentatively assigned these two features to
different conformers.

to be suppressed to a negligible amount in the two-color experiments,
enabling a detailed investigation of the variation in ionization
potential with degree of aggregation to be definitively established.

It is well known that the Stark effect leads to a shift in ionization
potential when measured in an electric field and correction is necessary
to account for shifts in the order of 50 cm$^{-1}$.  The ionization potentials
are found to vary with the square root of the electric field present in
the region of ionization in accordance with expectations and findings of
others (12).  Extrapolation to zero field is readily accomplished in view
of the linear dependence and the fact that various cluster systems
display lines of identical slopes in these weakly perturbed rare-gas
aggregates.

The shifts in ionization potential of p-xylene in the rare gas aggregates
is shown in Figure 3 for clusters with one to six argon atoms.  The shift
in relative ionization potential is observed to display a broadly linear
dependence on the number of argon atoms.  The largest deviations from
this trend are observed for the dimer and pentamer.  The observed total
shift of about 750 cm$^{-1}$ for the hexamer is to be contrasted with the
matrix isolated value which is about 6000 cm$^{-1}$ for a similar molecule

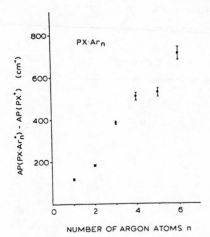

Fig. 3. Relative appearance potentials. Field ionization of $PX \cdot Ar_n$ (n=0-6) in a 150V/cm dc field. $AP(PX \cdot Ar_n^+)$ increases with the coordination number n.

(benzene) in an argon matrix (13). Evidently, the "local environment" with which a molecule interacts is relatively large in such a matrix and the observed shift is far from the expected bulk value. This is in interesting contrast to the metal systems which, when corrected for ion image potential effects, show that the ionization potential of aggregates comprised of only a few atoms display nearly the bulk ionization potential of the polycrystalline metallic system (14).

## 4. Intracluster "Penning Ionization" and Evidence for Slow Ionization Processes in Clusters

Clusters provide interesting systems for comparing ionization and con-comitant electron transfer processes for bimolecular processes in the gas phase, including Penning ionization, with analogous ones in the condensed phase. Toward this goal, ionization of clusters comprised of p-xylene (PX) bound to $NH_3$ and $N(CH_3)_3$ were studied following the absorption of photons through the perturbed $S_1$ state of p-xylene. An interesting comparison is provided by results of studies involving adducts of p-xylene bound to $NH_3$ and trimethylamine since the ionization of p-xylene is less than that of ammonia but greater than that of trimethylamine. In the case of $PX \cdot NH_3$, ionization by adsorption of a second photon which is adsorbed by the perturbed $S_1$ state of p-xylene begins near the ionization threshold of p-xylene and leads to the expected cluster ion $PX \cdot NH_3^+$. Two other channels are possible at higher photon energies, namely the formation of $NH_4^+$ at 0.1 eV above the ionizaton potential of p-xylene and $NH_3^+$ at 1.8 eV above; $NH_4^+$ is observed in the two-color experiments at high fluence of the ionizing laser where two photons are absorbed by the $S_1$ state.

By contrast, absorption into high Rydberg states of p-xylene below its ionization potential in $PX \cdot N(CH_3)_3$ leads to the production of pre-dominantly $N(CH_3)_3^+$ with $H^+N(CH_3)_3$ as a minor product. No $PX \cdot N(CH_3)_3^+$ ion is detectable. One conclusion is that photoexcitation of p-xylene leads to an intercluster ionization process bearing analogy to Penning ionization where the perturbed high Rydberg states of p-xylene interact with the partner molecule $N(CH_3)_3$. A second, and more startling observaton, was the finding of a slow ionization process as evident in the time-of-flight peak shapes shown in Figure 4. Since the laser interacts with the molecules in the first of a two-field acceleration region, a long tail is only possible when the ionization process is slow.

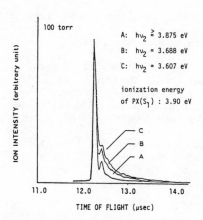

A: $h\nu_2 \gtrsim 3.875$ eV
B: $h\nu_2 = 3.688$ eV
C: $h\nu_2 = 3.607$ eV

ionization energy
of $PX(S_1)$ : 3.90 eV

Fig. 4. Ion mass peaks at different two-photon energies. Broadenings of $TMA^+$ ion peaks as a function of the ionization energy. The broadenings in B and C correspond to time constants of $160 \pm 20$ and $200 \pm 20$ ns, respectively. The peaks corresponding to $TMA \cdot H^+$ are also observable.

Fragmentation leads to a knee in the peak shape and not a long tail as observed in the figure (15). Interestingly, the process is substantially slower with a decrease in the energy of the ionizing photon. Questions arise whether the slow step is associated with the proton transfer channel (i.e., the $(CH_3)_3NH^+$ product) or an electron transfer process (i.e., the $(CH_3)_3N^+$ product). Careful measurements with deuterated species reveal that the tail is largely associated with the electron transfer process. Interestingly, Hatano (16) has found that orientational effects in the liquid phase, where motion is restricted, can lead to a significant reduction in the rate of Penning ionization. Likewise, Harris (17) has evidence for long delays in ionization of $NH_3$ on silver electrode surfaces. Whether there is some analogy to the foregoing observations is currently unknown. A plausible explanation for the present findings is that a large geometry change is involved in the formation of the trimethylamine ion.

## References

1. A. W. Castleman, Jr., in: Electronic and atomic collisions (J. Eichler, I.V. Hertel and N. Stolterfoht, Eds), Elsevier Science Publishers, Amsterdam, pp. 579-590 (1984).
2. A. W. Castleman, Jr. and R. G. Keesee, Chem. Rev. 86, 589 (1986).
3. A. W. Castleman, Jr. and R. G. Keesee, "Clusters: Properties and Formation," Ann. Rev. of Phys. Chem., in press.
4. A. W. Castleman, Jr. and R. G. Keesee, "Clusters: Bridging the Gas and Condensed Phases," Accts. Chem. Res., in press.
5. A. W. Castleman, Jr. and T. D. Mark, in: Gaseous Ion Chemistry/Mass Spectrometry (J. H. Futrell, Ed.) John Wiley and Sons, pp. 259-303 (1986).
6. M. F. Vernon, D. J. Krajnovich, H. S. Kwok, J. M. Lisy, Y. R. Shen, and Y. T. Lee, J. Chem. Phys. 77, 47 (1982); R. E. Miller, R. D. Watts and A. Ding, Chem. Phys. 83, 155 (1984); P. M. Dehmer and S. T. Pratt, J. Chem. Phys. 76, 843 (1982).
7. S. Leutwyler, U. Even and J. Jortner, J. Chem. Phys. 79, 5769 (1983).
8. S. Basu, Advan. Quantum Chem. 1, 145 (Eq. 46) (1964).
9. P. D. Dao, S. Morgan, and A. W. Castleman, Jr., Chem. Phys. Lett. 111, 38 (1984).
10. A.-M. Sapse (personal communication)

11. P. D. Dao, S. Morgan, and A. W. Castleman, Jr., Chem. Phys. Lett. 113, 219 (1985).

12. K. H. Fung, H. L. Selzle and E. W. Schlag, Z. Naturforsch 36a, 1257 (1981).

13. J. Jortner, in: Vacuum Ultraviolet Radiation Physics (E. E Koch, R. Haensel, and C. Kunz, Eds.) Pergamon Press, Oxford, p. 291 (1974).

14  M. M. Kappes, M. Schar, P. Radi, and E. Schumacher, J. Chem. Phys. 84, 1863 (1986); A. W. Castleman, Jr. and R. G. Keesee, "Metallic Ions and Clusters: Formation, Energetics, and Reactions," Zeitschrift fur Physik, in press.

15. P. D. Dao and A. W. Castleman, Jr., J. Chem. Phys. 84, 1435 (1986).

16. Y. Hatano (personal communication); see also T. Wada, K. Shinsaka, H. Namba, and Y. Hatano, Can. J. Chem. 55, 2144 (1977).

17. C. Harris (personal communication) Univ. of California, Berkeley.

## Acknowledgments

Support by the Department of Energy, Grant No. DE-ACO2-82-ER60055, and the U. S. Army Research Office, Grant No. DAAG29-85-K-0215, is gratefully acknowledged.

*Inst. Phys. Conf. Ser. No. 84: Section 6*
*Paper presented at RIS 86, Swansea, Wales, 7–12 Sept. 1986*

# Xylene isomer analysis multiphoton resonance ionisation spectroscopy

T G Blease, R J Donovan, P R R Langridge-Smith, T Ridley and
J P T Wilkinson

Department of Chemistry, University of Edinburgh, West Mains Road,
Edinburgh EH9 3JJ

Abstract. A method is described for the selective ionisation of
individual xylene isomers (ortho, meta and para) in the presence of an
excess of the other isomers. The method is based on laser resonance
ionisation spectroscopy and involves the use of a pulsed supersonic
jet to cool the molecules.

## 1. Introduction

The resonance ionisation technique is now well established as a means for
detecting very low atomic concentrations. However, the use of this
technique for the detection of low molecular concentrations remains in
its infancy. Two major problems are encountered in extending the
resonance ionisation technique from atoms to molecules: (i) the high
density of rovibronic states associated with molecules makes their
spectra more complex and also dilutes the states accessible for
excitation with monochromatic radiation (ie the population is spread over
a large number of states and only a few states can interact with monochromatic radiation), under normal laboratory conditions; (ii) the intermediate states involved in the excitation process may undergo rapid decay
via a variety of non-radiative channels.

The first of these problems can be overcome to a large extent by cooling
the molecules to ca. 10°K, using a pulsed supersonic jet. Under these
conditions, complex molecular spectra can be reduced to a few sharp
atomic-like lines. The second problem is less easily overcome and intermediate states that rapidly dissociate, undergo internal conversion or
intersystem crossing, should be avoided if high sensitivity is required.
Some choice of excitation scheme is normally possible but optimisation
requires a prior knowledge of the photophysics of a given molecule.
Increasing the laser intensity to increase the rate of ionisation is the
only other option but there are clear limitations to this approach.

In the present communication, we describe a spectroscopic study of the
three isomers of xylene (ortho, meta and para, see below) using the
resonance ionisation technique, together with a pulsed supersonic jet.

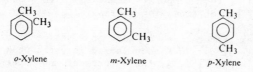

o-Xylene          m-Xylene          p-Xylene

We show that each of the three isomers can be selectively ionised and that this provides a basis for a sensitive and selective method for the analysis of xylene mixtures.

## 2. Experimental

The experimental arrangement involved the use of an excimer pumped dye laser, in conjunction with a pulsed supersonic jet. The laser system was of conventional design (Lambda Physik EMG102+FL2002, bandwidth = 0.3 cm$^{-1}$). For wavelengths below 335 nm a KDP doubling crystal was used. Experiments were also performed in the ionisation source of a quadrupole mass spectrometer (Vacuum Generators SX200), with the sample at ambient temperature.

The pulsed supersonic jet was produced by a modified automobile fuel injection valve (pulse width = 600 μs) having a 1 mm orifice (Behlen, Mikami and Rice 1979). Xylene samples were seeded into 1-6 atmospheres of helium carrier gas and expanded into a vacuum chamber where the jet was crossed (ca. 2 mm from the nozzle orifice) by the focussed beam from the tunable dye laser. Photoions were detected with a pair of nickel electrodes biased at 90 volts and located approximately 2 cm downstream from the nozzle. The electrodes were connected to a differential input operational amplifier which provided an amplification of 10$^7$. Signals were processed by an EG&G Brookdeal 9415/9425 gated integrator and recorded on a two channel chart recorder.

Absolute wavelength calibration was provided by neon optogalvanic lines from a hollow cathode lamp and accurate interpolations were made using the transmission fringes from a solid quartz etalon.

## 3. Results and Discussion

The effect of cooling on the spectrum of para-xylene, in a supersonic jet, is shown in Figure 1.

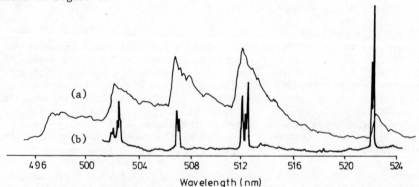

Fig. 1  Two-photon resonant (S$_1$) four-photon ionisation spectra of para-xylene obtained (a) at room temperature and (b) with a supersonic jet, illustrating the effect of expansion cooling.

The spectrum at room temperature consists of a series of broad over-lapping peaks, while at ca. 10 K (supersonic jet) the peaks are sharp and there is very little overlap. As the spectra of the three isomers all lie in the same region and overlap extensively with each other at 300 K, jet cooling is essential for selective ionisation.

Two ionisation schemes were investigated. The first and most sensitive scheme was resonant two-photon ionisation, i.e.

$$Xy(S_0) \xrightarrow{h\nu_1} Xy*(S_1) \xrightarrow{h\nu_1} Xy^+ + e^-$$

Wavelengths in the region of 270 nm are required for this scheme and the doubled output of the dye laser was therefore used. The resonant two-photon ionisation spectra of the three isomers are shown in Figure 2.

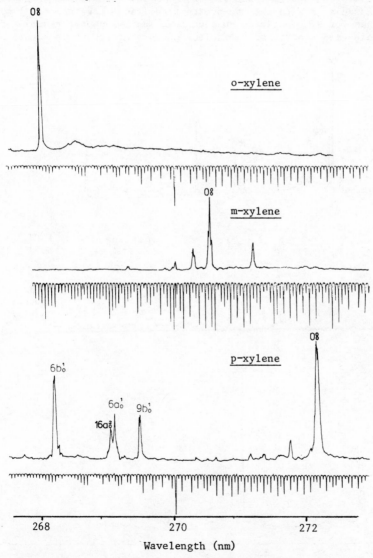

Fig. 2 One photon resonant (S$_1$) two-photon ionisation spectra of the jet-cooled xylene isomers in the region of the origin bands. The lower traces are a combination of optogalvanic lines and etalon fringes.

The observed resonances correspond with transitions to the first singlet state ($S_1$) of the xylene isomers (Blease 1985).

The second ionisation scheme involved two-photon resonant four-photon ionisation, i.e.

$$Xy(S_0) \xrightarrow{2h\nu_2} Xy^*(S_1) \xrightarrow{2h\nu_2} Xy^+ + e^-$$

This scheme has the advantage of directly employing the visible output of the dye laser (<u>ca.</u> 500 nm), but suffers from the disadvantage that cross-sections associated with the two-photon transitions are inherently much smaller than for single photon absorption.  The two-photon resonant four-photon ionisation spectra of the xylene isomers are shown in Figure 3.

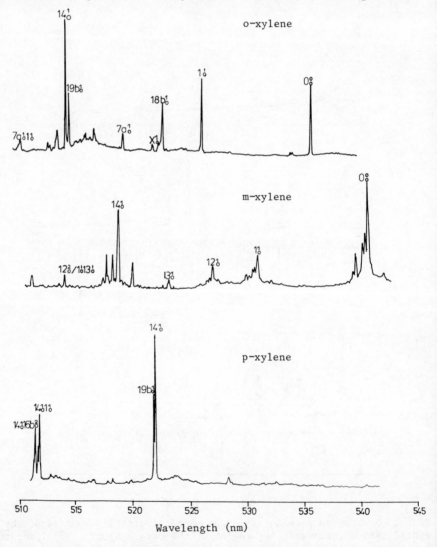

Fig. 3  Two-photon resonant four-photon ionisation spectra of the xylene isomers in the region of the origin bands.

It should be noted that the origin band is absent in the two-photon resonant spectra of para-xylene, as it is symmetry forbidden. Extensive vibronic structure was observed for all three isomers to shorter wavelength and will be discussed elsewhere.

Both ionisation schemes clearly allow the selective ionisation of individual xylene isomers and experiments were carried out to establish the limits for detecting one isomer in the presence of other isomers. The resonant two-photon ionisation scheme was chosen for this, due to the higher signal/noise ratio achieved with this approach. Mixtures of ortho- and para-xylene were examined and a limit for detecting para-xylene in such mixtures was determined as 0.5% (i.e. one part of para-xylene could be detected in the presence of 200 parts of ortho-xylene). This limit could be improved with further work and we note that Tembreull and Lubman (1984) were able to detect one of the cresol isomers in the presence of a 300-500 fold excess of a different isomer. The absolute detection limit for the xylene isomers with the present experimental arrangement was estimated as <u>ca.</u> $10^{10}$ xylene molecules $cm^{-3}$: again it should be possible to improve on this significantly with further work.

Experiments were also carried out employing a simple commercial quadrupole mass spectrometer modified to allow the laser beam to pass through the ion source region. At low laser intensities (10 MW $cm^{-2}$) only the parent and one fragment peak $(C_7H_7^+)$ were observed. However, at higher intensities $(10^2 - 10^3$ MW $cm^{-2})$ fragmentation increased (Figure 4).

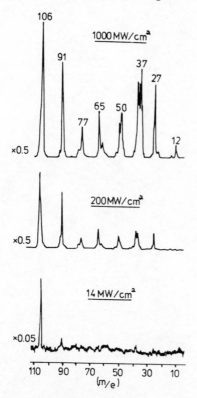

Fig. 4 Laser intensity dependence of the ortho-xylene resonant two-photon ionisation mass spectrum (268 nm).

The observed fragmentation patterns were similar to those described for other aromatic systems. In general, multiphoton ionisation produces different fragmentation patterns to those observed using electron impact: the fragmentation following multiphoton ionisation changes markedly with laser intensity and at high intensities substantial yields of small fragments, down to $C^+$ are observed (Gobeli, Yang and El-Sayed 1985).

4. <u>Acknowledgements</u>

We thank Shell Research Ltd. and the SERC for the award of a CASE studentship to T.B., and Shell Research Ltd. for provision of laser equipment.

5. <u>References</u>

Blease T G 1985 PhD Thesis, University of Edinburgh.
Behlen F M, Mikami N and Rice S A 1979 Chem. Phys. Letters <u>60</u> 364.
Gobeli D A, Yang J J and El-Sayed M A 1985 Chem. Rev. <u>85</u> 529.
Tembreull R and Lubman D M 1984 Anal. Chem. <u>56</u> 1962.

*Inst. Phys. Conf. Ser. No. 84: Section 6*
*Paper presented at RIS 86, Swansea, Wales, 7–12 Sept. 1986*

# Resonance ionization and time-of-flight mass spectrometry: high resolution, involatile molecules

U. Boesl, J.Grotemeyer, K.Walter, E.W.Schlag
Institut für Physikalische und Theoretische Chemie,TU München
Lichtenbergstr.4, D–8046 Garching, Germany

Abstract we present here a novel method for investigation of large involatile molecules. Laser desorption and cooling of neutral molecules is followed by laser ionization and analyzation by a RETOF instrument.

## 1. Introduction

For investigation of large involatile molecules several new desorption and ionization methods have been developed in the last two decades such as field desorption, direct chemical ionization, secondary ion emission, fast atom bombardment, $^{252}Cf$-plasma desorption and laser desorption of ions. More detailed information can be found in proceedings edited by Morris (1981) and Benninghoven (1983).

We present here a new mass spectrometric method with the following features :

1.) Separation of the vaporizing and ionizing process
2.) cooling of the evaporized neutral molecules
3.) ionizing by resonance enhanced multiphoton ionization (REMPI)
4.) detection by a high mass resolution reflectron time-of-flight mass spectrometer

The advantages of these new features are: a.) mass spectra with intensive pure molecular ion signals without transfer, abstraction or addition species like hydrogen, sodium etc.; b.) controllable fragmentation from very soft (base peak = molecular ion) to very strong fragmentation; c.) high yield of metastable ions; d.) high mass resolution of the TOF-spectrum; e.) high sensitivity.

In the following we shall describe our technique and in more detail the four stages mentioned above. As in the scheme of our instrument to be seen this is build up by three separated vacuum chambers, the laser desorption-,

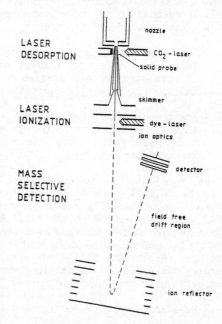

LASER DESORPTION

nozzle
$CO_2$ - laser
solid probe

skimmer

LASER IONIZATION

dye - laser
ion optics

MASS SELECTIVE DETECTION

detector

field free drift region

ion reflector

Fig.1 Experimental Setup

the laser ionization— and the mass spectrometer chamber. A solid probe is mounted near the nozzle of a pulsed valve. A low powered $CO_2$ laser desorbs ions as well as neutrals, which suffer many collisions in the nearby supersonic beam and then are cooled and transported to the skimmer. A clean molecular beam of argon and desorbed molecules enters the ionization region, while due to the construction of the ion source the desorbed ions are rejected. The ionization takes place by multiphoton absorption within the focus of a frequency doubled dye laser. The so formed ions are accelerated into the field free drift region, corrected for energy differences in an ion-reflector and detected by a tandem-channelplate detector. For more experimental details see Grotemeyer et al. (1986a).

## 2. Description of the experiment

### 2.1 Laser Desorption of neutral molecules

Most desorption techniques are used as an ion source of involatile molecules. However, it is well known that in these techniques neutrals are emitted in greater abundance than ions as demonstrated by van Breemen et al. (1983). We use the emission of neutrals by laser desorption in our technique and thus are able to separate desorption and ionization process. One major advantage is the possibility of cooling the desorbed molecules before ionization.

### 2.2 Cooling of the desorbed neutral molecules in a supersonic beam

A very efficient way of cooling is the cooling in a supersonic molecular beam. For an introduction see Levy (1981).Best cooling can be reached for the translational degrees of freedom, in comparison to rotational and vibrational motions. Translational temperatures of 10 K and better can be reached in our molecular beam.
The advantages of cooling are:

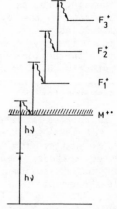

1.) low rotational and vibrational excitation leads very often to simplified structured optical spectra and enables state-selective and species-selective excitation by REMPI.

2.) low translational temperatures include narrow initial velocity distribution and make high mass resolution in a time-of-flight mass spectrometer possible.

Even in a RETOF the so called turn around times, due to initial velocities may spoil the mass resolution. Especially for desorption like SIMS, FAB and laser desorption the final kinetic energy distribution of the desorbed particles is typically in the 1eV region, which would lead in our instrument to mass resolutions of 1000 or less. With kinetic energy distributions in the 1meV region (= 10 K) we get a mass resolution of better than 10000.

Fig.2 Mechanism of REMPI and the following stepwise absorption -fragmentation.

### 2.3 Resonance enhanced Multiphoton Ionization (REMPI)

The third step after desorption and cooling is the REMPI. For fragile compounds and large molecules REMPI has many fascinating advantages. Boesl et al. (1981) could show that efficient soft ionization of organic molecules is possible with REMPI. The same authors (1982a) gave an explanation for the fragmentation following REMPI, which is valid for most organic molecules. Boesl et al. (1982b) also could demonstrate the enhanced production of metastable ions by REMPI, an important feature for mass spectrometry. Thus REMPI shows up a whole list of advantages, like high ion yields at fairly low laser power, state selective and species selective ionization, soft ionization and fragmentation, tuneable from production of mainly metastable ions to production of mainly elementary ions. In opposite to other ionization techniques like electron impact, REMPI is a multistep process. This process is dependent on the n-th power of the intensity with n the step number. This leads to a drastic change of the mass spectra with changing the laser intensity as demonstrated in Fig. 2 and Fig. 3.

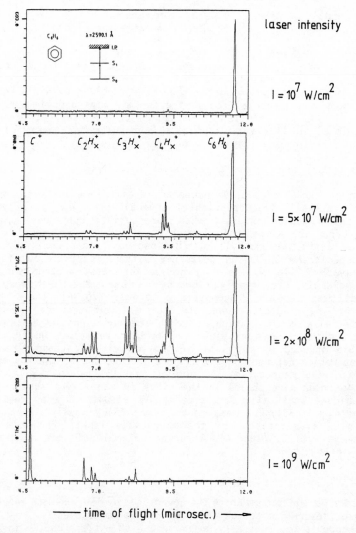

Fig. 3 Tunable fragmentation of benzene by varying the laser intensity. The mass spectra are taken with a low resolving linear TOF-Instrument.

## 2.4 Detection by a high resolution Reflectron-TOF mass spectrometer

The ideal mass spectrometer to be combined with a REMPI ion source is a TOF instrument. Both are pulsed techniques. The advantages of a TOF are unlimited mass range, high transmission and registration of all ions formed with one laser pulse. The advantage of a laser ion source are beneath those mentioned above short time and small spatial characteristics. With a Reflectron-TOF considerable mass resolutions can be achieved. However, even for a Reflectron instrument exist some limitations. Below the main contributions to the total width of a time-of-flight peak are listed:

$\Delta t$(laser)        5 - 8 nsec        commercial dye laser

$\Delta t$(turn around)   5 nsec           for room temp.;
                                           700 V/cm extraction field

$\Delta t$(Coulomb)       3 nsec           $10^4$ ions within
                                           $(0.15)^2 * 0.6$ $mm^3$ laser focus

By using a pulse cutting system for the laser, cooling within a supersonic molecular beam and a special focus size for minimizing Coulomb repulsion, Walter et al. (1986) could reduce $\Delta t$(laser) to 1.5 nsec, $\Delta t$(turn around) and $\Delta t$(Coulomb) to less than 1 nsec. This results in mass resolution of better than 10000 (50% valley) measured for p-Xylene.

## 3. Results : Mass spectra of large involatile molecules

Grotemeyer et al. (1986 a,b,c) succeeded in efficient and soft ionization as well as partially fragmentation of several large involatile molecules of biological and medical interest like Chlorophylls, Porphyrines, Tripeptides and the decapeptide Angiotensin I with mass 1295. As an example the mass spectra of native Chlorophyll **a**, obtained by methanolic extraction of the cyano-bacterium Spirulina geitlerie without further purification, are displayed in fig. 4. The soft ionization mass spectrum shows three different molecular ions, namely Chlorophyll **a** at mass 892, 10-Hydroxy-chlorophyll **a** at mass 908 and Phaeophytin **a** at mass 870. All three compounds are part of the original probe. Structural information can be obtained in the partial hard ionization mode. In the corresponding mass spectrum no longer the molecular ions yield in the base peak but the fragment ion at mass 615 induced by the loss of the phytyl side chain. The signals in the mass region between m/z 350 and m/z 500 are due to the fragmentation reactions of the macrocycle. The ion at m/z 481 can be explained by a loss of the polar side chains in the rings IV and V. The subsequent loss of the residual small aliphatic side groups results in the fragmentation pattern below m/z 481. The mass peak at m/z 414, finally, is the result of extensive fragmentation of the macrocycle itself. For more details see Grotemeyer et al. (1986 a). A further example of soft ionization is displayed in Fig.5.

## 4. Conclusion

In conclusion we want to summarize the advantages of MPI-TOF-Mass Spectrometry with a laser desorption source for free neutral molecules :
The mass range is theoretically unlimited. With Reflectron-TOF instruments mass resolutions of better than 10000 can be reached. Soft, as well as species selective ionization is possible. Tunable fragmentation promises

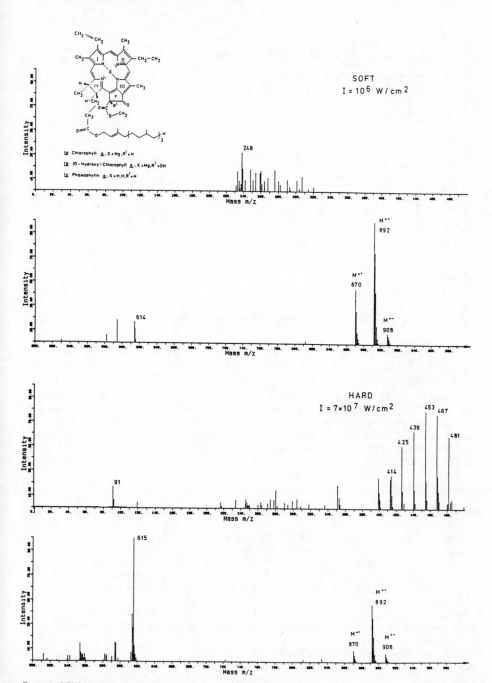

Fig.4 REMPI—mass spectra of Chlorophyll **a**.
By simply rising the laser intensity a medium fragmentation pattern
could be achieved. For low laser power soft ionization takes place
with the molecular ion being by far the dominating mass peak. The
unpurified probe contains two additional compounds (see the insert).

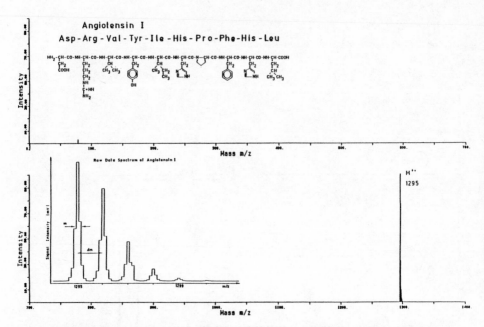

Fig.5 Soft laser ionization of Angiotensin I (m/z 1295). The insert displays the isotopic pattern of the molecular ion with a mass resolution of nearly 6000 (measured with a 5nsec laser pulse).

new insights concerning structural information as well as ion kinetics. We are sure, that the sensitivity, which is at the moment in the fmol range, can be improved by a few orders of magnitude. A great advantage is the very low background level, also called "chemical noise". At last, we should mention easy sample preparation, MS/MS capability and fast total mass spectrum recording.

## 5. References

A.Benninghoven 1983, "Ion Formation from Organic Solids", Springer Verlag, Berlin.
U.Boesl, H.J.Neusser, E.W.Schlag 1981 Chem.Phys. 55,193.
U.Boesl, H.J.Neusser, E.W.Schlag 1982a Chem.Phys.Lett. 87,1.
U.Boesl, H.J.Neusser, R.Weinkauf, E.W.Schlag 1982b J.Phys.Chem. 86,4857.
R.B.van Breemen, M.Snow, R.Cotter 1983 Int.J.Mass Spectrom Ion Proc. 49,35.
J.Grotemeyer, U.Boesl, K.Walter, E.W.Schlag 1986a Org.Mass Spectrom. in press.
J.Grotemeyer, U.Boesl, K.Walter, E.W.Schlag 1986b J.Am.Chem.Soc. 108,4233.
J.Grotemeyer, U.Boesl, K.Walter, E.W.Schlag 1986c Org.Mass Spectrom. in press.
D.H. Levy 1981 Science 214,263.
H.R. Morris 1981 "Soft Ionization Mass Spectrometry" Heyden&Son Ltd., London.

Acknowledgement: This work has been supported by a grant from the Bundesministerium für Forschung und Technologie (13N5307) in collaboration with Bruker-Franzen Analytik GmbH, Bremen.

*Inst. Phys. Conf. Ser. No. 84: Section 6*
*Paper presented at RIS 86, Swansea, Wales, 7–12 Sept. 1986*

229

# Resonant two photon ionization studies of amide groups in the UV

E. Benedetti,
Dipartimento di Chimica, Università di Napoli (Italy),
E. Borsella,
E.N.E.A., Via E.Fermi 27, 00044 Frascati (Italy),
R. Bruzzese, I. Rendina, A. Sasso, and S. Solimeno,
Dipartimento F.N.S.M.F.A., Pad.20 Mostra d'Oltremare, 80125 Napoli (Italy).

Abstract. We present in this paper resonant two-photon ionization studies at different UV laser wavelengths (193 nm, 248 nm, 308 nm, and 351
nm) of a series of simple molecules presenting primary, secondary, and
tertiary amide groups. In particular, ionic fragmentation patterns at
different laser wavelengths and fluences are reported and compared to
electron impact mass spectra. The power law for the total ion yield is
also discussed. The above studies have been carried out on samples in
gas phase at low pressure ( $10^{-6}$ mbar) by using a time-of-flight technique.

## 1. Introduction.

In the last ten years, the technique of multiphoton ionisation (MPI) of a-
toms and molecules has shown to be extremely efficient in a number of important research fields (Johnson, 1980; Antonov et al, 1984). In particular, laser ionization coupled to mass spectrometry has been extensively
used in the analysis of polyatomic molecules of chemical (Parker, 1983)
and biological interest (Pratesi and Sacchi, 1980).

In this paper we report MPI studies of a series of simple molecules pre-
sentig primary, secondary, and tertiary amide groups. The interest of
studying these molecules stems from the importance and prevalence of the
peptide bond in biological systems. In fact, the amide unit in addition
to being the building block of biologically relevant molecules such as peptide and proteins, is also a constituent of a variety of biologically active small molecules as penicillins and cephalosporins.

The experimental characterization of the lower energy excited states of
the peptide linkage has proven to be exceedingly difficult (Larson et al,
1974), and very little is known about their energy, intensity and, possibly, orbital-excitation nature. Thus, the problem of energy deposition
in variuos excited states of the peptide linkage is surely one of the primary events in radiation biology.

In what follows we shall describe two-photon ionisation studies, carried
out at different UV wavelengths, of N-methylacetamide and N,N-dimethyl-

acetamide molecules. In particular, we shall present ionic fragmentation patterns, obtained at different laser wavelengths and fluences, and compare them to mass spectra obtained by electron impact (EI) by Gilpin in 1959. The power law for the total ion yield will also be discussed.

The above studies have been carried out by using an experimental apparatus which is based on the time-of-flight (TOF) technique for the detection of parent and fragment ionic species.

2. Experimental setup.

A schematic view of the experimental apparatus including the laser, the vacuum chamber ($10^{-7}$ mbar background pressure), and the data acquisition system is shown in Fig.1. The vacuum system consits of a main chamber, pumped by a turbomolecular pump, and of a differentially pumped ($10^{-8}$mbar) side arm into which the TOF mass spectrometer is located. The sample, in gas form, is introduced into the vacuum chamber through a needle injector fed by a  molecular leak valve.

The excitation source is a commercial excimer laser (Lambda Physik, mod. E MG 103) operating at different UV wavelengths: 193 nm (ArF), 248 nm (KrF), 308 nm (XeCl), and 351 nm (XeF). The laser, whose pulse duration is about 15 ns (FWHM), was normally operated at a repetition rate of 10 pps. The laser beam enters the main chamber through a suprasil quartz window after passing a variable attenuator (set of different neutral density transmission filters) and a spherical focussing quartz lens, with f=25 cm. The focal intensity is, thus, in the range of several tens of MW/cm$^2$ to several GW/cm$^2$. The

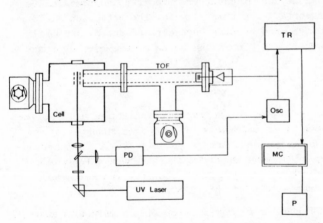

Fig.1 : Experimental setup.

laser intensity was measured, during the experiments, after an uncoated 45° quartz beam-splitter, by a photodiode (PD).

The ions generated in the focus of the laser beam are pushed by a repeller ( 600 V) before drifting through the field-free region of the TOF. Ions are detected  by a tandem channel-plate detector and the resulting signal is preamplified and transferred to either a fast oscilloscope (Osc), or a fast transient recorder (TR), (Tektronix, mod.7612D), which can store the complete mass spectrum for each laser shot. Data are transferred to a minicomputer (MC), and, finally, fed into a peripheral unit (P). The results presented in the following section have been obtained by averaging the mass spectra of several tens of laser shots (50 to 100). The detector resolution is 1 amu throughout the entire spectrum of parent and fragment ions, and its response is proportional to the relative abundances.

Finally, the amide molecules used in the experiments were the purest available commercially (>99.5 %).

## 3. Experimental results.

By using the apparatus described above we have carried out MPI studies of amide molecules. In particular, we shall describe here the results obtained on molecules presenting secondary (N-methylacetamide: $CH_3CONHCH_3$) and tertiary (N,N-dimethylacetamide: $CH_3CON(CH_3)_2$) amide groups.

The only fragmentation patterns of N-methyl- and N,N-dimethylacetamide molecules appeared thus far in the literature were the ones obtained by EI by Gilpin (1959), which have provided valuable information for the elucidation of molecular structure. We show in Fig.2 these EI mass spectra for the tertiary (case a) and secondary (case b) amide, for sake of comparison with our MPI fragmentation spectra. As for N-methylacetamide (b), whose molecular weight is 73.15 amu, the most intense peak corresponds to the parent ion (m/e= 73). Other mass spectral features are the intense m/e 30 ($CH_4N^+$) and 43 ($CH_3CO^+$) peaks. They can be correlated with rearrangement ions resulting from cleavage of the carbonyl carbon-nitrogen bond. In the case of the N,N-dimethylacetamide (a), with a molecular weight of 87.18 amu, the most intense peak in the mass spectrum (m/e 44 ion) corresponds to a cleavage similar to that proposed in the case of the m/e 30 ion of the secondary amide, i.e., cleavage of the nitrogen-carbonyl carbon bond.

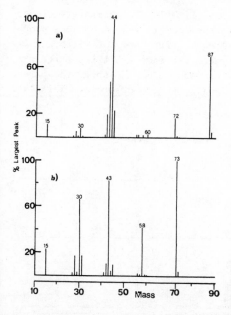

Fig.2 : EI fragmentation patterns of $CH_3CON(CH_3)_2$, case a), and of $CH_3CONHCH_3$, case b).

The MPI mass spectra were obtained in the gas phase with the pressure ranging between 1-2 x$10^{-5}$ mbar, and at a temperature of 20°C for the N,N-dimethylacetamide and 50°C for the N-methylacetamide.

Since the ionization potentials (IP) for N,N-dimethylacetamide and N-methylacetamide molecules are 9.70 eV and 9.20 (Baldwin et al, 1977) eV, respectively, two UV photons are needed to excite a molecule into the ionisation continuum. The overall process is characterized as a resonant one-photon excitation to an intermediate electronic singlet state from the ground state, followed by an incoherent one-photon excitation to the continuum. We have studied the MPI process by using four different UV laser wavelengths, already reported. The laser pulse energies reported in the following are all energies measured just before the 25 cm focal length fo-

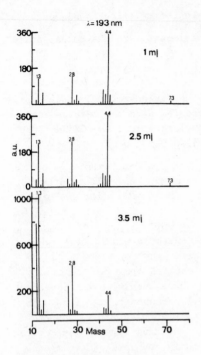

Fig.3 : MPI mass spectra of $CH_3CON(CH_3)_2$ at $\lambda$ =193 nm, for different laser pulse energies.

cussing lens.

A common feature of all the recently reported laser-ionisation mass spectrometry experiments on polyatomic molecules (Bernstein, 1982) is the extensive fragmentation of the molecules into smaller and, energetically, most costly ions, some requiring a minimum energy equivalent to 9 UV photons. We have observed the same feature, as shown in Fig.3 where MPI fragmentation patterns of N,N-dimethylacetamide are reported for different laser pulse energies, at the wavelength of 193 nm. In the case of 3.5 mJ, the m/e 13 peak is the most intense, and,moreover, at this laser wavelength we were unable to observe the parent ion (87 amu) even at very low fluence. At smaller energies one of the features characteristic of the EI spectra, i.e., the appearence of the most intense peak at m/e=44 is again observed.

In Fig.4, fragmentation patterns of the same molecule at the laser wavelength of 248 nm are shown. In this case we observe relevant differences with respect to $\lambda$=193 nm. Firstly, for comparable laser energies the overall number of ions produced is much smaller (see the different values of the ordinates in Figs. 3 and 4, where the same arbitrary units are used). Secondly, the percentage of heavier ions is much higher, the parent ion is clearly observed, and a peak at m/e=102 is also observed. This last

molecular ion is probably due to a fast reaction between the parent molecule and a methyl ion $(CH_3^+)$. Moreover, we also observe higher number of molecular ionic fragment peaks.

In Fig.5 we report the MPI mass spectra of N-methylacetamide at 193 nm. These spectra are similar to those of Fig.3, and to the EI mass spectra, even though the overall ionic yield is considerably smaller. In fact, we observe the predominance of the same molecular peaks, though the relative abundances are somehow different.

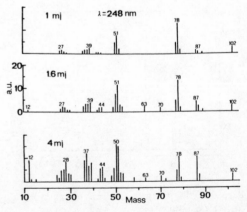

Fig.4 : MPI mass spectra, at $\lambda$=248 nm, of $CH_3CON(CH_3)_2$.

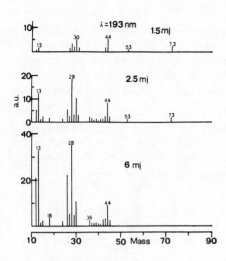

Fig.5 : MPI mass spectra of
CH₃CONHCH₃ at 193 nm.

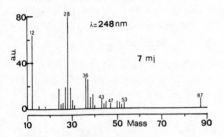

Fig.6 : MPI mass spectrum of
CH₃CONHCH₃ at 248 nm.

The results relative to N-methylaceta-mide at 248 nm are reported in Fig.6, for 7 mJ so to show the fragmentation pattern for a higher laser fluence. There is a number of features similar to those observed for N,N-dimethylace-tamide (Fig.4) at the same wavelength. Firstly, the number of molecular ionic peaks is much higher than at 193 nm. Secondly, we observe the heaviest peak at 87 amu, i.e., with a difference of 15 amu with respect to the parent ion. On the other hand, contrary to what is seen in Fig.4, no ion peak correspon-ding to the parent molecule (72 amu) is found, and the ion yield is gene-rally higher at any laser fluence.

We have also carried out the same mea-surements at 351 and 308 nm. We were unable to observe ionisation of the molecules, even at high laser enrgy, in the focussing condition of our ex-periment. This is expected on the ground of energetic considerations, since the energy of two UV photons at the above wavelengths is smaller than the IP of both molecules under study. Thus, a three-photon ionisation pro-cess would be required in this case. Moreover, the cross section of such an ionisation process is surely very small, since at 351 nm and 308 nm the-re is no intermediate resonance that could enhance the process. Much higher fluences are surely needed to ob-serve MPI in the above conditions.

Finally, we have studied the dependence of the total ion yield on laser power at 193 and 248 nm. As well known, by plotting on a log-log graph the above quantities one obtains straight lines the slopes of which bear information on the number of photons involved in the ionisation process ( Lambropoulos, 1976). For both molecules we have obtained slopes in the range 1.7 - 2 ± 0.2, thus confirming the two-photon character of our MPI scheme. It must be noted that the above values are an average between the different values characterizing the power laws of the various ionic frag-ments (smaller fragments are characterized by higher power dependences). However, the law indeces of the power laws suggest that the first step, one photon excitation to the intermediate state, is the rate-limiting step, whereas the overall fragmentation foollowing the ionisation is relatively rapid. This is in good agreement with the results reported in the lite-rature (Armenante et al, 1985).

## 4. Discussion and conclusions.

Regarding the mass spectra shown in the previous section, a first point which worth commenting upon is their differences with respect to EI ionisation spectra. In particular, characteristic features of the MPI fragmentation patterns are the almost complete absence of the parent ion, even at low laser fluences, and the extensive production of smaller fragments, particularly at high fluences.

These characteristics seem to confirm a well known feature of the MPI spectra of polyatomic molecules, i.e., the fact that as soon as one reaches the threshold laser fluence, which is needed for starting the MPI process, further absorption of laser photons of the ionic species produced takes place with great easy.

The other interesting feature of our results is the strong wavelength dependence of the mass spectra. The spectra of the two molecules obtained at the same wavelength are similar, and this is particularly true at 193nm.

The amide-group absorption spectrum in the UV region of our interest is characterised by a $\pi-\pi^*$ transition centered around 190nm and by a $n_o-\pi^*$ transition at about 220nm. In particular, for N-methylacetamide and N,N-dimethylacetamide in aqueous solution the absorption maxima lie, respectively, at 187 and 195 nm. Thus, the laser wavelength of 193nm is almost resonant with the absorption maximun of N,N-dimethylacetamide, and this can explain the high overall ionic yield at this wavelength.

On the other hand, the wavelength of 248nm is in the wing of the 220nm band and in this case we observe mass spectra which are completely different if compared to those at 198nm, though they show similar features for the two molecules (see Figs. 4 and 6).

The above discussion of our preliminary results, although still qualitative at the time of writing, shows that the different behaviour, at different laser wavelengths, of the simple molecules presenting amide groups analysed, can give interesting information on the problem of energy deposition in the electronic excited states of the peptide linkage.

## References.

Antonov V.S., Letokhov V.S. and Shibanov A.N. 1984 Sov.Phys.Usp. 27 81
Armenante M., Bruzzese R., Solimeno S., Spinelli N. and Vanoli F. 1985
    Jour.Opt.Soc.Am. B2 1088
Baldwin M.A., Loudon A.G. and Webb K.S. 1977 Org.Mass Spectr. 12 279
Bernstein R.B. 1982 J.Phys.Chem. 86 1178
Gilpin J.A. 1959 Anal.Chem. 31 935
Johnson P.M. 1980 Acc.Chem.Res. 13 20
Lambropoulos P. 1976 Adv. At. Mol. Phys. vol.12 (N.Y.: Ac. Press) p.87-164
Larson D.B., Arnett J.F., Seliskar C.J. and Mc Glynn S.P. 1974 J.Am.Chem.
    Soc. 96 3370
Parker D.H. 1983 Ultrasensitive Laser Spectroscopy (N.Y.:Ac.Press) p.233
Pratesi R. and Sacchi C. 1980 Lasers in Photobiology (Berlin: Springer).

*Inst. Phys. Conf. Ser. No. 84: Section 6*
*Paper presented at RIS 86, Swansea, Wales, 7–12 Sept. 1986*                                                      235


# Detection of trace amounts of actinides and technetium by resonance ionization mass spectrometry

H Rimke, P Peuser, P Sattelberger, N Trautmann, G Herrmann

Institut für Kernchemie, Universität Mainz, D-6500 Mainz, F.R.G.

W Ruster, F Ames, H-J Kluge, E W Otten

Institut für Physik, Universität Mainz, D-6500 Mainz, F.R.G.

Abstract. Laser resonant ionization mass spectrometry has been used for the determination of trace amounts of actinides and technetium. A sensitivity of less than $10^8$ atoms in the sample and an extreme selectivity can be achieved by three step photoionization followed by time-of-flight mass spectrometry.


## 1. Introduction

For studies of the ecological behaviour of radionuclides, particularly of plutonium, americium, curium and technetium, sensitive detection methods are required. An unambiguous element and isotope assignment is also desirable in order to determine the origin of the samples.

In the experiments described here, laser photo-ionization in combination with time-of-flight spectrometry has been tested as a novel technique for trace analysis. High sensitivity and selectivity can be achieved by excitation and ionization of the atoms via absorption of three resonant photons, followed by mass determination of the photo-ions.


## 2. Experimental

The experimental set-up is shown in Fig. 1. The laser system consists of three dye lasers (Lambda Physics, Mod. 2001 E), which are simultaneously pumped by a copper vapour laser (Oxford Lasers, Mod. CU40). The latter delivers about 30 W average power in an unstable resonator configuration at a pulse repetition rate of 6.5 kHz.

The three dye laser beams are deflected into the time-of-flight spectrometer, containing a Re-filament, on which the sample has been deposited by a specially developed electrolysis procedure. At filament temperatures of about 1500°C, evaporation of the atoms occurs. The photo-ions are accelerated by two grids to an energy of 2.9 keV and reach a channel-plate detector after a drift length of 2 m.

The pump laser emits light at two wavelengths of 510.6 and 578.2 nm with a pulse width of 30ns. With the dye lasers operated in the oscillator-amplifier configuration a band width of about 5 GHz and conversion efficiences between 10 and 25% have been obtained covering a spectral range between 530 and 850 nm.

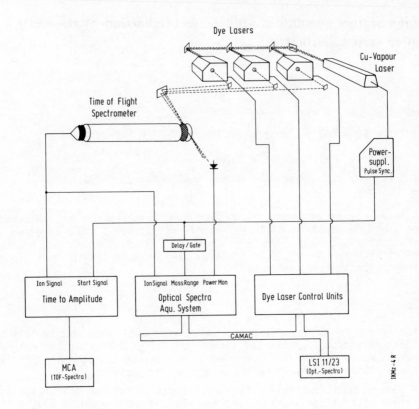

Fig. 1: Resonant Ionization Mass Spectrometer

For data aquisition two independent systems are in operation: i) The time-of-flight spectra are recorded by means of a time-to-amplitude converter, followed by a multichannel analyzer in pulse-height analysis mode. ii) The ion count rate is measured as a function of the laser wavelength by use of a LST-11 micro-computer, which also controls the wavelengths of the three dye lasers.

## 3. Results

In experiments with plutonium samples containing $10^{10}$-$10^{12}$ atoms (Peuser 1985) two dye lasers were used for excitation and a part of the yellow pump laser beam for the ionization step. A detection efficiency of $10^{-7}$ was achieved with two laser beams in resonance yielding a negligable background, i.e., $10^8$ atoms of plutonium in the sample should be detectable.

For further improvements of the detection system, gadolinium was used as a test element. By applying a pulsed acceleration voltage on the first grid triggered from the copper vapour laser pulse a mass resolution of

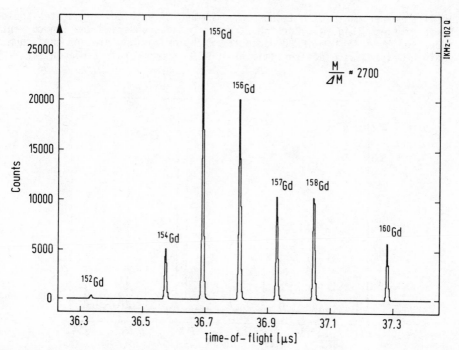

Fig. 2: TOF-Spectrum of a Gd-Sample

better than 2500 for a gadolinium sample (Fig. 2) has been obtained. The ionization probability was increased by two orders of magnitude by exciting autoionizing states near the ionization threshold with a third dye laser (Fig. 3). Saturation for the ionization step was obtained with $760 \, mW/cm^2$. A detection limit of $10^6$ atoms is also expected for plutonium by ionization via autoionizing states.

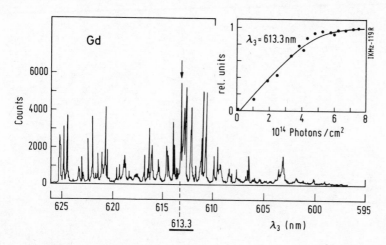

Fig. 3: Autoionizing Resonances of Gd

In first experiments with technetium a number of autoionizing states could be identified. (Fig. 4). For the strongest states observed the ion current increased by a factor of 200 compared with ionization into the continuum. Again we expect a detection limit of $10^6$-$10^7$ atoms for this element.

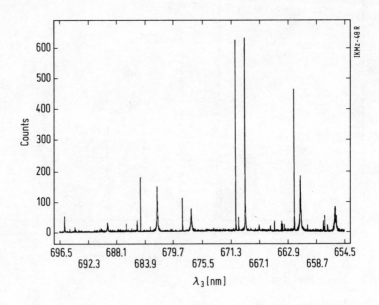

Fig. 4: Autoionizing Resonances of Tc

Peuser P, Herrmann G, Rimke H, Sattelberger P, Trautmann N, Ruster W, Ames F, Bonn J, Krönert U and Otten E W 1985 Appl. Phys. B38, 249

*Inst. Phys. Conf. Ser. No. 84: Section 6*
*Paper presented at RIS 86, Swansea, Wales, 7–12 Sept. 1986*

# Medical and biological applications of resonance ionization spectroscopy

L J Moore, J E Parks, E H Taylor, D W Beekman, and M T Spaar

Atom Sciences, Inc., 114 Ridgeway Center, Oak Ridge, Tennessee, 37830

## 1. Introduction

A compelling motivation for the measurement esthete using RIS goes something like this: If it is possible to form, transmit, and detect an ion of any element with nearly unit efficiency, what is left for the analyst to develop? The answer is elementary. However, analytical chemists need not despair immediately that retraining in elementary particle physics will be required to keep pace with rapidly developing measurement capabilities in RIS. As with any new technology, there is a significant gap between the conceptualization and realization of the analytical promise of RIS. We perceive that a significant portion of the RIS potential resides in its application to specific problems and analyses in medicine and biology. The purposes of this communication are to assess some of the technical impediments barring the realization of these applications, and to describe current development efforts on selected medical problems.

The analytical duality of RIS – sensitivity and selectivity – provides the bases for the extension of elemental and isotopic analysis to previously unexplored measurement frontiers. The ability to ionize an element selectively in the presence of many atoms of other elements permits the simplification of chemical separations and the commensurate reduction of the analytical blank that would otherwise limit much of conventional technology. Thus the selectivity of RIS reinforces the practicality of realizing the 'few-atom' sensitivity potential. From this perspective, it is useful to re-examine the requirements for chemical separation in elemental analysis. It is no longer necessary to think in terms of complex element-specific separation schemes, such as those typically required for thermal ionization mass spectrometry. Simple separation schemes that are neither element-specific nor quantitative may be ideally suited for RIS applications using isotope dilution methodology. Thus the labor savings involved in the elimination or reduction of chemical processing can be realized. The direct analysis of solids using ion sputtering and RIS has been successfully applied usually to conducting solids such as steel, other metal-based alloys, and 'high tech' electronic materials (Parks 1983, 1984). Direct analysis of more complex biological matrices has so far been less successful. Although direct, spatially resolved elemental analysis in biological materials with RIS sensitivity remains a goal worthy of pursuit, more immediate goals can be realized by employing a minimum of sample preparation that is mutually consistent with the demands of RIS and isotope dilution methodology.

Given the adoption of this philosophy, what are some potential application areas in medicine and biology? Elemental and isotopic studies can be performed using sample sizes that approximate entities perhaps responsible for the control of physiological functions. There are approximately fifteen trace elements essential to animals (Table 1), whose deficiency has been related to numerous conditions or disorders, or whose presence is necessary to sustain metabolic functions. Most of these elements are rather difficult to assay with presently available techniques, especially at the low concentrations encountered in biological samples.

Table 1

| Essential Element | Associated Metabolic Disorders/Functions |
|---|---|
| Zn | Sickle cell disease, liver disease, gastrointestinal disorders, impaired wound healing, genetic disorders (acrodermatitis enteropathica) |
| Cu | Wilson's disease, Menkes disease, anemia |
| Cr | Diabetes mellitus, cardiovascular disease, impaired glucose tolerance, elevated serum cholesterol |
| Se | Keshan disease (Se-responsive cardiomyopathy), atherosclerosis, muscular dystrophy, cystic fibrosis |
| Mn | Skeletal abnormalities, ultrastructural abnormalities, blood cholesteral level |
| Mo | Enzymatic reactions, severe bifrontal headache, night blindness, nausea, lethargy, disorientation, coma |
| Co | Part of Vitamin B-12, interacts with iron |
| I | Hyperthyroidism, cretanism |
| | Essentiality demonstrated for animals only --- |
| Ni | Depressed hematocrit, ultra-structural liver abnormality |
| As | Depressed birth weight, impaired fertility, elevated hematocrits |
| Si | Aberrant connective tissue and bone metabolism, atherosclerosis, hypertension, aging process |
| V | Cardiovascular disease, renal disease, kwashiorkor |
| Cd, Pb, Sn | Growth factor ... unknown |

These are but a few examples of the potential applicability of RIS in the medical and biological area. Generic arguments can be advanced for other metals and metallo-organic species whose intra- and inter-cellular functions play important roles in nutrition, toxicity, and disease etiology and related biological dysfunctions.

## 2. Atomization Modes

The analytical goal of atomization is to change the analyte bound within a complex matrix to an atomic form suitable for resonance ionization and mass analysis. Two atomization modes are being employed in this research – thermal vaporization and ion sputtering. Each of these modes has an interlocking set of advantages and disadvantages. Atomization through thermal vaporization is rooted historically in the vaporization process associated with thermal ionization mass spectrometry. The application of RIS to the (largely neutral) atomic vapor has become known as resonance ionization mass spectrometry (RIMS). In this process a drop of solution containing the analyte is added to a metal ribbon substrate, typically one of several metals that have been successfully employed for thermal ionization – rhenium, tantalum, tungsten, or platinum. Collectively these substrates limit ultimate sensitivity for RIS applications. Although the analyte thermal ions can be suppressed, the background impurities in the metal can be a limitation to the required sensitivity.

For example, there is a great deal of interest in the micro-determination of iron. It is extremely difficult to find a substrate of sufficient purity that still possesses the mechanical and refractory requirements for atomization. Across the periodic table there is a complex series of trade-offs from this perspective alone.

Due to the low pulse repetition rate and short pulse length of most lasers, the temporal characteristics of thermal atomization must be closely matched to utilize the atoms efficiently. Thermal atomization substrates thus far have been pulsed to produce atom plumes with a FWHM temporal distribution of 0.1-1ms (Fassett 1984, 1986). Thus the achievement of overall sample utilization efficiencies better than $10^{-3}$ will require further developments in this capability. Thermal atomization efficiency is also closely tied to the (analyte atom)/(analyte-containing molecule) ratio in the vapor phase. As noted in earlier work, approximately 80% of the elements have a metallic character, and can be placed on a metal substrate, as a metal, or metal oxide (Moore 1984). Alternatively, it is possible through in-situ reduction to manipulate the species emitted from the hot surface such that the atomic species is preferred over the oxide (Moore 1984). Other approaches are required for those strongly electronegative elements, such as the halides, that form simple compounds with metals, e.g., AgI. In this case the molecule can be photo- or thermally dissociated prior to resonance ionization.

Ion sputtering appears to offer many technical advantages over thermal atomization, but suffers from other effects that may be equally disadvantageous. Sputtering atomizes the sample essentially independently of the analyte's vapor pressure, thereby enabling broad elemental coverage. However, it produces secondary neutral species with a relatively large kinetic energy distribution, thus requiring a more sophisticated ion optics system. Thermal vapor sources produce a relatively well-characterized vapor phase that is in thermodynamic equilibrium with the condensed phase, whereas ion bombardment produces secondary species whose energy distribution is less well characterized. However, remarkable progress has been made in this area over the last few years (e.g., Kimock 1984). If the two atomization modes are equally inexpensive, the choice of atomization mode for RIS reduces to an appraisal of the potential interferences for each element or group of elements, as well as the sensitivity requirements. Given the availability of both approaches, often the most practical expedient is simply to try each.

### 3. Current Applications Development

Current efforts in biomedical applications at Atom Sciences are oriented toward the determination of trace and ultra-trace elements in blood serum samples. Of particular interest are the elements copper, molybdenum, and vanadium. Simplified methods of sample preparation have been devised for these elements. Typically, serum sample sizes ranging from 10 to 100 µl were prepared and chemically deposited for RIMS (Figure 1) or SIRIS (Figure 2).

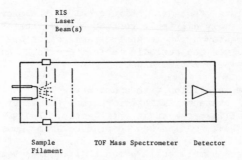

Figure 1  A schematic diagram of the time-of-flight thermal atomization RIMS system used for the $^{100}$Mo spectra of Fig. 5.

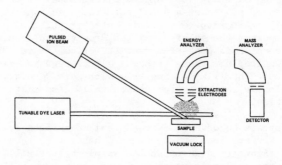

Figure 2  A schematic diagram of the sputter-initiated resonance ionization spectrometer (SIRIS) system used for the trace element determinations reported here.

## 4.   Results and Discussion

### 4.1   Copper

Data for the isotope dilution determination of copper in bovine serum (RM 8419) and human serum samples are tabulated in Table 2.   The SIRIS isotope dilution value for bovine serum is in good agreement with the suggested value derived from several atomic spectroscopic techniques utilizing 2-3 ml sample sizes (Veillon 1985).   The sample size used for SIRIS was 250 μl.   The reproducibility of the SIRIS measurements is illustrated by the replicate analyses of sample F.   A one gram sample of F was processed and split into three aliquots.   Isotope dilution analyses of the aliquots produced concentrations of 1.308±0.029 (std. dev., internal to an analysis), 1.304±0.047, and 1.408±0.057 μg Cu/g.   A separate analysis, one week later with a 100 μl sample, produced a concentration of 1.18 μg Cu/g, in good agreement.   We estimate the uncertainty of each determination to be ±15%, which includes contributions from chemical blank (<10% of the total Cu per sample), stoichiometry and impurities of the $^{65}$CuO, and isotope ratio accuracy and precision.   Typical isotope ratio precisions for copper, internal to an isotopic analysis, range from 0.9-4.0%, relative standard deviation (RSD), with errors of the mean down to 0.25%.   Thus the methodology described here would also be useful to determine isotopic enrichments in small samples for metabolism studies.

Table 2

Copper in Serum, µg/g

| | Suggested Value | SIRIS[a] | Notes |
|---|---|---|---|
| Bovine Serum (RM 8419) | 0.73±0.1[b] | 0.763 | [a]Estimated uncertainty ±15%. |
| Human Serum sample F (1g) | --- | 1.31[c] | [b]The value from (Veillon 1985) was |
| | | 1.30[c] | listed orginally in mg Cu/l; we |
| | | 1.41[c] | have converted to a weight basis |
| Human Serum sample F (0.1g) | --- | 1.18 | here for comparison using a density |
| | | | of 1.03 g/ml |
| | | | [c]Aliquots from the same solution. |

## 4.2 Molybdenum

Molybdenum in bovine and human serum occurs at levels of 16 and
$\sim$1 ng/ml, respectively. In a 100 µl sample, the amount of Mo to be
measured is $<10^{12}$ atoms, or $\sim$100 pg. The sensitivity of SIRIS
demonstrated thus far for Mo is illustrated in Figure 3 for a 2 ng sample
of $^{100}$Mo. The $^{100}$Mo/$^{98}$Mo ratio of 102, observed with larger $^{100}$Mo
samples, is slightly distorted here toward
a natural ratio. Assuming a 100% deposition
efficiency, there is a total of 57 pg
natural Mo, including about 11 pg of $^{98}$Mo.
A significant portion of this contamination
appears to come from reagent backgound.
The concentration of Mo in the reagents was
determined by isotope dilution, using
nominal 50 g samples: $H_2O$, 13 pg/g; $HClO_4$,
30 pg/g and $HNO_3$, 44 pg/g. Since
approximately 1 g of 10% $HNO_3$ solution
was used to deposit the 2 ng sample of
$^{100}$Mo, a total of $\sim$17 pg natural Mo would
be expected. Further work with system
blanks will be required to elucidate the
difference. SIRIS sensitivities for Mo
are currently at the few pg level; it
appears that this could readily be
extended to the femtogram range, although
a commensurate reduction in chemical
blanks would have to be achieved to permit
applications. A preliminary isotope
dilution analysis of Mo in a 250 µl bovine
serum (RM 8419) sample yielded 19 ng/ml,
compared to the suggested value of
16±4 ng/ml. In Figure 4 is illustrated
a $^{98}$Mo$^+$ profile produced by translating
the sample stage across the primary ion
beam. The RIMS system of Figure 1 was
used to produce the $^{100}$Mo spectrum shown
in Figure 5. The sensitivity illustrated is
for 4 ng $^{100}$Mo. We believe that the sensitivity
for RIMS analysis of Mo can easily be extended to the picogram range.

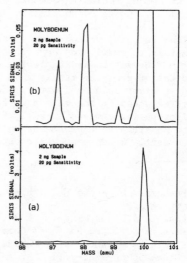

Figure 3  a) A mass scan over
a portion of the molybdenum mass
region, M/E 97 thru 100,
utilizing a molybdenum sample of
less than 2 nanograms. b) An
expanded scale (x78) of the
above spectrum. The peaks at
$^{97}$Mo and $^{98}$Mo illustrate
sensitivities of approximately
20 and 30 picograms, respectively.

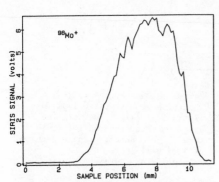

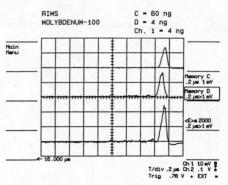

Figure 4   A plot of $^{98}Mo^+$ ion current versus the translational position of the deposition circle on gold foil relative to the primary argon ion beam.  The relative uniformity of the chemically deposited molybdenum, is typical for most samples.

Figure 5   An illustration of the sensitivity for $^{100}Mo$ using the thermal atomization RIMS system of Fig. 1.

Extrapolation of this methodology to vanadium in small samples (serum concentration $\sim 2$ ng/ml) appears straightforward, particularly since the V background levels in reagents are quite low (typ. $\leq 1$ pg/g) (Fassett 1985).  Several components of the vanadium methodology have been completed, including demonstration of deposition and ionization feasibility, but no concentration determinations have been attempted to date.

Thus far, the methodology development required to access ultra-trace metals in small samples has been successful, and analogous extrapolations to other metals and biological systems appear to be achievable.  Each metal or group of metals must be carefully examined to determine sources of contamination and error.  However, once developed, the RIS technology is expected to become a powerful tool to determine ultra-trace metals and isotope enrichments in small biological samples.  Consequently, it will be possible to explore the role of metal and metallo-organic transport at currently unachievable microscopic levels.

## References

Fassett J D, Moore L J, Shideler, R W and Travis J C 1984 Anal. Chem. <u>56</u> pp 203-206
Fassett J D and Kingston H M 1985 Anal. Chem. <u>57</u> pp 2474-2478
Fassett J D 1986 Personal Communication
Kimock F M, Baxter J P, Pappas D L, Korbin P H and Winograd N 1984 Anal. Chem. <u>56</u> pp 2728-2791
Moore L J, Fassett J D, and Travis J C 1984 Anal. Chem. <u>56</u> 2270
Moore L J, Machlan L A, Lim M O, Yergey A C and Hansen J W 1985 Pediatric Research <u>19</u> pp 329-334
Parks J E, Schmitt H W, Hurst G S and Fairbank W M 1983 Thin Solid Films <u>108</u> pp 69-78; patent no. 4,442,354
Parks J E, Schmitt H W, Hurst G S, and Fairbank W M 1984 Inst. Phys. Conf. Ser. No. 71:167-174
Veillon C, Lewis S A, Patterson K Y, Wolf W R, Harnly J M, Versieck J, Vanballengerghe L, Cornelis R and O'Haver T C 1985 Anal. Chem. <u>57</u> pp 2106-2109

Acknowledgements:  The methodology component of this research (excluding work with human subjects) was supported in part by NIH/SBIR contract 1R43 HD21831-01.

*Inst. Phys. Conf. Ser. No. 84: Section 7*
*Paper presented at RIS 86, Swansea, Wales, 7–12 Sept. 1986*

# Atomic vapor laser isotope separation using resonance ionization

B J Comaskey, J K Crane, G V Erbert, C A Haynam, M A Johnson, J R Morris, J A Paisner, R W Solarz, E F Worden

Lawrence Livermore National Laboratory, P.O. Box 5508, Livermore, California 94550

Abstract. Atomic vapor laser isotope separation (AVLIS) is a general and powerful technique. A major present application to the enrichment of uranium for light–water power–reactor fuel has been under development for over 10 years. In June 1985, the Department of Energy announced the selection of AVLIS as the technology to meet the nation's future need for enriched uranium. Resonance photoionization is the heart of the AVLIS process. We discuss those fundamental atomic parameters that are necessary for describing isotope–selective resonant multistep photionization along with the measurement techniques that we use. We illustrate the methodology adopted with examples of other elements that are under study in our program.

## 1. Introduction

Atomic vapor laser isotope separation (AVLIS) is a general process for enriching an atomic–vapor stream in one or more of its isotopes. The heart of the process is the selective multistep photoionization of the isotope of interest. In order to fully describe this selective photoionization, one must measure or otherwise determine the values for several atomic parameters of the particular element and isotope. A list of these key atomic parameters includes: radiative lifetimes, energy levels and ionization potentials, quantum numbers, hyperfine structure and isotope shifts, transition strengths and ionization rates, and electric– and magnetic–field effects. A large portion of the values comprising this list can be found in literature sources and more general spectroscopic compilations such as the NBS monographs. In this paper we describe some of the techniques that we use to measure those values, including some not found in the literature, and illustrate these techniques with results from elements we are currently investigating in our program.

Work performed under the auspices of the U.S. Department of Energy by the Lawrence Livermore National Laboratory under Contract W–7405–Eng–48.

## 2. Measurements

The measurement of the essential atomic parameters and the development of the necessary tools and techniques for these fundamental measurements have played a key role in the success of AVLIS. Ionization provides a very sensitive means of signal detection and consequently is used in many of the measurements. These ionization techniques typically use tunable pulsed dye lasers for resonant stepwise excitation to the level of interest followed by ionization by one of several possible techniques. Ions can be detected with biased parallel plates, microchannel plate detectors, or Channeltron detectors. Oftentimes these detectors are part of a mass spectrometric system employed for distinguishing individual isotopes. Using the basic technique of resonant excitation followed by ionization and adapting this technique to the particular measurement of interest, many of the atomic parameters necessary for designing an AVLIS process can be determined. Then by complementing this technique with standard methods such as absorption and fluorescence spectroscopy, the entire set of atomic parameters can be determined for the element of interest.

Perhaps the simplest measurement to make and one of the most important for the initial evaluation of a particular photoionization process is the radiative lifetimes of the various levels in the photoionization ladder. One technique that we use for determining radiative lifetimes is time-delayed photoionization (Carlson 1977). The atom is excited to the level of interest with one or more laser pulses, then photoionized with a final sequence of laser pulses that is varied in time. Ion signal is plotted as a function of delay time. In a more traditional fashion we also measure fluorescence decay using photomultipliers and fast transient digitizers.

An important technique that employs ionization for signal detection and which has yielded a wealth of atomic physics information over the years is scanning laser ionization spectroscopy. In this technique the atom of interest is stepwise excited to a high—lying level with 1, 2, or 3 lasers. This excited state can then be ionized by one of several methods including collisions with neutrals or electrons, field ionization, or photoionization. All of these techniques employ a scanning dye laser for probing levels just below or slightly above the atom's ionization threshold.

Figure 1 shows some data taken in dysprosium (Worden 1978) where the third—step laser was scanned from slightly below the ionization potential, up to and beyond the continuum. This scan gives the location of the onset of photoionization as well as information on different Rydberg series including level locations and principal quantum numbers, which are found by fitting the data to standard formulas.

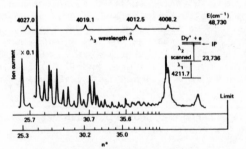

Fig. 1. Photoionization threshold of dysprosium and autoionizing Rydberg series converging to the first excited state of the ion.

In order to selectively photoionize one isotope of an element, a photoionization ladder must be chosen such that the isotope of interest is clearly separated by virtue of its isotope or hyperfine shifts with respect to the other isotopes. Consequently we must measure these shifts for any photoionization sequence of promise in a particular element. For first–step hyperfine structure, traditional spectroscopic techniques such as fluorescence or absorption are generally used. For the second– and third–step transitions in a three–step photoionization scheme, resonance ionization detection is used where the transition of interest is excited with a narrow–band (~1 MHz) cw dye laser and scanned over the bandwidth of the pulsed dye lasers, which are wide enough in frequency to excite a large portion of the hyperfine structure in a single shot. Figure 2 shows the hyperfine structure of each transition in a three–step photoionization scheme in Gd. The second– and third–step transition hyperfine spectra were obtained in the manner outlined above.

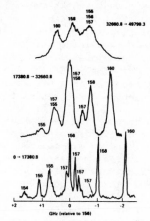

Fig. 2. Hyperfine structure for transitions of 3–step photo–ionization ladder in Gd.

In situations where small hyperfine splittings and isotope shifts are encountered such as for the odd isotopes of gadolinium, polarization selection rules can be exploited to achieve photoselectivity. In fact, polarization selection rules in resonant photoionization provide a convenient tool for unambiguously determining the angular momentum or J quantum numbers of previously unassigned levels.

The behavior of atomic levels in the presence of external electric and magnetic fields is another area studied extensively in our program. In Fig. 3 we illustrate some dramatic Stark effects that can be observed for high–lying valence levels even at extremely modest electric fields. The high–resolution spectra were taken with a cw dye laser scanned across the $J = 0$ autoionizing resonance at 49799 cm$^{-1}$ in $^{160}$Gd using a field ionization and collection arrangement. As the electric field is increased, the line breaks up into a multitude of separate lines. The autoionizing level is thought to be coupled to a high–lying (n* ~40) Rydberg level via the electric field. The multiple lines correspond to transitions to parabolic angular momentum states that are degenerate in a field–free environment.

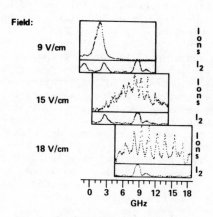

Fig. 3. DC field effect on the 32660–49799 cm$^{-1}$ transition in $^{160}$Gd.

Perhaps the most important spectroscopic parameters in the AVLIS process
are the transition oscillator strengths or dipole moments. The resulting
optical cross sections determine the laser fluences needed for effective
atomic photoionization. A method used extensively in our laboratories for
accurately measuring transition oscillator strengths relies on observing
Rabi flopping of atomic populations. An example of data obtained for

atomic gadolinium using
this technique is shown in
Fig. 4. The $^{160}$Gd was
photoionized in a stepwise
fashion by three resonant
laser pulses. The first-
and second-step excitations
were derived from pulse-
amplified cw dye-laser
systems allowing selective
excitation and detection of
$^{160}$Gd without a quadrupole
mass filter. The second and
subsequent photoionizing
steps were delayed in time
from the first-step
excitation pulse.

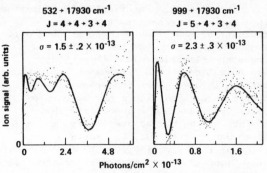

Fig. 4. Cross-section measurement using
Rabi oscillations in $^{160}$ Gd.

In Fig. 4 the photoion signal is plotted as a function of fluence of the
first-step laser. The dramatic oscillations observed in the photoion
signal arise from the coherent evolution of the atom between the manifold
of magnetic sublevels in the ground and first excited state. Since each
magnetic sublevel has its own dipole moment, the signal is not purely
sinusoidal as a function of field amplitude of the first-step laser, but
is instead periodic. Since the first-step laser fluence is accurately
measured, the only parameter varied to fit the data is the transition
oscillator strength. This method has been used in our laboratory to
measure transition oscillator strengths for bound-bound transitions
originating on both low- and high-J excited states.

In addition to the above technique, which requires a very narrow-bandwidth
pulse-amplified cw dye laser, oscillator strengths can be determined from
the branching ratio and lifetime of the level. We have already discussed
means for determining the radiative lifetime of an atomic level; we employ
two methods for obtaining transition branching ratios. In the first
method the branching ratio is measured using the time-delayed
photoionization technique described earlier. The upper state of the
transition of interest is populated with a pump laser. This level is
allowed to decay radiatively for a time $\tau$, whereupon a probe laser
photoionizes the lower level. By comparing the ion signal for a $\tau$
equivalent to several lifetimes with the ion signal when the lasers are
coincident, we obtain a value for the branching ratio of the transition.
Another means for obtaining transition branching ratios is to perform
large survey emission scans of discharge lamps containing the element of
interest. These types of scans have been done at the Kitt Peak National
Observatory using their high-resolution Fourier transform spectrometer.

A very accurate technique for determining relative oscillator strengths
uses absorption spectroscopy in a hollow-cathode discharge. The cathode
is constructed from the material of interest, usually in the form of a bar
with a deep slot. The discharge is operated with a few Torr of neon or a
heavier rare gas that ionizes and sputters the cathode material. By

running at a high–discharge–current density, high densities of ground– and low–lying excited states can be produced. With this device we can measure the relative absorption between two transitions with a common lower level and determine the relative oscillator strengths of the levels.

The key to any promising AVLIS process is the ionization step. In some cases the final step is to an autoionizing level that has been chosen from the photoionization scans described earlier. To determine the oscillator strength of a transition to an autoionizing level, we measure the Lorentzian line width of the transition and the peak photoionization cross section. To measure line width, we simply scan the photoionizing dye laser over the transition of interest and plot ion signal vs frequency. To obtain the photoionization cross section, the fluence of the photoionizing laser is varied to yield a plot of ion signal vs laser fluence.

Figure 5 shows the results for an autoionizing transition in hafnium. The points representing ion signal vs laser fluence are fit to a saturation curve that is derived from a simple rate equation model. Knowledge of the fluence of the photoionizing laser in the region of the intersection of the atomic beam with the

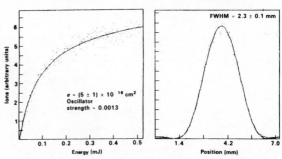

Fig. 5. Autoionization cross–section measurement of the 34991.64 → 55190 cm⁻¹ transition of Hf.

excitation lasers is vital for determining an accurate value for cross section. The plot on the right is the beam profile of the photoionizing laser fit to a curve representing the intensity profile of a circular Fraunhofer pattern. The curve for the saturation fit used in the left–hand plot shown in this same figure is generated by integrating the intensity–dependent photoionization rate equation over both the spatial and temporal bounds of the photoionizing laser pulse. Finally, the value for oscillator strength is determined from the product of oscillator strength and line width.

3.  Conclusions

In the case of oscillator–strength measurements, as well as the other atomic parameters necessary for designing an AVLIS process, there are advantages to having more than one technique available for the measurement. By making a particular measurement using different methods, we have a cross check of our result. This philosophy ensures the highest possible accuracy in our final values. Beyond this is the flexibility afforded by having a variety of available measurement techniques. This versatility becomes more important as we attempt to expand the AVLIS process to other elements.

References

Carlson L R, Johnson S A, Worden E F, May C A, Solarz R W and Paisner J A
    1977 Opt Comm. 21 116
Worden E F, Solarz R W, Paisner J A, Conway J G  1978 J. Opt Soc Am. 68 52

*Inst. Phys. Conf. Ser. No. 84: Section 7*
*Paper presented at RIS 86, Swansea, Wales, 7–12 Sept. 1986*

# The application of atomic vapor laser isotope separation to the enrichment of mercury

J. K. Crane, G. V. Erbert, J. A. Paisner, H. L. Chen, Z. Chiba, R. G. Beeler, R. Combs, S. D. Mostek

Lawrence Livermore National Laboratory, P.O. Box 5508, Livermore, California 94550

Abstract. Workers at GTE/Sylvania have shown that the efficiency of fluorescent lighting may be markedly improved using mercury that has been enriched in the $^{196}$Hg isotope. A 5% improvement in the efficiency of fluorescent lighting in the United States could provide a savings of $450 million dollars in the corresponding reduction of electrical power consumption. We discuss the results of recent work done at our laboratory to develop a process for enriching mercury. The discussion centers around the results of spectroscopic measurements of excited–state lifetimes, photoionization cross sections, and isotope shifts.

## 1. Introduction

Atomic vapor laser isotope separation (AVLIS) is a useful and economically attractive technique for the large–scale enrichment of uranium. In addition to uranium, for which the demand is greater than 1000 metric tons per year, there are smaller–scale applications for several other isotopically enriched elements. Researchers at GTE/Sylvania (Maya et al., 1984) have shown that the efficiency of standard fluorescent lamps used for lighting may be improved by approximately 5% by using mercury that has been enriched in isotope 196. The primary radiation source in a fluorescent lamp is emission from the electronically excited $6^3P_1$ state to the ground state. This emission at 2537 Å is strongly trapped on the other mercury isotopes (198, 199, 200, 201, 202, and 204) that comprise from 6.8% for $^{204}$Hg to 29.3% for $^{202}$Hg in natural mercury. Natural mercury contains only .15% of $^{196}$Hg and, as a result, does not trap radiation at 2537 Å emitted from these atoms at the operating densities of a fluorescent lamp. By increasing the portion of $^{196}$Hg up to 3%, an additional channel for untrapped radiation is available, providing an overall increase in lamp emission and efficiency of ~5%. Enriching the mercury beyond this point provides little additional improvement due to an increase in resonant energy transfer among isotopes. The motivation for providing isotopically enriched mercury is to provide more efficient lighting. Since the United States consumes $1.8 \times 10^{11}$ kWh at 0.05$/kWh annually in powering fluorescent lamps, a 5% increase in efficiency would yield a savings of $450 million annually.

## 2.   Isotope-Selective Ionization of Mercury

The design of an AVLIS process in any element inevitably centers around
the ionization and associated atomic physics. The photoionization of
mercury represents a departure from most of the schemes we have studied
for the lanthanides and other transition elements. At 10.4 eV, the
ionization potential (IP) of mercury is several volts higher than the IP
of the lanthanides or actinides. With mercury's filled shells and
relatively simple electronic configuration, there are fewer levels and
consequently fewer choices for a suitable photoionization ladder.
Absorption spectroscopy measurements dating back to the 1930s reveal
autoionizing levels above the $5d^{10}6s \, \epsilon l$ continua that belong to
Rydberg series converging to the lowest excited ion states at
35,514 $cm^{-1}$, 50,552 $cm^{-1}$, and 51,488 $cm^{-1}$.

Dyer (1985) photoionized mercury in three resonant steps to the
$5d^96s^26p \, (^3P_1)$ autoionizing level where $6^3P_1$ and $8^1S_0$ were
the intermediate levels. We have excited mercury to this state using
three-step ladders that first pump mercury to the $6^3P_1$ state, then to
one of the following levels:    $6^3D_1$, $6^3D_2$, $7^3D_1$, $7^3D_2$, $7^1D_2$,
$8^1S_0$, or $8^3S_1$. The final step pumps mercury to the autoionizing level
$5d^96s^26p(^3P_1)$ at 88,760 $cm^{-1}$.   Figure 1
shows data for the relative
photoionization cross section vs
energy above the ground state where
the upper state of the second step in
our ladder is $6^3D_2$.   The relative
cross section is given by:

$$\sigma_R = \frac{S_i}{E \cdot S_f} \quad ,$$

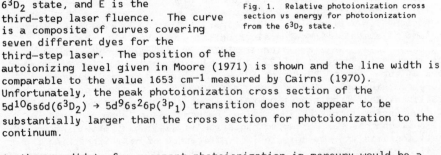

where $S_i$ is the ion signal, $S_f$ is
the fluorescence signal from the
$6^3D_2$ state, and E is the
third-step laser fluence.   The curve
is a composite of curves covering
seven different dyes for the
third-step laser.   The position of the

Fig. 1.  Relative photoionization cross
section vs energy for photoionization
from the $6^3D_2$ state.

autoionizing level given in Moore (1971) is shown and the line width is
comparable to the value 1653 $cm^{-1}$ measured by Cairns (1970).
Unfortunately, the peak photoionization cross section of the
$5d^{10}6s6d(6^3D_2) \rightarrow 5d^96s^26p(^3P_1)$ transition does not appear to be
substantially larger than the cross section for photoionization to the
continuum.

Another candidate for resonant photoionization in mercury would be a
two-step scheme:   $5d^{10}6s^2 \, (6^1S_0) \rightarrow 5d^{10}6s6p \, (6^3P_1) \rightarrow 5d^{10}6p^2 \, (6^3P'_0)$.
We generate the two ultraviolet transitions by Raman shifting in $H_2$ the
output of our YAG-pumped dye lasers.   For the first step transition at
2537 Å, we take the first anti-Stokes order of the frequency-doubled
dye-laser output at 5673 Å.   For the transition to the autoionizing
level at 1974 Å, we take the eighth anti-Stokes order of the dye-laser

output at 5733 Å. The lower trace in Fig. 2 shows ion signal vs second—step laser frequency, revealing a strong resonance. The upper trace in this figure is the signal from the pyroelectric detector used to measure laser energy. The dips in laser energy that produce corresponding decreases in ion signal are due to the $\nu''= 0 \to \nu'= 2$ Schumann—Runge absorption band in $O_2$. Our measured value of 90,099 $cm^{-1}$ for the $6p^2$ ($6^3P'_0$) level differs from the previous listing of 90,096 $cm^{-1}$ for this level (Moore, 1971).

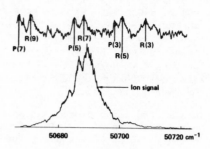

Fig. 2. Mercury autoionizing transition: $5d^{10}6s6p(6^3P_1) \to 5d^{10}6p^2$ ($6^3P'_0$); upper trace, laser intensity, shows absorption band in $O_2$: $X^3\Sigma g^- \to B^3\Sigma u_{\overline{1}}\nu'' = 0 \to \nu' = 2$.

To determine the photoionization cross section for the transition $5d^{10}6s6p$ ($6^3P_1$) $\to 5d^{10}6p^2$ ($6^3P'_0$), we measured ion signal vs laser fluence and fit the resulting saturation curve to a rate—equation model. The value we have measured for the peak photoionization cross section for this transition is $3 \pm 2 \times 10^{-15}$ $cm^2$. To fully illuminate a large cross—sectional area of vapor in a mercury separator, we would need several watts of average power at 1974 Å for this two—step photoionization scheme. We are investigating methods for generating this coherent UV radiation.

An alternative to resonant photoionization to an autoionizing level is photoionization to the continuum. The threshold photoionization cross section to the continuum increases as a function of principal quantum number. One clearly obtains an advantage by first exciting the atom to a high—lying state, then photoionizing to just above the threshold. A slight variation of this concept would be to choose an efficient high—average—power IR laser for photoionization, then excite the atom to the lowest level from which the atom can be photoionized.

We have designed a mercury enrichment process based on a three—step photoionization ladder using a 1.06—μm solid—state laser for the final transition to the continuum from one of three 7d states. In the remainder of this talk, we will discuss the existing data base for this particular photoionization scheme.

We have measured radiative lifetimes for the $7^1D_2$, $7^3D_1$, and $7^3D_2$ states using two—step laser—induced fluorescence. These measurements have been made in a low—pressure static cell. We send two laser pulses, the first tuned to the $6^1S_0 \to 6^3P_1$ transition and the second tuned from $6^3P_1$ to one of the three accessible states of the $5d^{10}6s7d$ configuration. We detect fluorescence from the excited state with an RCAC31034 photomultiplier tube and a fast transient digitizer.

Figure 3 shows results for these measurements along with values reported by other authors. We have also used this technique for measuring the radiative lifetime of the $6^3P_1$ state, corroborating a value of ~120 ns measured by several authors (Halstead et al., 1982 and King et al., 1974).

We have measured the photoionization cross section for the 1.06-μm radiation for each of the three 7d states using the saturation technique described earlier. Figure 4 shows a plot of ion signal vs 1.06-μm laser fluence. From this measurement we obtain a value of 1.35 $\pm$ 0.21 X $10^{-17}$ cm$^2$ for the photoionization cross section of the $7^3D_1$ state using 1.06-μm photons.

Measurements of the hyperfine structure and isotope shifts of the first step, $6^1S_0$ to $6^3P_1$, have been made by numerous authors including ourselves (Crane, 1985). Gerstenkorn (1977) gives values for the isotope shifts for the $6^3P_1$ to 7d states that we use in our photoionization data base. Finally, using the measured values for cross sections, radiative lifetimes, oscillator strengths, etc. that we have compiled in our photoionization data base, we can model photo—ionization vs laser fluence by integrating the density matrix equations relevant to our atomic system. Based on these predictions we are currently working on the design for a laser system for enriching mercury.

| STATE | LIFETIME (ns) | |
|---|---|---|
| | This work | Others |
| $7^1D_2$ | 41.5 ± 2.6 | 40.0 ± 1.0[a] |
| $7^3D_1$ | 13.4 ± 0.6 | 14 ± 3[b] |
| $7^3D_2$ | 17.5 ± 0.8 | |

a. Faisal, 1980
b. Anderson, 1973

Fig. 3. Excited—state lifetime measurements in mercury using laser—induced fluorescence.

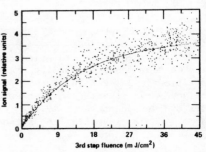

Fig. 4. Typical data of photoion signal vs third—step fluence for $7^3D_1 \rightarrow$ continuum at 1.06 μm. Solid curve represents the nonlinear least—squares analysis.

## References

Anderson T, Sorensen G 1973 J. Quant Spectrosc Rad Transfer 13 269
Cairns R B, Harrison H, Schoen R I 1970 J. Chem Phys 53 96
Crane J K, Erbert G V, Mostek S D, Kerlin R C, Paisner J A 1985 Proc. 1st Intl Laser Conf.
Dyer P, Baldwin G C, Sabbas A M, Kittrell C, Schweitzer E L, Abramson E, Imre D G 1985 J. Appl Phys 58 2431
Faisal F H M, Wallenstein R, Teets R 1980 J. Phys B 13 2077
Gerstenkorn S, Labarthe J J, Verges J 1977 Physica Scripta 15 167
Halstead J A, Reeves R R, 1982 J. Quant Spectrosc. Radiat. Transfer 28 289
King G C, Adams A 1974 J. Phys B 7 1712
Maya J, Grossman M·W, Lagushenko R, Waymouth J F 1984 Science 226 435
Moore C E, 1971 Atomic Energy Levels, NBS 35 3 195

*Inst. Phys. Conf. Ser. No. 84: Section 7*
*Paper presented at RIS 86, Swansea, Wales, 7–12 Sept. 1986*

255

# Observation and measurement of the high excited states of Sr atoms

Lu Jie   Hu Sufen   Zhang Sen   Qiu Jizhen   Sun Jiazhen
Department of Physics, Zhejiang University, Hangzhou,
People's Republic of China

ABSTRACT

The field of highly excited states of atoms is an active subject recently, because it is very important for many scientific and technological fields, such as in the isotope separation, plasma process astrophysics and in the new type lasers.

By using the Resonance Ionization Spectroscopy (RIS) approach and atomic beam technique, the 5sns and 5snd Rydberg series and $(5p_{\frac{1}{2}}ns)_{J=1}$, $(5p_{\frac{1}{2}}nd)$ $J=1,3$ autoionizing series of Sr have been observed.

In this experiment the 5sns and 5snd Rydberg series of Sr were selectively excited from the $5s5s's_0$ ground state, using two pulsed tunable dye lasers pumped by the third harmonics of the same Nd:YAG laser. (The pulse energy of the Nd:YAG laser is 700 mJ/pulse, the pulse duration is 9 ns). The Rydberg atoms were photoionized by the third pulsed laser. The photoionization signals were detected by an electromultiplier, were ac coupled to the Boxcar integrator, and were plotted by the x-y recorder.

Then, the first and second lasers were circularily polarized in the same sense by a linear polarizer and Frenel rhomb and the process was repeated again. In the second case, the polarization scheme allowed the population of the $(5snd)_{J=2}$ series only. Comparing the spectra of these two cases, the spectra of 5sns Rydberg series could be found. The wavelength calibration was achieved by using a Fabry-Perot etalon with a photomultiplier and known marker lines of Sr.

The energy levels and quantum defects of 5sns and 5snd for more than seventy Rydberg states of Sr have been measured (5sns's$_0$ n=10-20, 5snd'D$_2$ n=9-50 and some of the 5snd$^3$D$_2$).

In the observation of $(5p_{\frac{1}{2}}ns)_{J=1}$ and $(5p_{\frac{1}{2}}nd)_{J=1,3}$ of Sr, the Sr atoms are excited stepwise by means of three dye lasers. All three lasers are linearly polarized in the same direction. By sweeping the wavelength of the third dye laser we are able to record the excitation spectrum of the autoionizing states.

The energy levels and quantum defects of the $(5p_{\frac{1}{2}}ns)_{J=1}$ (n=10-26) and $(5p_{\frac{3}{2}}nd)_{J=1,3}$ (n=17-25) autoionizing series of Sr have also been measured.

It shows that the quantum defects of $(5p_{\frac{1}{2}}ns)_{J=1}$ are nearly constant. It indicates that the $(5p_{\frac{1}{2}}ns)_{J=1}$ series of Sr(n=10-26) has less coupling with other series. The single strong peak is observed in the $(5p_{\frac{1}{2}}ns)_{J=1}$ series at the Sr$^+$ 5s-5p$_{\frac{1}{2}}$ wavelength. And two strong peaks are observed in the $(5p_{\frac{1}{2}}nd)_{J=1,3}$ (n=17-25). These spectra are explained using a quantum-defect theory approach which shows that the cross section for photoexcitation $\sigma$ is proportional to a product of the spectral density of the autoionizing state and the overlap integral from the initial bound Rydberg state.

REFERENCES
(1) G S Hurst, M G Payne, et al; Phys.Rev.Lett. 35, 82 (1975)
(2) P Esherick; Phys.Rev. A, 15, 1920 (1977)
(3) W E Cooke, T F Gallagher et al; Phys.Rev.Lett. 40, 178 (1978)
(4) R Kachru et al; Phys.Rev.A. 31, 700 (1985)

*Inst. Phys. Conf. Ser. No. 84: Section 7*
*Paper presented at RIS 86, Swansea, Wales, 7–12 Sept. 1986*

257

# Nuclear isomer separation

P. Dyer

MS D449, Los Alamos National Laboratory, Los Alamos, NM  87545, USA

Abstract. Pure specimens of nuclear isomers are required for gamma-ray lasers. We have selectively photoionized atoms containing isomeric nuclei of $^{197}$Hg. The isomers were produced by the $^{197}$Au(d,2n)$^{197}$Hg reaction and distilled. Three pulsed dye lasers were used to selectively ionize mercury atoms by doubly resonant three-step photoionization. Other isomer separation techniques and their limitations are discussed.

## 1. Introduction

Our scheme for a gamma-ray laser involves two-step pumping (Baldwin 1981, 1986), in which a long-lived nuclear isomer would be produced by a nuclear reaction, separated, and implanted into a crystalline host (for supporting the Mossbauer effect and Borrmann modes). The crystal fiber would be irradiated by a beam of photons to transfer energy (probably through the electrons) to "tickle" the nucleus into a nearby short-lived nuclear state that would then decay to the upper lasing level. First nuclear superradiance would be measured by observing the time dependence and angular distribution of emitted gamma rays from $10^{10}$ to $10^{14}$ excited nuclei implanted in a low-Z substrate about a micrometer in diameter and several millimeters long. The first step in making such a gamma-ray laser is to provide a nuclear population inversion. However, when nuclear excited states are produced by a nuclear reaction, many more ground-state nuclei are usually formed. A separation step is generally required. The work reported here demonstrates the feasibility of isomer separation; the particular nucleus for making a gamma-ray laser has not yet been chosen.

We also include here a summary of various possible isomer separation techniques, outlining advantages and disadvantages of each.

## 2. Resonance Ionization of $^{197m}$Hg

We have demonstrated isomerically-selective photoionization of $^{197m}$Hg (nuclear half-life 24 hours) via the atomic excitation sequence $6^1S_0$ - $6^3P_1$ - $8^1S_0$ - Hg$^+$ (Dyer 1985). Three collinear pulsed dye laser beams were used: 254, 286, and 696 nm, selectively exciting the first two transitions and ionizing through an autoionization state in the continuum.

First, gold target foils were bombarded by deuterons at the Los Alamos tandem Van de Graaff accelerator, to generate $^{197}$Hg by the (d,2n) reaction. These target foils were then heated in vacuum to distill

mercury onto a second gold "catcher" foil, which was then sealed in a
shielded capsule for transportation to the Massachusetts Institute of
Technology Laser Research Center.

At the Laser Center, optical excitation experiments were performed with
two vapor cells, one containing natural mercury and the other, mercury
enriched in $^{202}$Hg, for adjustment and calibration of the apparatus.  Upon
arrival of the radioactive sample at the Laser Center (sixteen hours
after the end of bombardment), the active catcher foils were introduced
into a clean irradiation cell and heated to expel mercury.  The Pyrex
irradiation cells, shown in Fig. 1, were 12 cm long and 15 mm in
diameter, with fused-silica Brewster windows at each end.  No materials
that had been exposed to natural mercury were used in constructing the
$^{197}$Hg cell.  Other materials to which mercury was exposed in the chamber
were limited to Teflon, Viton O-Rings, ceramic adhesive and clean iron;
all had been previously found, using $^{197}$Hg as a tracer, to have low
tendency to adsorb mercury.

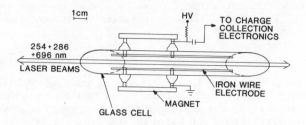

Fig. 1 Schematic diagram of the mercury vapor cell.

The collecting electrodes were a pair of magnetically supported Fe wires
on opposite sides of the laser beam.  A collecting potential of 400
volts was applied to the electrodes.  Currents from a phototube and from
the ion-collector were amplified, passed to a boxcar integrator, and
registered on a chart recorder.

The bandwidths of the 254- and 286-nm beams were about 3 and 2 GHz,
respectively.  For the 254-, 286- and 696-nm beams, the intensities were
about 1, 10, and 400 $\mu$J/pulse, respectively.  Beam diameters were about
2 mm.  The 20-nsec pulses were repeated 20 times per second.  Frequency
doubling was provided by KDP (286nm) and lithium formate (254 nm)
crystals.  The quantities of $^{197m}$Hg released into the cell were of the
order of 2 x $10^{12}$ atoms.  Ion-collection rates were of the order of
4 x $10^6$ s$^{-1}$.

For orientation, Fig. 2 shows a rough computer simulation of the ion
current as a function of the frequencies of the 254- and 286-ion beams.
Figure 3 shows the measured ionization current when the  254-nm laser
was fixed at the $^{197m}$Hg-c peak (selecting the F = 11/2 $6^3P_1$ hyperfine
state), the 286-nm radiation was scanned, and the ions that were created
in the final transition to the continuum were collected.  Figure 4 is
the same, but with the 286-nm laser fixed and the 254-nm laser scanned.
Peaks in the ionization current are observed at the expected positions.
The combined selection by both 254- and 286-nm radiations was sufficient
to achieve a clean separation of $^{197m}$Hg.

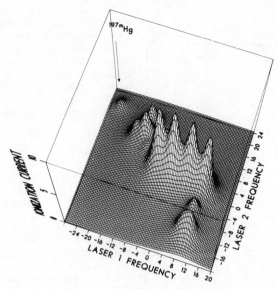

Fig. 2 Computer generated simulation of the
ionization current as a function of the
254-nm laser frequency (laser 1) and the
286-nm laser frequency (laser 2). The ¹⁹⁷ᵐHg
peak of interest is labeled. Most of the
other peaks are from natural mercury
contamination. Frequency scales are in GHz;
ionization current is in arbitrary units.

We did not, however, obtain an enriched sample outside the cell. An
attempt to measure enrichment by counting gamma rays from the positive
and negative electrodes failed. There was no significant difference in
isomeric enrichment between the two electrodes. Moreover, the total
number of radioactive atoms on the wire exceeded nearly 100-fold the
number of ions collected, estimated from the ionization current and
collection time. Presumably, the selectively ionized and collected
portion was greatly diluted by nonselective adsorption of $^{197}$Hg, despite
precautions to use clean Fe electrodes.

The isomer $^{142m}$Eu has been resonantly ionized by Alkhazov et al. (1985).
In this experiment, three laser beams intersected the atomic beam from a
mass separator on-line to a proton synchrocyclotron. Ionization rates
of $10^4$ s$^{-1}$ were achieved.

## 3. Comparison of Techniques

A number of laser techniques are available for isomer separation:
resonance ionization (where the ionization step may be performed by a
laser or by electric fields or collisions acting on Rydberg states), the
optical piston (light-induced drift in a buffer gas) (Gel'mukhanov 1979,
Werij 1984), photochemistry, radiation pressure, magnetic or electric
deflection of an optically pumped atomic beam (Zhu 1985). The optimum
choice will depend on atomic state energies, hyperfine structure, vapor
pressure, chemistry (especially surface), nuclear state lifetime, isomer
production rate, and initial enrichment factors. Various factors limit

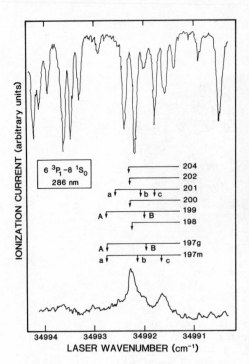

Fig. 3 Ionization current, as a function of 286-nm scanning frequency, with the 254-nm radiation fixed at the $^{197m}$Hg-c hyperfine component. The larger of the two peaks is contributed primarily by stable isotopes of mercury. The smaller peak is due to $^{197m}$Hg. The upper trace is an iodine comparison spectrum (at the fundamental frequency).

the efficiency of the separation, the enrichment achieved, and the time required to perform the separation. In general, separations in cells have high efficiency, but low resolution, whereas the opposite is true for separations in atomic beams.

Formation of an atomic beam is necessarily an inefficient process. In a cell there is the potential for a given atom to pass through the laser beam many times, but to achieve this, there must be little loss of the material to adsorption on the walls. A further factor limiting the efficiency for resonance ionization, particularly in cells, is space charge.

If the separation is performed in a cell, and the Doppler width is greater than the hyperfine splitting of the lines involved in discrete transitions, loss of enrichment results. Collisions in a cell also limit enrichment. In the case of resonance ionization, resonance charge exchange results in non-specific collection of ions. In the case of the optical piston or of photochemistry, inelastic collisions at high sample densities dilute the enrichment. Multi-photon ionization is also a source of non-selective background in the case of resonance ionization by photons, whether in a cell or an atomic beam.

There are various considerations involved in the choice of pulsed versus CW lasers. CW lasers offer narrower bandwidth and high duty factors. Pulsed lasers are more suitable for producing UV wavelengths by frequency doubling. They are also better suited to multistep processes, such as resonance ionization with the ionization step performed by photons.

The most widely applicable technique thus far appears to be that of resonance ionization. In the case of isomer separation for a gamma-ray laser, this technique offers the advantage that implantation into a crystal may be performed by the field that collects the ions. If far UV wavelengths are not required, if ionization is performed by electric fields rather than photons, and if the time constraints are not too severe, CW lasers may be used. Otherwise pulsed laser excitation is

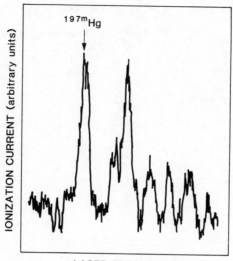

**LASER FREQUENCY**

Fig. 4. Ionization current, as a function of 254-nm scanning frequency, with the 286-nm radiation fixed at the $^{197m}$Hg-c hyperfine component. In this case, the largest peak is due to $^{197m}$Hg.

necessary. The time to separate the required number of isomers for a gamma-ray laser, once the sample is in the laser beam, can be much shorter than a second. Thus, the entire process of producing implanted isomers will be more limited by the time to produce the isomers in nuclear reactions and to transfer them from the reaction target to the laser beam, than it is by the laser ionization time.

It is difficult to scale up charged-particle-induced reaction rates to produce $10^{10}$ or more isomers in a second. There is, however, the possibility of producing such yields with a pulsed nuclear reactor. Such devices can emit over $10^{16}$ neutrons into a column of about 2000 cm$^3$ in 30 msec. If such a reactor, the lasers for isomer separation, and the photon source for transferring the nuclei from isomeric states to nearby short-lived states, can be located in one place, it may be possible to work with isomeric states as short-lived as 1 sec for a first gamma-ray laser demonstration.

## References

**Work supported in part by the Division of Advanced Energy Projects of the United States Department of Energy. Part of this work was performed while the author was a Visiting Scientist at the MIT Laser Research Center, which is a National Science Foundation Regional Instrumentation Facility.

Alkhazov G D, Barzakh A E, Berlovich E E, Denisov V P, Dernyatin A G, Ivanov V S, Letokhov V S, Mishin V I and Fedoseev V N, 1985 JETP Lett. 40 836
Baldwin G C, Solem J C and Goldanskii V I 1981 Rev. Mod. Phys. 53 687

Baldwin G C 1986 Proc. First International Laser Science Conference, Dallas, TX, Nov 1985, ed W Stwalley (New York:  American Institute of Physics) in press
Dyer P, Baldwin G C, Sabbas A M, Kittrell C, Schweitzer E L, Abramson E, and Imre D G 1985 J. Appl. Phys. $\underline{58}$ 2431
Gel'mukhanov F Kh and Shalagin A M 1979 JETP Lett. $\underline{29}$ 711
Werij H G C, Woerdman J P, Beenakker J J M and Kuscer I 1984 Phys. Rev. Lett. $\underline{52}$ 2237
Zhu X 1985 Chinese Physics $\underline{5}$ 375

*Inst. Phys. Conf. Ser. No. 84: Section 7*
*Paper presented at RIS 86, Swansea, Wales, 7–12 Sept. 1986*

# Optogalvanic investigation of hollow-cathode discharge of noble gases

Kenji Tochigi

Production Engineering Laboratory, Hitachi Ltd., 292 Yoshida, Totsuka-ku, Yokohama, 244 Japan

Shigehiko Fujimaki, Akihide Wada, Yukio Adachi and Chiaki Hirose

Research Laboratory of Resources Utilization, Tokyo Institute of Technology, 4259 Nagatsuta, Midori-ku, Yokohama, 227 Japan

Abstract   Optogalvanic spectroscopy has been applied to the investigation of hollow cathode discharge of noble gases. The topics include the observation of transient ionization wave triggered by a pulsed excitation of neutral neon atoms, the determination of threshold energy of sputtering from cathode surface, and the observation of autoionization spectrum of krypton.

## 1. Introduction

The high pressure of fill gas(generally more than 1 torr, 1 torr=133.322 Pa) and the cylindrical geometry of electrodes of hollow cathode discharge (HCD) make it difficult to apply many of the techniques used to probe normal glow discharges. The technique of optogalvanic spectroscopy(OGS) has been applied to the observation of transient ionization wave which is triggered by the optical excitation of neutral neon atoms in HCD by a pulsed dye laser. Laser optogalvanic spectroscopy using a ring dye laser provides a simple yet reliable method for measuring electric field, which in turn gives the ion density and the energy of ions impinging onto the cathode surface. The latter can be combined with the observation of sputtered substance to derive the energy of sputtering. OGS is also sensitive enough to make the observation by visible light of the optical transitions from the excited to the autoionizing levels of krypton in HCD.

## 2. Experimental Setup

The experimental setup of OGS is shown in Fig. 1. Output beam from a tunable dye laser is passed through HCD plasma and the laser-induced change of plasma impedance is picked up via a coupling capacitor. OG signal was processed by a lock-in amplifier or a BOXCAR integrator for spectroscopic measurements using a cw or a pulsed lasers, respectively, or, to measure the temporal behavior of the signal, by a programmable digitizer. The discharge tubes which were either home-made or purchased

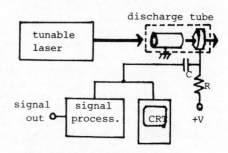

Fig. 1   Schematic layout of OGS experiment of HCD

from Hamamatsu Photonix Inc.(ST-type) had both ends of cathode cylinder cut off so that laser beam goes through

without being reflected back.    Tubes were operated by a  voltage-regulated
DC  source.    The  observation  of the OG signal requires that  the  inter-
electrode  voltage  is  stable within a few millivolts  and  the  discharge
condition for this was dependent on the kind and pressure of fill gas,  the
geometry and size of electrodes,  and the make of individual tube.   Commercial tubes were factory-sealed at 6 torr of fill gas while home-made  tubes
were  operated  in the pressure range of 1 to 6 torr(measured  by  Baratron
pressure transducer)  after several preparatory procedures of discharge-and-
refill.  Gas handling system consisted of an oil diffusion pump and a liquid
nitrogen trap.

### 3. Optogalvanic Observation of Ionization Waves

Ionization waves in normal glow discharges have been extensively  investigated  by many authors  as reviewed by Garscadden(1978) and the theory by
Pekarek(1962,1963)  was  extended by Lee et al.(1966),  Pekarek(1968)  and
Garscadden  et al.(1969),  but the study of the phenomenon in HCD  has  not
been reported to the authors' knowledge.    Smyth et al.(1978) reported that
in  neon-filled  HCD substantial change in ion concentration  results  from
exciting  neutral neon transitions by irradiation with a cw tunable  laser.
We  have succeeded in the optogalvanic observation of transient  ionization
waves  in neon HCD triggered by the pulsed dye laser excitation of the $2p_8$-
$1s_5$(in Paschen notation) transition(Tochigi et al. 1986).    The observation
has  been extended to the excitation of the $2p_4$-$1s_5$ transition  and a  comparison  has  been  made.   Fig. 2  shows the
shape of the used tube which was  filled by
6 torr of neon.   The positions of laser illumination  are  indicated  by the numbers.
Arrows and dots imply that the laser is directed parallel  or  perpendicular, respectively,  to the cathode axis. Typical output
power of the  pulsed laser was  20  $\mu$J/pulse
with  the duration time of 1 nsec.    The OG
signal was  picked  up  through a coupling
capacitor of 0.01 $\mu$F and  30  pulses  were
accumulated  by  a  programmable  digitizer
(Techtronix 390AD,the input impedance:1 M$\Omega$).

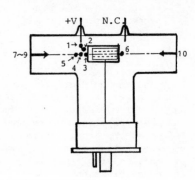

Damped oscillations appeared on the OG signal when the  frequency  of the  laser was
resonant with an optical transition of the          Fig.2    HCD tube used to
neutral neon.   The  results  which were ob-         observe the ionization wave
tained  by setting  the  wavelength of the
laser to the $2p_4$-$1s_5$ line at 594 nm are shown in  Fig. 3 by solid curves of
upper traces.   The positions and directions of the illumination are  indicated  in the figure.   Traces on the left are the  results  for  different
positions  of  illumination under the same discharge current of 3.5 mA  and
those on the right are for different discharge current at the same position
of illumination.    Similar results were obtained by the illumination at 633
nm,  the transition frequency of the $2p_8$-$1s_5$ line.   The laser-induced fluorescence from the $2p_8$-$1s_4$ transition  at 651 nm decayed  within  less  than
a  $\mu$sec in good accordance with the radiative lifetime of  the  $2p_8$  level
which  is about 20 nsec.    The oscillation was damped faster at lower  discharge current and disappeared below 1 mA.

The oscillating components were simulated by the equation,

$$\Delta V(t)=V_0 \exp[-t/\tau]\cos(\nu t+\theta) \tag{1}$$

where,  $\Delta V(t)$ is the laser-induced change in inter-electrode voltage at time
t  after the firing of the laser pulse,and $V_0$,  $\tau$,  $\nu$,  and $\theta$ are parameters.
The derived values of the parameters for various experimental conditions at

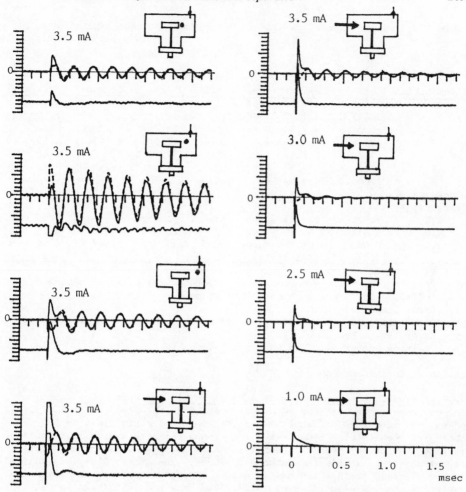

Figure 3.   Temporal response of pulse-excited OG signals at various positions of illumination under constant discharge current of 3.5 mA(left, full
scale:75 mV)   and   for different discharge current at the same position   of
illumination(right, full scale:100 mV)

the two laser wavelengths are listed in Table 1 and the calculated   signals
are   shown   by   the dotted curves and the residue signals of   the   observed
minus   the calculated signals are shown by the curves drawn below   the   two
curves.   The   initial   amplitude $V_0$ is highly sensitive to the position   of
illumination.   As it was difficult to achieve identical position of illumination at the two wavelengths because the laser had to be adjusted at   each
wavelength,   the   comparison   of $V_0$ for the two wavelengths will not be   of
much significance.   Similar situation seems to exist on the phase term $\theta$.
It   has   been   concluded that the observed oscillations are   due   to   the
ionization wave which is triggered by the pulsed excitation of neutral neon
atoms   from   metastable to upper excited levels and the signal   $\Delta V(t)$   was
related with the time-dependent change of ion density by the   consideration
of   conductivity   of the plasma(Tochigi et   al.   1986).   Unfortunately the
discharge tube of the present experiment was not designed to stand      quantitative analysis on the basis of plasma parameters, but we believe that we
have demonstrated the usefulness of OGS for extensive and detailed investi-

Table 1.   Derived Values of Parameters for the Transient Ionization Waves[a]

| Position of illumination[b] | $2p_4-1s_5$ | | | | $2p_8-1s_5$ | | | |
|---|---|---|---|---|---|---|---|---|
| | $V_0$ (mV) | $1/\tau$ (1/msec) | $\nu$[c] (kHz) | $\theta$[d] (deg.) | $V_0$ (mV) | $1/\tau$ (1/msec) | $\nu$[e] (kHz) | $\theta$[d] (deg) |
| 1 | 12.4±0.3 | .40±.04 | 4.88 | 65 | 9.7±.3 | 1.2 ±.1 | 5.10 | -78 |
| 2 | 57.0±1 | .47±.03 | 4.88 | 13 | 45.1±1 | 1.5 ±.1 | 5.10 | -44 |
| 3 | 19.3±1.3 | .47±.03 | 4.88 | 265 | 14.7±1 | 1.43±.02 | 5.10 | 36 |
| 4 | 1.1±0.1 | .45±.08 | 4.88 | 332 | 3.6±1 | 1.47±.02 | 5.10 | -4 |
| 6 | 0.75±0.1 | 1.5 ±.8 | 4.88 | 80 | 2.0±.1 | 1.47±.05 | 5.10 | 231 |
| 7 | 4.2±0.4 | .30±.5 | 4.88 | 250 | 18.0±1 | 1.54±.05 | 5.10 | 128 |
| 8 | 22.6±0.4 | .62±.04 | 4.91 | 195 | --- | --- | --- | --- |
| Discharge current(mA)[f] | | | | | | | | |
| 3.5 | 27.5±1 | .82±.01 | 4.95 | 210 | 22.7±1 | 1.54±.05 | 5.10 | 128 |
| 3.0 | 24.0±2 | 2.5±.1 | 4.67 | 210 | 18.0±1 | 1.82±.1 | 4.79 | 139 |
| 2.5 | 20.0±2 | 3.0±.3 | 4.10 | 185 | 13.8±1 | 2.86±.1 | 4.46 | 116 |
| 2.0 | 12.0±1 | 5.5±1 | 3.6 | 195 | --- | --- | --- | --- |

a)   Fit is made to Eq. 1 of the text.
b)   See Fig. 2 for the correspondence between the cited numbers and the positions and directions of illumination.
c)   Error limits are ±0.02 kHz.
d)   Cited values are for positive $V_0$ values and the error limits are ±2.
e)   Error limits are ±0.05 kHz.
f)   Position number of illumination is 9(see Fig. 2).

gation of the phenomenon.

## 4. Optogalvanic Determinations of Electric Field and Sputtering Energy

We have shown that the OG measurement of Stark effect of atomic lines makes the otherwise difficult determination of electric field in the cathode fall region of a HCD possible.   The experiment was carried out by letting the output beam of a single mode cw laser to pass through various radial positions inside the cathode cylinder.  The atomic lines are shifted by the electric field which exists near the cathode surface, and the radial variation of the electric field is derived from the shift(Nakajima et al. 1983).   New experiment has been done on the Kr 8d[3/2]2-5p[3/2]2 line by using a home-made discharge tube with the cathode of 4.5 mm ID and 20 mm length under various fill pressure(Fujimaki et al. 1986a).   Derived values of electric field at distance x from cathode surface, E(x), have been found to follow the relation,

$$E(x)=E_c(1-x/d) \qquad (0<x<d) \qquad (2)$$

$$V_c=E_c d/2 \qquad (3)$$

where, $E_c$ is the field strength at cathode surface, d is the distance corresponding to the width of the cathode fall region,  and $V_c$ is the voltage applied between the electrodes.  Linear decrease of the electric field with distance x implies that the charge density in the cathode fall region is uniform and is given by the relation $V_c\mathcal{E}/d^2$ where $\mathcal{E}$ is the dielectric constant of the plasma.  The mobility of electrons is more than an order larger than that of ions and we can assume that the derived charge density is mostly due to ions.

It has been shown by Little and von Engel(1954) that the average energy of the ions which impinge onto cathode surface is given by,

$$E_{ion}=eV_c[2\lambda_{+0}/pd-(\lambda_{+0}/pd)^2] \qquad 4$$

where p is the pressure of fill gas measured at 0 °C . The mean free path of the krypton ion at unit pressure of 1 torr, $\lambda_{+0}$, is estimated from the cross-section for charge-exchanging collisions (Brown, 1966) to be 5.65 ×10$^{-2}$ mm. The average ion energy can now be calculated for each discharge conditions for which the values of $V_c$ and d differ. The OG signal of sputtered copper atoms became visible for the discharge with 1 torr pressure of fill gas when the discharge current exceeded 5 mA. The plot of the signal intensity vs. the ion energy is shown in Fig.4. The value of 19.0±0.5 eV is derived from the plot as the threshold energy for sputtering of copper atoms from the brass surface by Kr ions(Fujimaki et al. 1986b). The detection of sputtered substances by absorption or emission spectroscopy will extend the application of the method to the investigation of sputtering under practical discharge conditions of HCD.

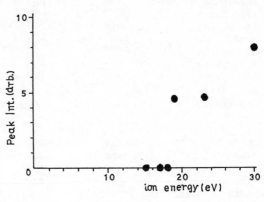

Fig. 4  Plot of OGS intensity of Cu line vs. average energy of Kr ions

## 5. Autoionization Spectroscopy Using Visible Light

The optical transitions to autoionizing levels have generally been studied by VUV and multiphoton spectroscopy, but the optogalvanic observation by Grandin and Husson(1981) of the one-photon transition from the excited to the autoionizing levels of Xe in rf discharge proved that OGS enables us to investigate the subject by using a visible light source. We describe about the extensive investigation of the autoionization spectrum of Kr.

In krypton, several of the levels converging to the $^2P_{1/2}$ sublevel of Kr$^+$ lie above the ground $^2P_{3/2}$ state of Kr$^+$, which is at about 5370 cm$^{-1}$ below the $^2P_{1/2}$ sublevel, and the atoms excited to these levels autoionize to the continuous levels of the system made up of an electron plus the [Kr$^+$]$^2P_{3/2}$ core. The optical transitions to these levels by visible light are possible from the levels higher than the 5p levels. In the present study, the optogalvanic autoionization spectrum has been observed in Kr HCD by using a cw dye laser and passing the laser beam near the cathode surface. The autoionizing lines are characterized by their broad widths ranging from 5 to 30 cm$^{-1}$ (FWHM), and the assignment was confirmed by optogalvanic double resonance(OGDR). Observed OGDR spectrum is shown in Fig. 5 where the curves (i), (ii) and (iii) are the spectrum obtained by pumping the 5p'[3/2]2, 5p'[1/2]1 and 5p'[3/2]1 levels, respectively. The spectrum shows several characteristic features. First, the spectral profiles are symmetric in contradiction to the expected presence of asymmetry due to the interference between the optical transition in question and the direct photoionization(Fano 1961). Secondly, the linewidth of the OGDR line at 16933 cm$^{-1}$ is dependent on the choice of the pumped level and this dependence cannot be explained by the consideration of the change in the number of transitions buried under the profile but the interaction between two autoionizing levels has to be taken into account. These features lead us to conclude(Wada et al. 1986a) that (i) the probability of direct photoioniza-

Fig. 5 Auto-
ionization
spectrum of
Kr observed
by OGDR.
Autoioniza-
tion lines
are marked
by *, and,
as for the
difference
of (i),(ii)
and (iii)
see text.

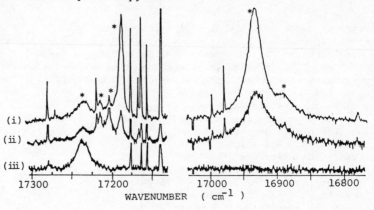

tion from the 5p' sublevels is negligibly small compared to that of the optical transitions to the autoionizing levels and (ii) the autoionizing 7d'[5/2]3 and 7d'[3/2]2 levels are mutually coupled by their configuration interaction with the continuous states. The analysis using a higher order perturbation theory has been carried out to derive the decoupled term values and lifetimes of the levels and the results are shown in Table 2. The fact

Table 2. Term Values and Lifetime of Autoionizing 7d' and 9s' Levels of Kr

| Level | Term value (cm-1) | Lifetime (psec) |
|---|---|---|
| 7d'[3/2]1[a] | 115 019.0 | 2.7 |
| 7d'[5/2]3 | 114 883 | 0.23 |
| 7d'[3/2]2 | 114 831 | 0.20 |
| 7d'[5/2]2 | 114 728 | 1.3 |
| 9s'[1/2]1[a] | 115 135.4 | 1.1 |
| 9s'[1/2]0 | 115 124 | 0.48 |

a) Yoshino K and Tanaka Y 1979 J. Opt. Soc. Amer. 69 159

that atoms in the metastable states can be raised to the autoionizing levels by two-step excitation with visible light and that the autoionization takes place in the order of picosecond will provide practical use in the processes which require temporally or spatially controlled ionization of gases(Wada et al. 1986b).

References

Brown C S 1966 Basic Data of Plasma Physics(Cambridge:The M.I.T. Press)p 73
Fano U 1961 Phys. Rev.124 1866
Fujimaki S Adachi Y and Hirose C 1986a submitted to Appl. Spectrosc.
Fujimaki S Adachi Y and Hirose C 1986b submitted to Appl. Surf. Sci.
Garscadden A 1978 Electrical Discharges ed M N Hirsh and H J Oskam (New York:Academic Press)Part 2.2
Garscadden A Bletzinger P and Simonen T C 1969 Phys. Fluids12 1833
Grandin J-P and Husson X 1981 J. Phys.B14 433
Lee D A Bletzinger P and Garscadden A 1966 J. Appl. Phys.37 377
Little P F and von Engel A 1954 Proc. Roy. Soc. London A224 209
Nakajima T Uchitomi N Adachi Y and Maeda S and Hirose C 1983 J. de Phys.44 suppl.C7 497
Pekarek L 1962 Czech. J. Phys.B12 450
Pekarek L 1963 Czech. J. Phys.B13 881
Pekarek L 1968 Sov. Phys. Usp.11 188
Smyth K C Keller R A and Crim F F 1978 Chem. Phys. Lett.55 473
Tochigi K Maeda S and Hirose C 1986 Phys. Rev. Lett.57 711
Wada A Adachi Y and Hirose C 1986a J. Chem. Phys. submitted
Wada A Adachi Y and Hirose C 1986b J. Phys. Chem. accepted

*Inst. Phys. Conf. Ser. No. 84: Section 7*
*Paper presented at RIS 86, Swansea, Wales, 7–12 Sept. 1986*

269

# RIS and competing processes in high concentration atomic vapors

Rainer Wunderlich* M. G. Payne, and W. R. Garrett

Chemical Physics Section, Oak Ridge National Laboratory,
Oak Ridge, Tennessee 37831, USA

Processes competing with the resonant ionization of an excited atomic state have to be considered in choosing proper resonance ionization spectroscopy (RIS) schemes in applications such as resonance ionization mass spectroscopy (RIMS), certain analytical work, or material separation where a high efficiency of the RIS process is required. Some of these competing processes, including the one described here, lead also to additional line broadening effects which are important in isotope selective resonance ionization where a Doppler–free two photon resonance transition is used (Clark et al 1984, Whitaker and Bushaw 1981).

The competing processes can be divided into coherent and incoherent processes. Straightforward examples for the latter case are the spontaneous transitions from the resonance state to some intermediate states. Their influence on the efficiency of the RIS process is described in the flux and fluence conditions (Hurst et al 1979) for saturation to the continuum and can be overcome by simply increasing the power of the ionization laser. An example for coherent processes is the suppression of the multiphoton ionization of Xe with a three–photon resonant transition (Miller et al 1980). This effect is observed at higher densities (p > 0.05 Torr) and it was shown (Payne and Garrett 1980, 1983) that the quenching of the ionization is due to a cancellation effect caused by the simultaneous interaction of the atoms with the driving laser field and the coherently generated third–harmonic polarization. This process prevents any significant buildup of population in the three–photon resonance state. Malcuit et al (1985) observed the suppression of amplified spontaneous emission (ASE) out of the Na 3d two–photon resonant state which was explained by a competition between four wave mixing and ASE. It was shown that under certain conditions the probability amplitude for the excited state vanishes. These are two examples of suppressed atomic transitions.

Here we report some experimental results on the three–photon ionization of Na out of the 4d two–photon resonant state. An ionization scheme of this type seems very promising in various applications because only visible photons are required and the resonance state has a photoionization cross section of $\sigma_I = 1.5 \times 10^{-17}$ cm$^2$ (Smith et al 1980). Figure 1 shows the relevant energy levels for this system. The 4d and 4p state can be ionized by the laser photons (not shown). This ionization process, referred to as Process I, can be compared to the three–photon ionization using the 3p single photon resonance transition, Process II.

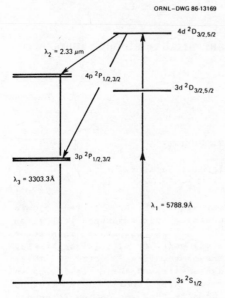

ORNL—DWG 86-13169

Fig. 1

In the latter case the single photon transition is easily saturated (and power broadened) while the effective two—photon ionization cross section of the 3p states is expected to be rather low.

Experiments were carried out with a Na atomic beam and in a Na heat pipe with an atomic vapor zone of 20 cm length. The heat pipe was equipped with a charge collecting wire. An excimer laser—pumped dye laser of 0.1 cm$^{-1}$ bandwidth, $4 \times 10^{-9}$ sec pulse duration and a maximum power of $I_p \sim 14$ MW/cm$^2$ was used for the experiments. Both two—photon and one—photon resonance transitions, i.e., 4d5/2 and 3p1/2,3/2 (5788.9 Å, 5897.5 Å, and 5891.6 Å, respectively) could be pumped exactly under the same

conditions with R6G dye. We observed in the atomic beam experiment that the ionization signal in process I (4d two—photon resonance) is a factor of $1.3 \times 10^3$ bigger than the ionization signal from the two—photon ionization of the 3p resonance state (process II). Modest focusing had to be used to obtain the signal from process II. The pump intensity in both cases was ~14 MW/cm$^2$. Process I proved to be saturated. Experiments in the heat pipe had the opposite result. At Na vapor pressures as low as $10^{-2}$ Torr, and above, the ionization signal due to process I was a factor of 10 smaller than the ionization signal from process II. The 4d ionization is suppressed by a factor of $10^4$ in the heat pipe experiment!

The heat pipe was filled with 200 Torr Xe to prevent a continuous gas discharge which is present even at low voltages on the charge collecting wire. If the ionization probability for process I is calculated without considering other processes e.g., stimulated Raman emission or four—wave mixing, we obtain an ionization probability of 70% which should yield large ionization signals even at low Na pressures. The two photon Rabi frequency $\Omega_2$ can easily be calculated because of the small detuning from the intermediate 3p state (305 cm$^{-1}$)

$$\Omega_2 = 500 \times I_p \text{ sec}^{-1}$$

The two photon transition rate $R_{01}$ is:

$$R_{01} = |\Omega_2|^2/\Gamma_L = 1.2 \times 10^{-5} I_p^2 \text{ sec}^{-1}$$

$\Gamma_L$ is the laser bandwidth ($\Gamma_L$ = 2 X 10$^{10}$ sec $^{-1}$). These numbers show that for $I_p$ = 10$^7$ W/cm$^2$ the two-photon transition to the 4d state is strongly saturated, and population leveling occurs early in the laser pulse. The only contribution to pump absorption should be ionization, and if the number of atoms in the beam, $N_T$, is larger than the number of photons in the laser pulse, $N_L$, the medium should be opaque. Given

$$N_T = 4 \text{ X } 10^{16} \text{ } P_{Na}$$
$$N_L = 8 \text{ X } 10^8 \text{ } I_p$$

This should be the case for $P_{Na}$ > 0.2 Torr, $I_p$ = 10$^7$ W/cm$^2$. Instead we observe just 20% absorption at $P_{Na}$ = 0.2 Torr. These results can only be explained by some very fast processes which dump the population from the resonant state into some lower levels and if some process suppresses the two-photon excitation at high concentrations. Consequently, we observed strong infrared (IR) and ultraviolet (UV) emission in the forward direction corresponding to the 4d -> 4p and 4p -> 3s transition, respectively. Figure 2 shows the emission vs pump laser detuning at $P_{Na}$ = 2.0 Torr and $I_p$ = 13 MW/cm$^2$. Under these conditions the conversion efficiencies were 1.5% for the IR and 3% for the UV in terms of power. Here 5% of the laser photons are converted into IR photons in the forward direction. The FWHM widths of the signals are 2.4 cm$^{-1}$ and 1.25 cm$^{-1}$ for the IR ($\lambda$ = 2.33 µm) and UV ($\lambda$ = 330.3 nm), respectively, compared to an 0.1 cm$^{-1}$ bandwidth of the laser. In Fig. 3 the FWHM width, with respect to the dye laser detuning, of the IR signal (upper curve) and the two-photon absorption is shown as a function of the pump intensity.

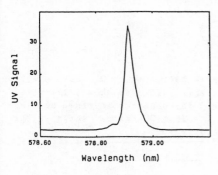

Fig. 2                                    Fig. 3

Considering that the ASE lineshape should be close to the two-photon absorption lineshape, the IR signal is taken to be stimulated Raman emission. A monochromator scan of the UV emission with the pump laser tuned exactly on resonance showed a slight displacement +~1.5 cm$^{-1}$ from the 3p - 3s transition and a width of 2.4 cm$^{-1}$, which is typical for a four-wave mixing signal. Also the UV beam has a very good beam quality and a convergence lower than that of the pump laser. This would suggest that the observed processes is a parametric four-wave mixing signal (Bokor et al 1980, Malcuit et al 1985) although our results can be very well understood in terms of strong stimulated transitions from the two-photon resonant state. A theory which <u>might</u> have some relation to our observation was given recently by Agarwal (1986). Experiments are under way to decide this question. The absorption cross section for the 4p – 4d is, in the case of weak light, $\sigma_R = 10^{-12}$ cm$^2$. The gain length for stimulated emission is then

$$\ell_R = \frac{1}{N_d \, X \, \sigma_R} = \frac{1}{N_d \, X \, 10^{-12} \text{cm}^2}$$

where $N_d$ is the concentration of atoms in the 4d state. With the rapid leveling of the populations between the 3s and 4d resonance state $\ell_R >$ 1 cm is readily achieved for $P_{Na} > 10^{-4}$ Torr. This process will limit itself when the population in the 4p state builds up. But, if there is a rapid depletion of the 4p state this process can be repeated several times. The upper limit for the number of IR photons produced, $N_R$, is then:

$$N_T/3 < N_R < N_T \, R_{01} \, X \, t_p \simeq N_T \, 4.$$

This process is shown in Fig. 4 where the $N_R/N_T$ is plotted as a function of pressure at a constant pump intensity of 13 MW/cm$^2$. The strong Raman signal causes an additional broadening of the resonance transition which eventually limits the two–photon absorption itself. The Rabi frequency for this transition is very large:

$$\Omega_R = 10^9 \, \sqrt{I_R} \, \sec^{-1}$$

$I_R$ is the intensity of the Raman emission in Watts/cm$^2$. $\Omega_2$ can easily exceed the laser bandwidth. We find that this influence should limit the two–photon absorption for $P_{Na} > 10^{-2}$ Torr. The power broadening of the 4d – 4p level due to $I_R$ is 2.4 cm$^{-1}$ for $I_R \simeq 200$ kW/cm$^2$ which is the observed width with respect to pump laser detuning.

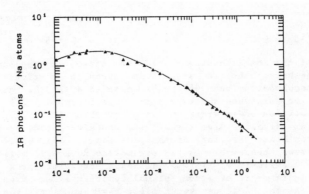

Fig. 4

It is instructive to compare the two-photon absorption spectrum in the heat pipe with the three-photon ionization signal from the atomic beam experiment ($I_p = 7$ MW/cm$^2$ unfocused beam) as shown in Fig. 5. In the latter case collective processes are absent due to the much lower density and spatial extent (~0.2 cm) of the interacting atoms. The width of the ionization signal is 0.042 Å close to $\sqrt{2} \times \Delta\lambda_L$ as predicted (Payne et al 1980) for the ionization linewidth with broad bandwidth lasers under saturating conditions. This demonstrates that the observed width of the two-photon absorption is due to the interaction of the additional waves with the atoms. With a 5% IR beam in each direction and two laser photons for each IR photon, we arrive just at the observed 20% absorption.

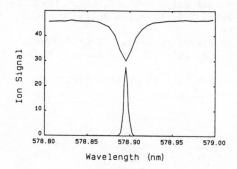

Fig. 5

In summary, we have demonstrated a striking example for the quenching of ionization via a two-photon resonance and given a qualitative explanation of the effect in terms of stimulated emission out of the two-photon resonance level. The qualitative relations allow an estimate the importance of this effect in other systems. Two conditions have to be met for this effect to occur. First, a strong two-photon transition is required so that a large fraction of atoms are in the resonant state, and second, a large cross section for stimulated emission from the resonance state to some lower state. Both factors lead then to a rapid depopulation of the resonance level. Our description of the system is by no means complete. As mentioned before, parametric four-wave mixing can also occur in this system. Further, if we take a situation at $I_p \sim$ 5 MW/cm$^2$ and $P_{Na} \sim 0.1$ where a strong Raman beam is present this beam can be treated as a separate laser interacting with the medium. Then the same principles apply which lead to the cancellation mention in the introduction (Payne and Garrett 1983) and no IR emission should occur at all! These questions will be the subject of further studies.

**Acknowledgment**

Research sponsored by the Office of Health and Environmental Research, U.S. Department of Energy under contract DE-AC05-84OR21400 with Martin Marietta Energy Systems, Inc. Ranier Wunderlich is a postdoctoral fellow under appointment through the University of Tennessee and Oak Ridge National Laboratory and supported in part by the Deutche Forschungsgemeinschaft-DFG through the Max-Planck Institute fur Kernphysik.

**References**

Agarwal G S 1986 Phys. Rev. Lett. 57 827–830
Bokor J, Freeman R R, Panoch R L, and White J C 1981 Opt. Lett. 36 182–184
Clark C W, Fassett J D, Lucatorto T B, and Moore L F 1984 in Resonance Ionization Spectroscopy (conference series No 71) eds G S Hurst and M G Payne (Bristol: The Institute of Physics) pp 107–117
Hurst G S, Payne M G, Kramer S D, and Young J P 1979 Rev. Mod. Phys. 51 767–805
Malcuit M S, Gauthier D F, and Boyd R W 1985 Phys. Rev. Lett. 55 1086–1089
Miller J C and Compton R N 1982 Phys. Rev. A 25 2056
Payne M G, Garrett W R, and Baker H C 1980 Chem. Phys. Lett. 75 408–472
Payne M G, Chen C H, Hurst G S, Kramer S D, and Garrett W R 1981 Chem. Phys. Lett. 79 142–147
Payne M G and Garrett W R 1983 Phys. Rev. A 28 3409–3429
Smith A V, Goldsmith J E M, Nitz D E, and Smith S F 1980 Phys. Rev. A 22 577–581
Whitaker T J and Bushaw B A 1981 Proc. SPIE 286 40–47

*Inst. Phys. Conf. Ser. No. 84: Section 8*
*Paper presented at RIS 86, Swansea, Wales, 7–12 Sept. 1986*

# A search for new elementary particles using sputter-initiated resonance ionization spectroscopy

W M Fairbank Jr., E Riis, and R D LaBelle
Physics Department, Colorado State University, Fort Collins CO  80523

J E Parks and M T Spaar
Atom Sciences Inc., 114 Ridgeway Center, Oak Ridge TN  37830

G S Hurst
Institute of Resonance Ionization Spectroscopy, One Pellissippi Center,
Box 22238, Knoxville TN  37933-0238

Abstract.  Sputter-Initiated Resonance Ionization Spectroscopy is being
used to search for new elementary particles which may exist at very low
concentrations in stable matter, perhaps as relics of the Big Bang.
Details of developments which have reduced backgrounds in the method to
the parts-per-trillion level are discussed in detail.  The latest
results in the search for heavy fractional- and integer-charged species
are reported.  Preliminary concentration limits of $2\times10^{-11}$ and $5\times10^{-12}$,
respectively, have been obtained.

## 1.  Introduction

It has been known for some time that Resonance Ionization Spectroscopy
(RIS) could provide the ultimate in analytical sensitivity, the ability to
detect and count single atoms of a specific species in a sample with near
100% efficiency.  While this has now been demonstrated for noble gases,
progress in using RIS for the more generally interesting and also more
difficult case of ultrasensitive solids analysis has not been as rapid.

One application which requires the ultimate sensitivity and selectivity
which RIS can provide is the search in stable matter for exotic atoms
which may contain a new elementary particles, perhaps as relics of the Big
Bang.  Our interest in this area was originally stimulated by the
observation of free fractional charges on niobium spheres heat treated on
tungsten (LaRue et al 1981).  Theoretical developments in the last decade
have also provided impetus for a search for heavy integer-charged
particles with a mass beyond the capabilities of current accelerators,
$\sim10^2$ GeV/C$^2$ ($\sim10^2$ amu).

In this paper we present our progress to date on searches for exotic atoms
using Sputter-Initiated RIS (SIRIS).  More detailed background information
on the method, motivating concerns, predicted exotic atom spectra, optimum
choice of element and sample, etc. can be found in a previous paper
(Fairbank et al 1984).  Our main emphasis here will be on experiments
which demonstrate important characteristics of the apparatus (e.g., mass
resolution and mass discrimination effects) and on a discussion of the
techniques we have used to reduce backgrounds in the method to the ppt
level.  Much of this experience is relevant also to ultrasensitive normal

atom analysis. In the last section our latest results from exotic atom searches are presented and compared to achievements of alternate methods.

## 2. The SIRIS Apparatus

The experimental apparatus which we have developed for the exotic atom searches is illustrated in Fig. 1. The salient features of the method

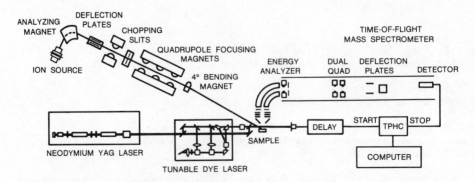

Fig. 1. Schematic diagram of the SIRIS apparatus in the configuration used for exotic atom searches.

include a pulsed ion beam for vaporizing the solid sample, a laser for resonant ionization of the atoms of interest, energy analysis of the created ions, and time-of-flight mass analysis. Each laser shot a time-to-pulse-height converter (TPHC) is triggered by the laser light after a variable delay. If an ion arrives at the detector within the specified time interval (e.g., 40 μsec or 80 μsec in heavy atom searches), a voltage pulse with amplitude proportional to the time of arrival is generated by the TPHC. The pulses are digitized and stored in a multichannel array in a computer according to arrival time. Later the time scale is converted to mass through a square root dependence.

## 3. Sensitivity and Background Reduction Considerations

For exotic atom searches interest begins at a concentration of $10^{-10}$, a level often quoted by cosmologists for unusual species surviving the Big Bang. System features which are desired for analysis at this level and below are: (1) high sample throughput, (2) high efficiency and (3) low background levels. In the present SIRIS system all three factors are barely adequate, and further improvements in each area are still desired.

Sample throughput is presently low because of low duty cycle (8.5 μsec current pulses at 30 Hz repetition rate) and moderately low peak current (3 μA). The low repetition rate results from the limits of our Nd:YAG laser system. The low peak current derives from alternate interests in depth profiling of semiconductors. The net result is an average sputter current of only 0.8 nA, or $5 \times 10^9$ argon ions per second. This problem is compounded by a useful yield (sputter coefficient times RIS efficiency) that is also somewhat lower than desired, e.g., 0.2% in the heavy atom experiments reported here. This arises from a placement of the laser beam far (4-5 mm) from the sample for reasons of background reduction, as discussed below. Thus the useful throughput in the present system is only about $10^7$ atoms per second. At this rate a concentration of $10^{-12}$ yields only about one count per day.  Clearly improvements are needed for practical ppt analysis, but initial studies can be undertaken if one is patient.

In the long run, improvements of $10^4$ to $10^5$ in throughput should be fairly easy to obtain.  Alternative lasers (copper vapor, 6000 Hz) and ion sources (>100 μA in 1 mm$^2$) are available commercially, and ion optics modifications should be possible which preserve background reduction features but allow a closer placement of the laser.  Then ppt analyses can be done in seconds with average useful currents approaching those of competitive techniques such as accelerator mass spectrometry (AMS).

As throughput is thus not a great concern, background reduction has been the major focus of our experimental program to date.  We are concerned, of course, with stray counts which arrive at the detector with long time delays after the laser pulse.  Potential sources of these background counts are:  (1) real heavy atoms or molecules ionized by the laser (SIRIS ions) or by the sputtering process (SIMS ions), (2) light SIRIS or SIMS ions delayed by scattering off parts of the apparatus, (3) light SIMS ions from any residual DC component of the sputter beam, and (4) other DC counts from dark current, ion pumps, etc.  Of these problems those involving the SIMS ions are initially the most severe but also the most amenable to solution.

In our system a great deal of SIMS discrimination is obtained by placing a potential hill of about 50 Volts in front of the sample.  Most of the SIMS ions do not have enough energy to make the top of the hill and are repelled back to the sample.  Those that do get over the hill acquire a greater energy than the SIRIS ions which are created part way down the back side of the hill.  The energy analyzer in the system (Fig. 1) removes these residual SIMS ions effectively (Fairbank 1984).  Scattering of SIMS and SIRIS ions in the system is minimized by controlling the paths which ions can take.  For example, apertures limit the solid angle of ions entering the energy analyzer, and a dual quadrupole in the time-of-flight tube ensures proper focussing of SIRIS ions on the detector.  Potential problems with detector saturation on initial light ions is avoided by directing the light ions downwards away from the detector with deflection plates during the delay time before the TAC is started.

In our system the sputter beam is pulsed by deflecting a DC argon ion beam vertically (out of the page in Fig. 1) across a pair of chopping slits. Unfortunately this beam has a weak halo (~100 pA) which still reaches the sample.  The residual SIMS ions from this beam represent a major source of heavy atom background.  We have at least partially alleviated this problem be deflecting the residual beam sideways on the sample after the main current pulse with raster plates located after the 4° bending magnet. This moves the beam out of the effective region of the extraction optics. This makeshift solution helps a lot, but there is probably room for improvement here.  Dark current contributions are negligible at the present level.

Problems with counts due to real heavy SIRIS or SIMS ions can be controlled somewhat by the choice of sample and the laser wavelengths and intensities.  Obviously samples containing low mass components (e.g., lithium metal) are desirable in searches for superheavy atoms.  Note also that the SIMS yield and the extent of molecular cluster formation are highly dependent on the chemical form of the sample surface.

4.    Diagnostic Tests on Normal Atoms

In exotic atom searches it is imperative, of course, that the response of
the system for these atoms is completely understood.    In most of the
interesting cases predictions of the exotic atom energy levels can be made
reliably by scaling of normal atom data (Fairbank et al 1984, 1985).
Similarly the response of the system can be predicted from diagnostic
tests with normal atoms.    For example, we have studied in this way the
system mass resolution (Fig. 2), the useful yield as a function of current
pulse length and laser time delay, and the mass dependence of the
sputtering process (Fig. 3).

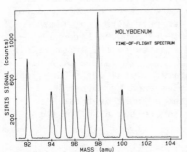 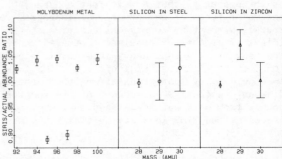

Fig. 2.    Time-of-flight
spectrum for molybdenum
isotopes.

Fig. 3.    Isotope ratio measurements which
test mass-dependent effects in SIRIS.

The system mass resolution has been improved substantially in recent
months by proper energy spread compensation and the use of focussing in
the flight tube.    Each peak in Fig. 2 is about 20 nsec wide.    This
corresponds to a mass resolution greater than 500 at mass 100.    From this
data we can argue that heavy atoms of a particular species should arrive
predominantly in one or two 40-80 μsec channels in our exotic atom
searches.

The results in Fig. 3 are extremely important to heavy atom searches and
are quite interesting in their own right.    It has been found in SIMS that
yields of heavy isotopes of an element can be reduced substantially
compared to light isotopes.    Dependences as large as $m^{-3/2}$ are observed,
although $m^{-1/2}$ is more typical.    Our recent tests (Fig. 3) with molybdenum
and silicon, two elements which have small isotope shifts and hyperfine
structures, indicate that mass dependent effects are completely absent in
SIRIS.    The unusual behavior of the odd isotopes in molybdenum is not
completely reproducible.    We understand it to be due to different
saturation characteristics of magnetic sublevels.    While correctible, we
have particularly emphasized this point as a caution to those attempting
precision isotope ratio measurements with RIS.

One important experiment with normal atoms which has not yet been
completed is a test of the velocity dependence of the detector response.
This information is needed before reliable concentration limits at high
mass can be extracted from the results of exotic atom searches.    We do
know that the detector has some response at 3000 amu because sputtered
tungsten clusters as large as $W_{17}$ have been detected through non-resonant
ionization by intense 1.06 micron light.

## 5. Results of Exotic Atom Searches

Sample results from our most recent searches for fractionally-charged atoms in niobium and tungsten and for superheavy isotopes of lithium are shown in Figs. 4 and 5. The vertical scale in these figures should not be taken too seriously, especially above mass 1000 amu, since the detector response remains uncalibrated. These graphs are therefore intended mainly to indicate the extent of our progress, rather than to establish rigorous limits.

The tungsten results (Fig. 4) date from a year ago, when the source current was higher (40 μA) but the background level was worse than at present. Typical backgrounds in Fig. 4 are two counts per channel, or $5 \times 10^{-12}$ concentration. The largest signals are less than $2 \times 10^{-11}$ for m > 600 amu.

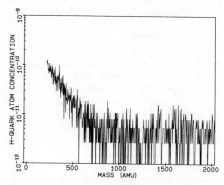

Fig. 4. Results of searches for heavy atoms with Z=2/3 in W metal. Comparable results are obtained with Nb metal samples.

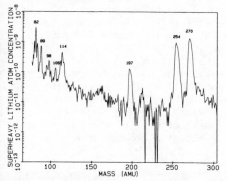

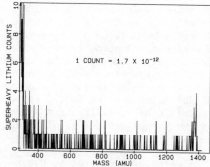

Fig. 5. Results of a search for superheavy Li isotopes in lithium metal. The low mass region is shown on the left; the higher mass region on the right.

In both our quark and heavy atom data the largest peaks at low mass appear at mass 254 and 270 amu. We attribute these peaks to nonresonant ionization of UO and $UO_2$. Uranium is a major system contaminant from previous work with uranium metal. The mass 197 peak in Fig. 5 is also systemic. We identify it as TaO from sputtering of the extraction electrode closest to the sample. None of these peaks exhibit the predicted wavelength dependence of exotic atoms. The broadness of these peaks in Fig. 5 could indicate that they are indeed higher energy atoms back-sputtered from the ion optics. Obviously further improvements in background reduction in the low mass region are desired.

In the lithium experiments (Fig. 5) the average background in the mass range 350-1350 amu is now only 0.3 counts per channel ($5 \times 10^{-13}$ concentration). We can set a tentative concentration limit in this region of $5 \times 10^{-12}$, the size of the largest peaks. Note that the corresponding limit on heavy -1 charge particles surviving the Big Bang is much lower than this (perhaps $10^{-16}$) due to the large cosmological enhancement factor for this case (Cahn and Glashow 1981, Dicus and Teplitz 1980). The bunching of counts near 1375 amu is interesting and warrants further investigation. One suspects saturation of the digitizer from the appearance of this spectrum, but such an effect has not been observed in previous experiments with this system.

6. <u>Conclusion</u>

As SIRIS is still a relative infant compared to more mature technologies such as AMS, we feel that the results reported here represent important milestones in the general development of the method. Nevertheless, our present limits for fractional charges are still well above those of Milner et al (1985) with AMS (e.g., $10^{-16}$ to $10^{-17}$ for Z=2/3 particles in W and Nb over the mass range 0.2-260 amu). Our Li results are superior to those ($10^{-10}$) obtained on sodium with the Photon Burst method (Dick et al 1985), but are not as good as the results ($10^{-16}$ to $10^{-18}$ without the enrichment factor) of Smith et al (1982) in the special case of hydrogen using a 130 keV mass spectrometer. To our knowledge no AMS searches for heavy integer-charged particles (m > 100 amu) have yet been reported, although experiments with a projected sensitivity of $10^{-16}$ are apparently underway (Elmore et al 1984).

We would like SIRIS to compete eventually on equal footing with the best of these technologies. To achieve this level about four orders of magnitude improvement is needed in both useful throughput and background reduction. We have seen that the throughput deficiency in SIRIS can probably be made up by instrumental changes. The path to a further $10^4$ background reduction is less clear. The installation of further redundancy in the system (e.g., energy analysis at the end of the flight tube) should help. Perhaps innovations which allow coincidence counting such as an installation of a photon burst detector (Fairbank et al 1984) or post-acceleration to 100 keV may be necessary. In many ways the capabilities of SIRIS and AMS complement one another. It would be nice for many applications if SIRIS could be extended to a comparable sensitivity.

This work was supported by National Science Foundation grants PHY-8210835 and PHY-8106763. We also acknowledge the collaboration of Robin Hutchinson and support from the Department of Energy on the isotope ratio measurements of Fig. 3.

<u>References</u>

Cahn R N and Glashow S L 1981 Science <u>213</u> 607
Dick W J, Greenless G W and Kaufman S L 1984 Phys. Rev. Lett. <u>53</u> 431
Elmore D et al 1984 Nucl. Instr. Meth. <u>B5</u> 109
Fairbank W M Jr., Hurst G S, Parks J E and Paice C 1984 Resonance
    Ionization Spectroscopy 1984 ed G S Hurst and M G Payne (Bristol:
    Institute of Physics) pp 287-296
Fairbank W M Jr., Perger W F, Riis E, Hurst G S and Parks J E 1985 Laser
    Spectroscopy VII ed T W Hansch and Y R Shen (Berlin: Springer-Verlag)
    pp 53-54
LaRue G S, Phillips J D and Fairbank W M 1981 Phys. Rev. Lett. <u>50</u> 1640
Milner R G, Cooper B H, Chang K H, Wilson K, Labrenz J and McKeown R D
    1985 Phys. Rev. Lett. <u>54</u> 1472
Smith P F, Bennett J R J, Homer G J, Lewin J D, Walford H E and Smith W A
    1982 Nucl. Phys. <u>B206</u> 333

Inst. Phys. Conf. Ser. No. 84: Section 8
Paper presented at RIS 86, Swansea, Wales, 7–12 Sept. 1986

# Solar neutrino spectroscopy

R D SCOTT
Scottish Universities Research and Reactor Centre
East Kilbride, Glasgow G75 0QU, UK

Note:  this unscheduled, off-the-cuff, introduction to Dr Hurst's talk is reproduced as nearly as possible in its original form; its purpose was merely to smooth the transition from atomic to nuclear matters.

The basic fusion process by which energy is produced in the sun may be written $4p \rightarrow \alpha + 2e^+ + 2\nu + 27$ MeV, and the neutrinos alone are capable of escaping and being observed. They are produced in three main ways:  (a) $p + p \rightarrow d + e^+ + \nu$; this is the rate-controlling reaction which proceeds through the weak interaction and is barely exothermic, producing a flux of some $6 \times 10^{10}$ cm$^{-2}$sec$^{-1}$ low energy neutrinos (a continuum extending to 0.42MeV).  (b) electron capture decay of $^7$Be; this gives a flux of $4 \times 10^9$cm$^{-2}$ sec$^{-1}$ at what we may call intermediate energy (0.86MeV). (c) positron emission from $^8$B; here the flux is only $6 \times 10^6$ cm$^{-2}$ sec$^{-1}$ and the neutrinos have a spectrum of energies extending to 14MeV.  Detection of the solar neutrinos in radiochemical experiments uses the inverse electron capture reaction $(A,Z - 1) + \nu \rightarrow (A,Z)^+ + e^-$ (threshold = $Q_{ec}$). The reaction rate is very low; the cross section is $\sim 10^{-45}$cm$^2$ so that the interaction rate per atom of target, $\sigma\emptyset$, is expressed in units of $10^{-36}$sec$^{-1}$ (= 1SNU, or solar neutrino unit) and the ensuing practical problem is that of detecting perhaps one event per day in a hundred tons of target.

Here, then, is the solar neutrino problem: in the only, justly famous, experiment to date, Raymond Davis Jr. has succeeded in observing the reaction $^{37}$Cl + $\nu \rightarrow$ Ar$^+ + e^-$, at a rate, above all known background processes, of $2.1 \pm 0.3$ SNU which has to be compared with the theoretical prediction, based upon the standard solar model, of 5–7 SNU for the capture of (mainly) type (c) neutrinos in this detector.

The table below gives a list of radiochemical detectors which, by virtue of their different reaction thresholds, are predominantly sensitive to different parts of the neutrino energy spectrum and together offer the possibility of a neutrino spectroscopy of the sun, both present and past.  Those in the column 'current' explore the present state of the sun whereas those headed 'geological' have long-lived products which integrate the sun's output over the past $10^6$ - $10^7$ years provided that the target material can be found in suitably deep geological deposits.  $^{81}$Br can, in principle, do both jobs since RIS obviates the need for detection of the electron capture decays of $^{81}$Kr.

| Neutrino Type | Current | Geological |
|---|---|---|
| a | $^{71}$Ga - $^{71}$Ge (11d) | $^{205}$Tl - $^{205}$Pb (3x10$^7$y) |
| b | $^{81}$Br - $^{81}$Kr (2x10$^5$y) | $^{81}$Br - $^{81}$Kr (2x10$^5$y) |
| c | $^{37}$Cl - $^{37}$Ar (35d) | $^{98}$Mo - $^{98}$Tc (4x10$^6$y) |

*Inst. Phys. Conf. Ser. No. 84: Section 8*
*Paper presented at RIS 86, Swansea, Wales, 7–12 Sept. 1986*

# Feasibility of a $^{81}$Br $(v,e^-)^{81}$ Kr solar neutrino experiment

G. S. Hurst,[||] Institute of Resonance Ionization Spectroscopy
University of Tennessee, Knoxville, TN 37996 USA

## 1. Introduction

Modern astrophysics, as described by the Nobel lecture of Chandrasekhar in 1984, has made remarkable progress in understanding the interior of stars and the mechanisms of energy production. From the early suggestion of Russell, Eddington followed up the search for an internal energy source (in addition to gravity) and speculated that the sun produces energy by the burning of hydrogen atoms. In 1939 Bethe showed that hydrogen atoms are fused to make helium atoms in two ways, the P-P chain and the C-N-O cycle.

It is believed that the sun and its internal energy source are understood well enough to make an accurate prediction of solar neutrino production. A review by Bahcall et al (1982) gives a critical evaluation of uncertainties in predicting capture rates in solar neutrino detectors, using standard solar models (Bahcall 1978,1979). However, a careful experiment has been done and the result is a neutrino flux that is significantly lower than predicted. Because of this problem, the calculations by Bahcall and associates, as well as experimental refinements by Davis and associates, have continued but have not resolved the problem. The origin of the discrepancy is still unknown; it may be due to a flaw in the standard solar model, or it could be due to a lack of knowledge in neutrino physics.

The standard theory of main-sequence stars assumes local hydrostatic equilibrium, energy transport by radiation and convection, and energy production by hydrogen burning. These assumptions are expressed by differential equations, but to solve them requires additional information. An equation of state is used, together with the radiative opacity (which depends on detailed composition of the sun) and cross sections for the various nuclear reactions. Models based on this standard theory, with reasonable assumptions of the initial composition, are referred to as standard stellar models.

[*]Research sponsored by the Office of Health and Environmental Research, U.S. Department of Energy under contract ACO5-84OR21400 with Martin Marietta Energy Systems, Inc.

[||]Consultant to the Oak Ridge National Laboratory, Oak Ridge, TN 37831, U.S.A.

## 2. Solar Neutrino Flux on Earth

Figure 1 shows a schematic representation of the standard solar model. Examples of non-standard models involve unusual assumptions of initial composition, black holes in the sun, fractional charges, etc.

**NEUTRINOS FROM THE SUN**

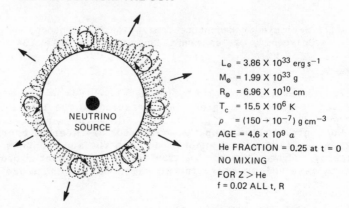

$L_\odot$ = 3.86 X $10^{33}$ erg s$^{-1}$
$M_\odot$ = 1.99 X $10^{33}$ g
$R_\odot$ = 6.96 X $10^{10}$ cm
$T_c$ = 15.5 X $10^6$ K
$\rho$ = (150 → $10^{-7}$) g cm$^{-3}$
AGE = 4.6 x $10^9$ $a$
He FRACTION = 0.25 at t = 0
NO MIXING
FOR Z > He
f = 0.02 ALL t, R

Fig. 1. Schematic representation of standard stellar model

The results of folding nuclear reaction cross sections, the solar constant, elemental abundance, mean opacity, equation of state, and solar age into the standard solar model are shown in Fig. 2. According to the currently accepted solar model of Bahcall, the fraction of energy produced by the P-P chain is 0.985 and only 0.015 for the C-N-O cycle.

The only solar neutrino flux measurement to date was made in the landmark experiment by Ray Davis, Jr. and his associates, using the reaction $^{37}$Cl$(\nu,e^-)^{37}$Ar. For a very interesting history of this classic experiment, see the account by Bahcall and Davis (1982). This difficult experiment has been done, refined, and repeated over the last two decades; for recent reviews, see Davis et al (1983) or Cleveland et al (1984). The total rate of $^{37}$Ar production was found to be just 0.38 atoms per day. After background correction, the measurement of 0.30 + 0.08 atoms per day corresponds to 1.6 + 0.4 SNU. Thus, the solar neutrino problem....the measured flux is a factor of 3.6 smaller than the value calculated from the standard model. These low values persisted in a series of experiments from 1971 until 1983.

The large discrepancy between predictions of the standard model and experimental results from the Cl experiment has set the stage for a number of interesting speculations. In very general terms, the problem must be in the standard model or in neutrino physics. In the standard model it is possible that some astrophysics or nuclear physics facts are missing, or that some uncertainties come into the transport calculations. In the area of neutrino physics, perhaps all types of neutrinos have rest mass. A difference in mass amongst $\nu_e$, $\nu_\mu$, and $\nu_\tau$ could lead to oscillations between neutrino types. A new mechanism in which electron neutrinos, $\nu_e$, are converted into $\mu$ neutrinos, $\nu_\mu$, was suggested by Mikheyev and Smirnov (1985). In this process, $\nu_e$ having energies greater than some minimum can be converted into $\nu_\mu$ in material of unusually high density. Bethe (1986) adopts the general features of this mechanism and argues that all of the

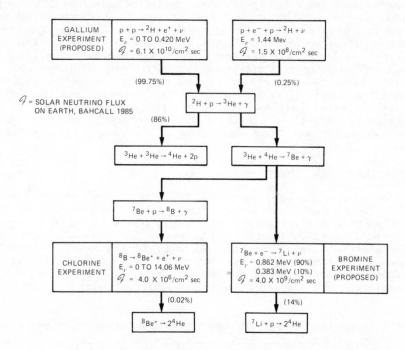

Fig. 2. Calculated flux of solar neutrinos on the earth, according to the standard stellar model of Bahcall

high energy component of $\nu_e$ starting at the core of the sun (where $\rho$ is about 150 g/cm$^{-3}$, see Fig. 1) will pass through a density region in which the conversion $\nu_e \rightarrow \nu_\mu$ is complete. From the flux measured with the chlorine experiment, combined with the calculation of Bahcall et al (1985), Bethe (1986) estimates the mass difference $m_2^2 - m_1^2 = 6 \times 10^{-5}$ eV$^2$, placing a limit of about 0.008 eV for the mass of $\nu_\mu$. Such small mass differences would be difficult to detect in earth-based neutrino oscillation experiments.

3. **Proposed Experiments**

It is not surprising that a number of new experiments have been proposed to try to resolve the solar neutrino mystery. Other radiochemical experiments, following the original chlorine experiment, are the gallium and the bromine experiments. With gallium, low-energy neutrinos could be measured since $^{71}$Ga$(\nu,e^-)^{71}$Ge has a threshold at 233 keV. The inverse decay, by electron capture, occurs with a half-life of 11.4 days; thus, detection of $^{71}$Ge decay is a measure of the neutrino flux provided cross sections are well known. This experiment was first developed at Brookhaven National Laboratory (see Davis et al 1983) with theoretical support by Bahcall (see Bahcall 1978, Bahcall et al 1982). Currently, an experiment is planned in West Germany (Kirsten 1984, Hampel 1985) and in the USSR (Barabanov et al 1985).

The bromine experiment would utilize the reaction $^{81}Br(\nu,e)^{81}Kr$ which has a threshold at 470 keV and would measure the neutrinos of intermediate energy, primarily the $^7Be$ source in the sun. Since $^{81}Kr$ has a half-life of $2.1 \times 10^5$ yr, a solar neutrino experiment using this reaction cannot be done by decay counting. It appears that RIS is the only feasible way to count small numbers of $^{81}Kr$ atoms. Scott (1976) first suggested the use of bromine as a geophysical experiment--here we are discussing a Davis type radiochemical solar neutrino detector. Approximately 65% of the total signal is due to the $^7Be$ neutrinos. Fortunately, the bromine experiment is dominated by a single solar-neutrino-flux component even if higher excited states contribute to the capture rate. The gallium experiment, in contrast, is dominated by the P-P neutrinos if excited states are neglected, but becomes increasingly sensitive to the $^7Be$ neutrinos as the effect of excited states increases. Results from a bromine experiment may thus be essential to interpret a gallium experiment.

## 4. RIS Counting of $^{81}Kr$ Atoms

With a detector size comparable to that of the present chlorine experiment, the rate of production of $^{81}Kr$ is about five atoms per day for $CH_2Br_2$, according to the standard solar model (see Table 1). Each run of six months will then contain about 900 atoms of $^{81}Kr$. Extraction of $^{81}Kr$ from a bromine-containing compound can be performed by using a helium purge system like that used for extracting $^{37}Ar$ from $C_2Cl_4$. The entire krypton extraction process using the 380-m$^3$ tank of $C_2Cl_4$ at Homestake has been demonstrated by Davis and Cleveland (unpublished data).

Table 1. Summary of proposed bromine solar neutrino experiment

---

Reaction:       $^{81}Br(\nu,e^-)^{81}Kr$, primarily $^7Be$

Facilities:     Davis radiochemical, like Cl except that atom
                counting requires RIS method

Compound:       380 m$^3$ (1000 tons) $CH_2Br_2$

Signal levels (atoms $^{81}Kr$ per day): Consistent model ----- 2
                                         Standard solar model - 5

Noise/signal examples: Sudbury (6000 hg/cm$^2$) ---- 0.1%
                       Homestake (4000 hg/cm$^2$) -- 2%
                       Gran Sasso (3800 hg/cm$^2$) - 3%

---

Background effects have to be considered for any proposed solar neutrino experiment (Rowley et al 1980). These arise from the penetrating cosmic-ray muons, from $\alpha$ decay in the target itself, and from neutrons generated by fission decay or $(\alpha,n)$ reactions in surrounding rock. Cosmic-ray muons can create protons by the photonuclear process and the reaction $^{81}Br(p,n)^{81}Kr$ leads to a background. At the depth of the Homestake mine (410 kg/cm$^2$ or 4100 hg/cm$^2$, where hg = hectogram), this background is reduced to about 0.07 atom of $^{81}$Kr per day, with the assumption of a volume of 380 $m^3$ for the bromine-rich organic solution. Alpha particles from the decay of uranium or thorium in the target would initiate $^{81}$Br$(\alpha,p)^{81}$Rb followed by $^{81}$Br$(p,n)^{81}$Kr. Furthermore, the alpha process $^{78}$Se$(\alpha,n)^{81}$Kr leads to background if $^{78}$Se is an impurity. The total alpha-induced background is about 0.03 atom of $^{81}$Kr per day, assuming impurity levels to be the same as in the chlorine solution. Similarly, $^{81}$Br$(n,p)^{81}$Se followed by $^{81}$Br$(p,n)^{81}$Kr is a neutron-induced background. And $^{84}$Sr$(n,\alpha)^{81}$Kr leads to $^{81}$Kr if $^{84}$Sr is an impurity (unlikely). These neutron-induced reactions would produce about 0.1 atom of $^{81}$Kr per day in the target; however, the neutron flux can be easily reduced by a water shield around the tank. Thus, we are left with a total background rate of about 0.1 atom per day, considerably less than the expected rate of about 5.0 per day due to solar neutrinos.

Any krypton from air leaks will contain the atmospheric abundance of $^{81}$Kr, viz., $1.6\times10^7$ atoms of $^{81}$Kr per cubic centimeter of krypton (Loosli and Oeschger 1969. The measurements made of krypton extraction from the 380-$m^3$ $C_2Cl_4$ tank give an upper limit of $10^{-6}$ $cm^3$, and that will not be a serious source of background. Excessive $^{82}$Kr could interfere with the RIS detection of $^{81}$Kr; thus, one step of isotopic enrichment could be necessary to reduce the number of $^{82}$Kr atoms due to air contamination before doing the RIS counting.

As reported in detail (Hurst et al 1985), the Maxwell demon (based on RIS) would be used to count $^{81}$Kr. In Session III, Thonnard reported on the superb progress made at Atom Sciences, Inc. to improve on the RIS method for counting noble gas atoms, and Lehmann discussed important uses of the method for dating polar ice caps and groundwater.

References

Bahcall J N 1978 Rev. Mod. Phys. 50 881
Bahcall J N 1979 Space Sci Rev 24 227
Bahcall J N, Cleveland B T, Davis R Jr, and Rowley J K 1985 Astrophys. J. 292 L79
Bahcall J N and Davis R Jr 1982 In Essays in Nuclear Astrophysics eds C A Barnes, D D Clayton, and D N Schram (Cambridge: Cambridge University Press) pp 243-285
Bahcall J N, Huebner W F, Lubow S H, Parker P D, and Ulrich R K 1982 Rev. Mod. Phys. 54 767
Barabanov I R, Gavrin V N, Golubev A A, and Poomansky A A 1985 Bull. Acad. Sci. USSR, Phys. Ser. 37 45
Bethe H A 1986 Phys. Rev. Lett. 56 1305
Cleveland B, Davis R Jr, and Rowley J K 1984 in Resonance Ionization Spectroscopy (conference series No 71) eds G S Hurst and M G Payne (Bristol: The Institute of Physics) pp 241-250
Davis R Jr, Cleveland B T, and Rowley J K 1983 in Science Underground (AIP conference proceedings No 96) eds M M Nieto, W C Haxton, C M Hoffman, E W Kolb, V D Sandberg and J W Toevs (New York: American Institute of Physics) p 2
Hampel W 1985 in Proceedings Conference on Solar Neutrinos and Neutrino Astronomy eds M L Cherry, W A Fowler, and K Lande (New York: American Institute of Physics)
Hurst G S, Payne M G, Kramer S D, Chen C H, Phillips R C, Allman S L, Alton G D, Dabbs J W T, Willis R D, and Lehmann B E 1985 Rep. Prog. Phys 48 1333

Kirsten T 1984 in Resonance Ionization Spectroscopy  (conference series
  No 71) eds G S Hurst and M G Payne (Bristol: The Institute of Physics)
  pp 251–161
Loosli H H and Oeschger H 1969 Earth Planet. Sci. Lett. 7 67
Mikheyev S P and Smirnov A Yu 1985 paper Tenth International Work-
  shop on Weak Interactions, Savanlinna, Finland (unpublished)
Rowley J K, Cleveland B T, Davis R Jr, Hampel W, and Kirsten T 1980
  Geochim. Cosmochim. Acta (Suppl. 13) 45
Scott R D 1976 Nature 264 729

*Inst. Phys. Conf. Ser. No. 84: Section 8*
*Paper presented at RIS 86, Swansea, Wales, 7–12 Sept. 1986*

289

# An application of resonant ionisation spectroscopy to accelerator based high energy physics

K W D Ledingham, J W Cahill, S L T Drysdale, C Raine, K M Smith, M H C Smyth, D T Stewart and M Towrie

Department of Physics and Astronomy, University of Glasgow, Glasgow G12 8QQ, Scotland

C M Houston

Department of Chemistry, University of Glasgow, Glasgow G12 8QQ, Scotland

Abstract. The simulation of charged particle tracks by pulsed UV lasers is now used extensively in the calibration of multiwire drift chambers. The identity of the trace quantities of low ionisation potential impurities responsible for the laser induced ionisation in conventional chamber gases has caused much discussion. Using two photon resonant ionisation spectroscopy (R2PI) two of the major sources of ionisation in proportional counters have been identified as phenol and toluene.

## 1. Introduction

CERN is constructing a Large Electron–Positron storage ring (LEP) which is expected to begin operation in 1989. This will allow intense beams of electrons and positrons to collide at very high energies to investigate some of the outstanding questions in our understanding of the electroweak forces. It is hoped that new information on the $Z^0$ and $W^{\pm}$ will be forthcoming as well as evidence of the top quark and other hitherto unseen particles such as Higgs bosons or supersymmetric particles.

ALEPH, one of the four LEP experiments, will use a very large time projection chamber (TPC, shown in Fig. 1) to detect the charged particles produced when the electron–positron beams collide. It has been shown that the tracks of these charged particles can be simulated by using UV laser induced ionisation. The laser tracks produced are straight in space and hence will be used to discover any global distortions in either the electric or magnetic fields of the ALEPH TPC drift volume. An excellent review of the present status and range of applications of laser ionisation in high energy physics has been given recently by Hilke (1986). The source of the ionisation has been the subject of much debate but it is now believed that two photon absorption in some low ionisation potential impurity is the most likely explanation. The source of the impurity could be outgassing from the materials of which the chamber or gas flow system is constructed and large variations in this 'background' ionisation have been reported (Ledingham et al. 1985; Towrie et al. 1986a).

It was felt therefore that an investigation of the wavelength dependence of the two photon 'background' ionisation spectra might produce information leading to the identity of the impurities. This is important to the ALEPH program but it is also important to identify the 'background' if the ultimate sensitivity of RIS as an analytic technique is to be realised, since the same impurities are likely to exist in a typical resonant ionisation spectroscopy (RIS) system.

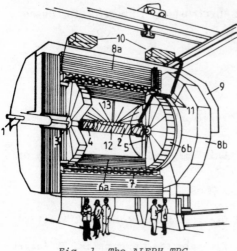

1. Beam pipe
2. Inner tracking chamber
3. Luminosity monitor
4. Beam pipe cone
5. TPC
6a. e-γ calorimeter (barrel)
6b. e-γ calorimeter (end cap)
7. Superconducting solenoid
8a. Hadron calorimeter (barrel)
8b. Hadron calorimeter (end cap)
9. Muon chambers
10. Yag laser (265nm)
11. Mirrors
12. Prisms attached to inner
      field cage
13. Laser tracks inside the
      TPC volume

Fig. 1   The ALEPH TPC

## 2.  Experimental Arrangement

The laser system used was a Lumonics TE-861M-3 (XeCl filling) which pumped a Lumonics EPD-330 dye laser to provide a pulse of light of about 6 ns duration at a repetition rate of a few hertz. Three dyes, DCM, Rhodamin 6G and Coumarin 153 gave a wavelength coverage from 521 nm to 686 nm. The output of the dye laser was then frequency-doubled using Inrad frequency-doubling crystals KDP'B' and KDP'R6G' to give a UV wavelength range from 262 to 330 nm. The laser wavelengths were checked at regular intervals using a calibrated monochromator. The frequency-doubled laser beam, with a cross-sectional area 1 mm by 1 mm, defined by a collimating aperture, passed into a single wire proportional counter. The counter was filled with conventional 90% argon/10% methane gas mixture and operated in static or flowing mode. Calibrations were carried out using an $^{55}$Fe source. The electronic arrangement has been dealt with in greater detail elsewhere (Ledingham et al. 1985) and is described only briefly here. Part of the laser beam was reflected from a quartz beam splitter just before the dye laser into a photodiode to provide a timing reference signal. Signals from the proportional counter and a Molectron joulemeter were measured simultaneously using voltage sensitive ADC's, gated by the photodiode signal. The digitized signals were stored event by event on floppy disk. For each wavelength setting some 500 events were recorded and the mean values for the ionisation and joulemeter signals were determined. For selected wavelengths it was shown that the laser induced ionisation was a quadratic function of the laser fluence. The pulse to pulse variation of the laser fluence amounted to between 10 and 20%, and therefore the dependence of the ionisation signal on the laser fluence can be determined, over a limited dynamic range, for each wavelength setting.

## 3.  Results

The laser induced 'background' ionisation in a proportional counter as a function of laser wavelength is shown in Fig. 2. The laser was stepped by increments of 1 nm. The laser cross-sectional area was 1 mm by 1 mm with a fluence of 8.1 μJ/mm². The ionisation signal has been normalised

quadratically to a laser fluence of 1 µJ/mm² so that comparisons with the
work of other authors can readily be made. It can be seen that, as the
wavelength decreases from 325 to 278 nm, the ionisation increases
monotonically by around four orders of magnitude. Below 278 nm the
fluctuation of the ionisation signals suggests the presence of fine
structure. The region between 278 and 266 nm was examined with greater
precision, the wavelength being stepped by increments of 0.025 nm, to give
the data shown in Fig. 3 (spectrum A).

*Fig. 2*

*Variation of 'background'*
*ionisation with wavelength*
*in a proportional counter.*
*All data are normalised*
*quadratically to a laser*
*fluence of 1 µJ mm$^{-2}$*

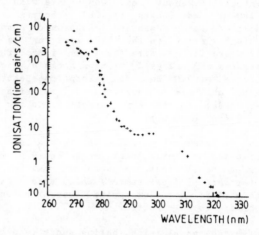

The first step from the ground state to an excited state in a two-photon
process is precisely that involved in single electronic absorption
spectroscopy and hence the two-photon ionisation spectrum of an organic
molecule is likely to be similar to its vapour phase electronic absorption
spectrum.

*Fig. 3*

*A comparison of the laser*
*induced R2PI spectrum using*
*a proportional counter with*
*Ar/CH₄ mixture plus traces*
*of phenol (spectrum A) with*
*the single-photon UV*
*absorption spectrum of*
*phenol (spectrum B) showing*
*an identical wavelength*
*dependence*

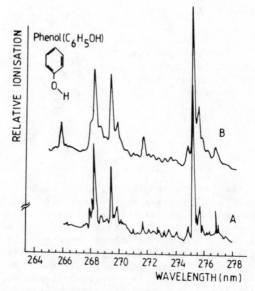

The UV absorption spectra of organic molecules are well documented (e.g.
Murrell (1963) and UV Atlas of Organic Compounds (1966)). Organic

molecules which absorb in the near UV wavelength range used in this study
are typically of two types. Those with chromophores containing an atom
with a lone pair of electrons generally have absorption energies of about
4 eV (310 nm). The other class of molecules consists of those containing
conjugated double bonds. These often contain a chain of carbon atoms with
alternate single and double bonds. In particular, simple 6-membered
aromatic ring compounds like benzenes and substituted benzenes exhibit
electronic absorption in the same wavelength region as that shown in
Fig. 3 (spectrum A). Furthermore benzenes have, in general, ionization
potentials of about 9 eV, compatible with two-photon ionization in the
260-280 nm wavelength region. Several substituted benzenes were
identified which possessed electronic absorption spectra (in methanol
solution) similar to the R2PI spectrum of Fig. 3 (spectrum A). Quartz
cells were filled with the room temperature vapour pressure of these
substances and analysed using a Beckman spectrophotometer UV 5270 with a
resolution of 0.05 nm. The absorption spectrum of the hydroxy-substituted
benzene molecule (phenol) shows fine structure which is identical, in
first order, to the R2PI spectrum in the wavelength range 266-278 nm
(Fig. 3 (spectrum B)).

Phenol is commonly used in the manufacture of plastics and resins as a
plasticiser or antioxidant (Mark and Gaylord (1969)). By adding small
known concentrations of phenol it can be estimated that in a typical
counter (Fig. 3 (spectrum A)) which uses plastics as electrical insulators
or as part of a vacuum system, concentrations of between 0.1 to 1 ppm of
phenol are to be expected.

After extensive cleaning, baking under vacuum and removal of all plastics
the background was again taken (Fig. 4) and found to be an order of

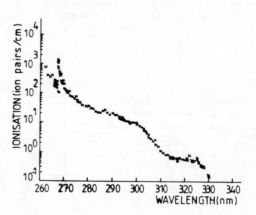

*Fig. 4*

*Laser induced 'background'*
*ionisation as a function of*
*wavelength in a proportional*
*counter after extensive*
*cleaning and baking under*
*vacuum*

magnitude lower than in Fig. 2 and without the fine structure attributed
to phenol. There was still some fine structure at 268 nm however. As
previously, a high resolution resonant two photon ionisation spectrum was
taken between 262 and 269 nm (Fig. 5 (spectrum A)). This was again
compared to the single photon absorption spectra of several substituted
benzene samples and the spectrum of toluene (Fig. 5 (spectrum B)) was
found to match closely the R2PI structure of Fig. 5 (spectrum A). Toluene
is a common impurity in proportional counters especially if methanol and
ethanol have been used as cleaning agents and by calibrating with known
quantities of toluene it has been estimated that 1 ppb of toluene can be
present in a very clean counter system.

*Fig. 5*

*The continuous curve shows
a single UV spectrum of
toluene obtained using a
Beckman spectrophotometer.
The experimental points (+)
show the R2PI spectrum of
trace quantities of toluene
in a proportional counter
operated with A/CH₄ mixture*

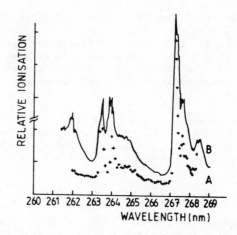

## 4. Conclusions

Phenol and Toluene have been shown to be present in proportional counters filled with conventional gases. These molecules or other substituted benzenes are likely to be a major source of laser ionisation in a typical TPC since the preferred laser for producing the ionisation is a quadrupled Nd:YAG ($\lambda$ = 266 nm). Although this study is an application of RIS to high energy physics it is relevant to RIS as an analytic technique.

Hurst et al. (e.g. 1977) have shown that single Cs atom detection is possible in conventional counter gas using R2PI at 455 nm. This corresponds to a detection sensitivity of parts in $10^{18}$. However if the laser wavelength is reduced below 300 nm the background (Fig. 2) is likely to increase several orders of magnitude. Resonant ionisation wavelengths in this region are required for the detection of several elements (Travis et al. 1984) and hence the sensitivity is unlikely to be as high as parts in $10^{18}$. The background can of course be reduced when mass spectrometers are used to detect the positive ions. However especially when saturation spectroscopy conditions exist, a fragmentation spectrum in molecules (Fig. 6) is likely to be a major influence on the ultimate sensitivity of RIS (Towrie et al. 1986b)

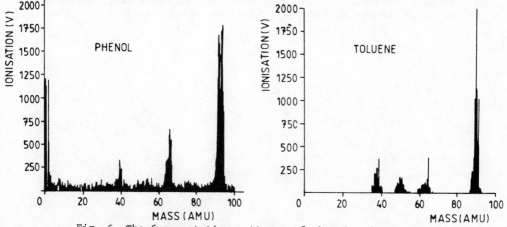

*Fig. 6 The fragmentation patterns of phenol and toluene*

The data shown in Fig. 6 were taken using a conventional quadrupole mass spectrometer operated at a pressure of about $10^{-6}$ torr and modified to permit a laser beam to ionise the gas previously ionised by an electron beam. The laser was operated at a wavelength of 266 nm and a fluence of about $3mJ/mm^2$. If the laser fluence is reduced only the peaks corresponding to the total phenol and toluene masses remain.

Detection schemes 2, 4 and 5 (Hurst et al., 1979) are likely to be particularly affected by such 'background' fragmentation patterns especially if real environmental or biological samples, containing quantities of hydrocarbons are to be analysed.

## Acknowledgements

Four of us (SLTD, CR, MT and CMH) wish to thank SERC while JWC and MHCS acknowledge support from the Wolfson Foundation and DENI respectively.

## References

Hilke H J 1986 Proc. of the Vienna Wire Chamber Conference, Feb. 1986 – To be published

Hurst G S, Nayfeh M H, Young J P 1977 Phys. Rev. A 15 2283

Hurst G S, Payne H G, Kramer S D and Young J P 1979 Rev. Mod. Phys. 51 767

Ledingham K W D, Raine C, Smith K M, Smyth M H C, Stewart D T, Towrie M and Houston C M 1985 Nucl. Instrum. Meth. A241 441

Mark H F and Gaylord N G (eds.) 1969 Encyclopedia of Polymer Science and Technology Vol. 10 (New York:Wiley) pp 1–110

Murrell J N 1963 The Theory of the Electronic Spectra of Organic Molecules (London:Methuen)

Towrie M, Cahill J W, Ledingham K W D, Raine C, Smith K M, Smyth M H C, Stewart D T and Houston C M 1986a J. Phys. B At. Mol. Phys.

Towrie M, Cahill J W, Ledingham K W D, Raine C, Smith K M, Smyth M H C, Stewart D T and Houston C M 1986b – to be published

Travis J C, Fassett J D and Moore L J 1984 Resonant Ionization Spectroscopy (Institute of Physics, Bristol, England) pp 97–106

UV Atlas of Organic Compounds 1966 ed. Photoelectric Spectrometry Group London and Institut fur Spektrochemie und Angewandte Spektroskopie, Dortmund (London:Butterworths; Weinheim:Verlag Chemie)

*Inst. Phys. Conf. Ser. No. 84: Section 8*
*Paper presented at RIS 86, Swansea, Wales, 7–12 Sept. 1986*

295

# The on-line resonance ionization mass spectroscopy of short-lived isotopes

G. Bollen[1], A. Dohn[1], H.-J. Kluge[1, 2], U. Krönert[1] and K. Wallmeroth[1]

[1]Institut für Physik, Universität Mainz, Postfach 3980, D-6500 Mainz, Fed. Rep. Germany
[2]CERN, ISOLDE, CH-1211 Geneva 23, Switzerland

**Abstract.** On-line resonance ionization mass spectroscopy (RIMS) is discussed in order to study short-lived isotopes produced at on-line isotope separators. We report on a first application of RIMS to a determination of the isotope shift and hyperfine structure of $^{185-189}$Au and the I = 11/2 isomer of $^{189}$Au in the 6s $^2S_{1/2}$ → 6p $^2P_{1/2}$ ($\lambda$ = 268 nm) transition. The Au atoms were obtained as daughters of mass-separated Hg isotopes produced at the ISOLDE facility at CERN.

## 1. Introduction

The use of tunable lasers for nuclear physics research is growing rapidly. The most important application of lasers in this field is the study of hyperfine structure (HFS) and isotope shift (IS) in optical transitions. Such experiments yield information on nuclear spins, moments, and changes of nuclear charge radii, which are key input parameters for nuclear-model calculations. The use of lasers is especially efficient when the investigations are performed at on-line isotope separators where long chains of isotopes are available with high purity and intensity. Such a facility is the on-line isotope separator ISOLDE, which uses the 600 MeV proton synchro-cyclotron at CERN, Geneva, as a dedicated injector. Just 10 years ago, the first on-line laser experiments were performed by Kühl et al (1977). These experiments yielded a drastic nuclear-shape staggering in very-neutron-deficient Hg isotopes, and demonstrated for the first time the outstanding power that results from the combination of lasers with on-line isotope separation.

Now, 10 years later, some 300 short-lived isotopes have been investigated, and the results obtained have produced quite an important impact on nuclear structure research. A comprehensive overview can be found in Hyperfine Interactions, Vol. 24 (1985). In most experiments, resonance fluorescence spectroscopy was applied to atoms confined in resonance cells, or to thermal atomic beams or fast atomic or ion beams. Recently, Alkhazov et al (1983) reported the first application of resonance ionization spectroscopy (RIS) to a study of $^{141-150}$Eu. In this contribution, which comprises the talks given by K. Wallmeroth and H.-J. Kluge, we report on the first on-line application of RIMS.

## 2.  The Status of Laser Spectroscopy at On-Line Isotope Separators

Laser spectroscopy programmes are being carried out at the on-line isotope separators at Daresbury (Eastham et al 1986), at GSI/Darmstadt (Ulm et al 1985), at ISOLDE/CERN (Kluge 1985a 1986), at Leningrad (Alkhazov et al 1983), TRISTAN/Brookhaven (Schuessler 1985), and at UNISOR/Oak Ridge (Bounds et al 1985). Most of the work has been done at ISOLDE because of the large variety and high intensity of the available radioactive beams. Hence, the discussion will be restricted to the actual ISOLDE programme. Table 1 lists the on-going programme and the experiments that are in preparation.

Today, collinear laser spectroscopy (Neugart 1985) is the most general, sensitive, high-resolution method for optical spectroscopy of radioactive beams delivered by on-line isotope separators. This can easily be seen from Table 1, which lists nine different collinear laser experiments. Still, this method needs a minimum ion-beam intensity of about $10^4$–$10^6$ atoms per second, depending on the strength and multiplicity of the optical transitions. This is due to the low detection efficiency of the photons from the resonance fluorescence process and to the background detected by the photomultiplier. A big improvement can be expected if, instead of the photons, high-energy particles are detected. This has been done or is being planned for the experiments 4, 5, 7–9 listed in Table 1, where the nuclear radiation or the ions after resonant ionization are used to determine the optical resonance. It is obvious from this table that, in the near future, RIS or RIMS will play an increasingly important role in laser spectroscopy of short-lived nuclei. Furthermore, the sensitivity of conventional collinear spectroscopy with photon detection might be enhanced. This can be achieved through an increase of the signal-to-background ratio with the help of a pulsed ion source. Such a device (Kluge et al 1985b) has recently been realized by Andreev et al (1986) for the case of Sr. A pulsed ion source using resonance ionization will have important applications also in other research areas at isotope separators, because it allows a very efficient reduction of isobars. The isobaric contamination is quite often a major problem when nuclei very far away from the valley of $\beta$ stability are studied. The RIS technique may also increase the sensitivity when spectroscopy of thermal atomic beams is performed, as will be discussed below.

## 3.  The Motivation to Study Short-Lived Au Isotopes

The nuclei near $Z = 80$ and $N = 104$ have attracted considerable attention following the discovery by Bonn et al (1972, 1976) that there is a big change in the ground-state mean-square charge radius between $^{187}$Hg and $^{185}$Hg, which was interpreted as the onset of strong deformation at $A \leq 185$. This observation, obtained from a measurement of the optical IS, was completely unexpected, since at that time the unanimous view was that nuclei near closed shells, such as the isotopes of Hg with $Z = 80$, are only weakly deformed. This view and our knowledge of the region around $^{185}$Hg have since changed dramatically. It is becoming evident that coexisting nuclear shapes occur widely throughout the region below $Z = 80$.

A measurement of the isotope and/or isomer shift represents the most straightforward and model-independent signature for shape transitions or shape coexistence. In recent experiments, the ISs in the isotopic chains of Tl (Bounds et al 1985) and Au (Streib et al 1985) have been determined. No indication of a shape transition or coexistence was observed because the experiments were restricted to the study of heavier isotopes ($N \geq 108$), whereas the shape effects are expected to occur in nuclei with an approximately half-filled neutron shell.

**Table 1**

Present on-line techniques for optical spectroscopy at ISOLDE/CERN

| Basic technique | Method of sensitivity increase | Application | Status | Ref. [a] |
|---|---|---|---|---|
| COLLINEAR SPECTROSCOPY IN FAST ATOMIC AND ION BEAMS | | | | |
| 1. Observation of optical fluorescence | — | Rare earth, Ra, Rn, In, S, Fr $D_2$ line of RaII | Data-taking | [M203, P77] |
| 2. Observation of optical fluorescence with a frequency-doubled CW dye laser | — | | Data-taking | [IP-16] |
| 3. Observation of precession of optical polarization in a magnetic field | — | $g_I$ of Ra isotopes | Data-taking | [P76] |
| 4. Reionization by charge exchange after laser interaction | Ion detection after charge exchange | Rn, Xe | Tested | [M203] |
| 5. Resonant three-photon ionization with pulsed Cu vapour pump laser | Photoionization | Rare earth | In preparation | [P78] |
| 6. Observation of fluorescence with pulsed ion beam | $10^{-4}$ duty cycle of laser ion source | Rare earth | In preparation | (Kluge et al 1985b) |
| 7. Two-step photoionization with a CW laser via Rydberg states | Ion detection | In | Tested | [P77] |
| 8. One-step photoionization with a CW laser via Rydberg states | Ion detection | Fr, Cs | Tested | [IP-17] |
| 9. Implantation of optically pumped beams in a crystal, observation of $\beta$ asymmetry | Asymmetry of $\beta$ radiation | Li | Data-taking | [IP-15] |
| OPTICAL PUMPING IN RESONANCE CELLS | | | | |
| 10. Polarization by spin exchange of optically pumped Rb, observation of $\gamma$ anisotropy | Anisotropy of $\gamma$ radiation | Rn | Data-taking | [P79] |
| RESONANCE IONIZATION SPECTROSCOPY IN THERMAL ATOMIC BEAMS | | | | |
| 11. Resonant three-photon ionization with Nd:yag, mass measurement of photoions by TOF | Ion detection, TOF | Au, Pt | Data-taking | [IP-14] |
| 12. As above with pulsed thermal beam using laser desorption | Ion detection, TOF, duty cycle | Au | In preparation | [IP-14] |

a) References in square parentheses give the CERN filing number of the proposal. Copies can be obtained from the ISOLDE Secretariat, EP Division, CERN, CH–1211 Geneva 23.

## 4.  Experimental Technique and Performance

The present experiment was designed to investigate shorter- lived Au isotopes and to determine their ISs. As a by-product the magnetic hyperfine splitting is obtained. The Au isotopes were produced at the ISOLDE mass separator at CERN. Since no target and ion-source system exists for a direct production of Au isotopes, these nuclei have at present to be obtained as daughters of Hg isotopes, which can be produced with yields up to $10^{10}$ atoms per second and mass number by a Pb(p,3pxn)Hg spallation reaction (Kluge 1986). The RIS technique was used and combined with time-of-flight (TOF) mass spectroscopy. Because of the short half-lives and the small production rates of the Au isotopes far from stability, the investigations had to be performed on line with the ISOLDE facility. This contribution reports the first application of RIMS to an on-line study of short-lived isotopes.

Figure 1 shows the experimental set-up. The ISOLDE beam is focused into an atomic beam oven. After a suitable time has elapsed during which the Hg nuclei decay to the daughter isotope, a thermal atomic beam is formed by evaporating the radioactive sample. The Au atoms are excited and photoionized by a three-step resonant process: by the light from two tunable dye lasers pumped by a Nd:yag laser, and by the frequency-doubled output of the same Nd:yag laser (repetition rate = 10 Hz). The optical transitions used for this experiment are indicated in Table 2.

In order to increase the signal-to-background ratio, the photoions were detected and mass-selected by a TOF spectrometer. This consists of a two-stage acceleration region, a field-free drift tube, a 40° ion reflector, and a microchannel plate detector (MCP). With the exception of the reflector the spectrometer is of the same type as the one described in detail by Krönert et al (1985). The ion reflector is installed to allow an efficient shielding of the MCP against nuclear $\beta$ and $\gamma$ radiation

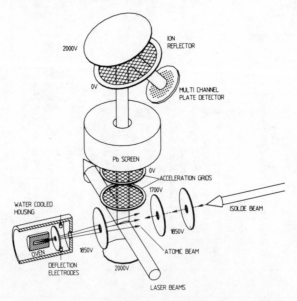

Fig. 1  Set-up for on-line resonance ionization mass spectroscopy (RIMS) on short-lived Au isotopes

**Table 2**

Optical transitions used for resonance ionization spectroscopy on short-lived Au isotopes,
and data on the different lasers and laser beams (SH = second harmonic)

| Excitation step | Optical transition | Wave-length (nm) | Light source | Dye width | Pulse energy ($\mu$J) | Pulse duration (ns) | Band-width (GHz) |
|---|---|---|---|---|---|---|---|
| | – | 535 | Molectron D16P[a)] | C485 | $10^3$ | 6 | $\geq 0.2$ |
| # 1 | $6s\,^2S_{1/2} \rightarrow 6p\,^2P_{1/2}$ | 268 | SH of 535 nm | – | 25 | 4 | $\geq 0.4$ |
| # 2 | $6p\,^2P_{1/2} \rightarrow 6d^2D_{3/2}$ | 406 | Lambda FL2001 | DPS | 150 | 6 | 6 |
| # 3 | $6d\,^2D_{3/2} \rightarrow$ continuum | 532 | SH of Nd:yag | – | $4 \times 10^4$ | 15 | 15 |

a) Modified as described by Bollen et al (1986).

from the oven and acceleration region where strong radioactivity is accumulated during the on-line experiment. Figure 2 shows the TOF spectrum for the measurement of $^{186}$Au. A background rate as low as 1 event per 1000 laser shots was obtained with a mass resolution of the TOF set-up of 250 and a time window of 70 ns for gating on the mass of the isotope under investigation.

The overall detection efficiency $\epsilon$ of our RIMS technique for Au was determined to $\epsilon = 10^{-8}$ (defined as the ratio of Au ions detected by the MCP to the number of atoms collected in the oven). The efficiency is mainly limited by the low repetition rate of the laser system and the small cross-section of the third excitation step, which cannot be saturated even with laser powers up to 5 MW/cm$^2$.

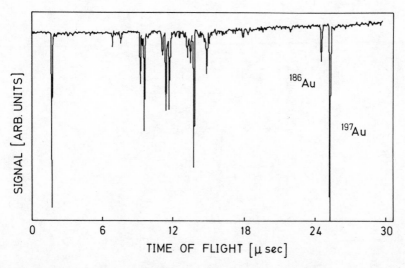

Fig. 2   Time-of-flight spectrum. The $^{186}$Au was collected in the atomic beam oven. Stable $^{197}$Au was added to serve as a mass marker. The medium-mass ions are due to photoionization of the rest gas and can be suppressed by an appropriate time window.

## 5.  Results

Measurements were performed on Au isotopes in the mass range $190 \geq A \geq 185$. The shortest half-life of the isotopes investigated was $T_{1/2} = 4.2$ min ($^{185}$Au), where $\sim 10^9$ atoms could be collected in the oven. Figure 3 shows the signals obtained for the measurement of $^{187}$Au. Here, the ion counting technique was applied. Similar spectra were obtained for the low-intensity cases $^{185}$Au and $^{186}$Au and for the isomeric state of $^{189}$Au. For the more abundant isotopes $^{188-190}$Au, the signals were digitized by an analog-to-digital converter and stored in the memory of the computer. A much better signal-to-noise ratio was obtained in these cases.

Nuclear magnetic moments and changes of mean-square charge radii ($\delta\langle r^2\rangle$) were deduced from the HFS splitting and the IS in the $D_1$ line. The $\delta\langle r^2\rangle$ values are plotted in Fig. 4 relative to $^{197}$Au.

In addition, the $\delta\langle r^2\rangle$ values of Hg are shown (Ulm et al 1986). The slope of the mean-square charge radii is very monotone and is similar to the slopes of Hg down to $^{187}$Au. Then, between $^{187}$Au and $^{186}$Au, a drastic change occurs, similar to the one in the Hg isotopic chain but i) two neutron numbers earlier, and ii) with a much less pronounced odd-even staggering than for the Hg

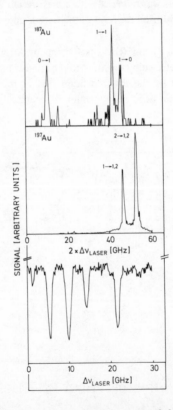

Fig. 3   Photoion yield as a function of the laser frequency of the first optical transition ($D_1$) for $^{187}$Au (top), and the simultaneously measured signals of the resonance fluorescence of stable $^{197}$Au in an atomic beam (middle) and of the absorption spectrum of iodine in a cell (bottom).

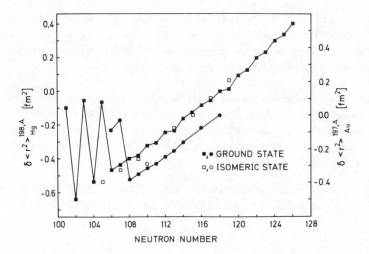

Fig. 4 Changes of the mean-square charge radii of Au (dots) and of Hg (squares; Ulm et al 1986) isotopes. $^{198}$Hg and $^{197}$Au are used as reference isotopes.

isotopes. As in the case of Hg, the break in the IS can be related to a change in the nuclear quadrupole deformation $\beta_2$. From the IS values and with $|\beta_2(^{197}\mathrm{Au})| = 0.11$ (Streib et al 1985), a smoothly increasing $|\beta_2| = 0.14$ to $0.16$ can be calculated for the isotopes $^{190}$Au to $^{187}$Au; $|\beta_2| = 0.25$ is obtained in the case of $^{186}$Au, and $0.24$ for $^{185}$Au. These results confirm earlier suggestions by Ekström et al (1980) and Porquet et al (1983) that the ground states of these nuclei might be strongly deformed. Clear evidence for a shape transition from weak (A $\geq$ 188) to strong (A $<$ 187) deformation in the neutron-deficient Au isotopes has now been observed by optical spectroscopy, providing a sensitive test of nuclear model calculations.

## 6. Future Developments

The present experiment made it possible to study just those very neutron-deficient Au isotopes where the nuclear shape changes. It would be interesting to extend the investigations to still shorter-lived isotopes further away from the valley of stability, and to study more isomers which have nuclear spins as large as I = 12. Such experiments would allow determination of the type of coupling of the particles or holes to the Pt or Hg cores. However, each mass number further away from stability corresponds to a decrease of available Au atoms by about one order of magnitude. Furthermore, the isomers might only be populated quite weakly in the decay of Hg to the Au daughter. Hence a substantial increase of the sensitivity of our RIMS technique has to be achieved. As briefly discussed in Section 4, the small detection efficiency is mainly due to i) the low repetition rate of the laser systems used (10 Hz), and ii) the small cross-section of the third excitation step.

### 6.1 Increase of duty cycle

Since the atoms traverse the region of interaction with the laser beams (D = 1 cm) within $10^{-4}$ s, the duty cycle is of the order of only $10^{-3}$. This factor can be regained by using a pulsed laser with a repetition rate of 10 kHz. The only laser with such a high rate is the Cu vapour laser as used by Rimke et al (1986) for the trace analysis of actinides. In our case, however, the first transition ($\gamma_1$ = 268 nm) is in the ultraviolet spectral region, which cannot be reached by frequency-doubling the output of a dye laser pumped by a Cu vapour laser.

An alternative way of increasing the duty cycle is to produce a pulsed atomic beam. Figure 5 shows the experimental set-up used for preliminary studies. About $10^{17}$ Au atoms were adsorbed on a graphite foil, and this target was irradiated by the 1064 nm output of a second Nd:yag laser. The resulting pulsed Au beam was probed by the laser beams (Table 2). The ions obtained in resonance were recorded as described above.

Changing the delay between the lasers used for desorption and for RIS, a signal is obtained as shown in Fig. 6. The solid line represents a fit to the experimental data by a Maxwell distribution

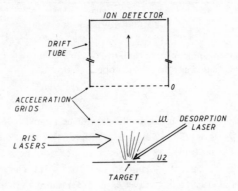

Fig. 5   Experimental set-up for RIMS on a pulsed atomic beam using laser desorption. The laser beams, as listed in Table 2, are used for RIS of a pulsed thermal Au atomic beam which is obtained by pulsed desorption with the help of a second Nd:yag laser.

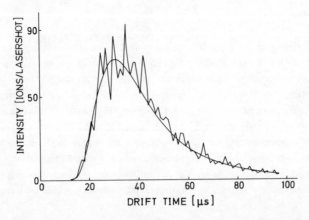

Fig. 6   Photoion yield as a function of the delay between the laser pulse used for desorption ($\lambda$ = 1.06 $\mu$m) and the pulses for RIS of Au. A graphite foil covered with $10^{17}$ Au atoms was used. Each data point represents the ion yield obtained for 30 laser shots. The solid curve represents a fit by a Maxwell distribution (T = 1600 K).

corresponding to a temperature of 1600 K. A gain in efficiency by a factor of 30 was observed, compared with that determined for continuous evaporation of the Au atoms. This is much lower than expected. Additional studies are now being performed with different target materials and experimental conditions. Furthermore, the efficiency has to be determined for ions implanted at an energy of 60 keV in the target material, and compared with that for surface-adsorbed Au atoms.

## 6.2 Use of autoionizing states

The cross-section is not known for the ($\lambda$ = 532 nm) transition from the 6d $^2D_{3/2}$ state into the continuum. A linear dependence of the ion yield on the laser intensity was observed up to the maximum available power of 5 MW. The ionization probability might be enhanced by using a transition from the 6d state to an autoionizing or a Rydberg state. A search for autoionizing states was performed by Lindenlauf (1986) in the wavelength ranges 430–439 nm, 460–510 nm, and 615–657 nm. Rather strong resonances were found only in the last wavelength region shown in Fig. 7. The resonance at 628.70 nm was known (Janitti et al 1979), whereas the resonance at $\lambda$ = 641.25 nm was found in this present work. Even the strongest transition to an autoionizing state found so far will not increase the detection efficiency. Only a small factor of 2–3 is gained, compared with the non-resonant transition (Fig. 7). Taking the conversion efficiency of the dye laser into account, the best detection efficiency is still obtained by the frequency-doubled output of the Nd:yag laser. A search for possibly stronger transitions to autoionizing or Rydberg states will soon be performed.

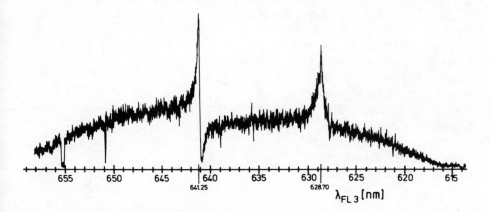

Fig. 7 Autoionizing resonances in the transition from the 6d $^2D_{3/2}$ state of Au. A dye laser was used to induce the third excitation and was scanned in the wavelength region 615–657 nm.

## 7. Conclusion

In the future, resonance ionization spectroscopy will play a major role in the determination of nuclear properties of unstable nuclei. We have applied resonance ionization mass spectrometry to determine the isotope shift and the hyperfine structure of short- lived Au isotopes. A nuclear shape transition was found which takes place between $^{187}$Au and $^{186}$Au. The detection efficiency of $10^{-8}$ might be enhanced by using pulsed laser desorption and transitions to autoionizing or Rydberg states.

This work has been supported by the Bundesministerium für Forschung und Technologie.

## BIBLIOGRAPHY

Alkhazov et al 1983 JETP Lett. **37** 274, and 1985 Opt. Commun. **52** 24

Andreev S V et al 1986 Opt. Commun. **57** 5

Bollen G et al 1986 J. Opt. Soc. Am. in press

Bonn J et al 1972 Phys. Lett. **38B** 302

Bonn J et al 1976 Z. Phys. **A276** 203

Bounds J A et al 1985 Phys. Rev. Lett. **55** 2269

Eastham D A et al 1986 Daresbury preprint EAST–86/263

Ekström C et al 1980 Nucl. Phys. **A348** 25

Janitti E et al 1979 Phys. Scripta **20** 156

Kluge H-J 1985a Hyp. Int. **24** 331

Kluge H-J et al. 1985b Proc. Workshop on Accelerated Radioactive Beams, Parksville, BC, 1985, eds. L Buchmann and JM D'Auria (TRIUMF report TRI–85–1, Vancouver), p. 119

Kluge H-J 1986 ISOLDE Users' Guide, CERN 86–05

Krönert U et al 1985 Appl. Phys. **B38** 65

Kühl T et al 1977 Phys. Rev. Lett. **39** 180

Lindenlauf F 1986 Diploma work, Mainz, unpublished

Neugart R 1985 Hyp. Int. **24** 159

Porquet M G et al 1983 Nucl. Phys. **A411** 65

Rimke H et al 1986, contribution to this conference

Schuessler H A 1985, private communication

Streib J et al 1985 Z. Phys. **A321** 537

Ulm G et al 1985 Z. Phys. **A321** 395

Ulm G et al 1986 Z. Phys. **A325** in press

*Inst. Phys. Conf. Ser. No. 84*
*Paper presented at RIS 86, Swansea, Wales, 7–12 Sept. 1986*                                                    305

# Reflection on the history of ionization phenomena†

Professor Frank Llewellyn-Jones
Department of Physics, University College of Swansea

Ladies and Gentlemen,

In responding to your invitation to give my reflections on early studies of ionization phenomena, it is best to do, as did Dylan Thomas of this city, that is "to begin at the beginning"! In this context, the effective beginning surely was the decade following with the discovery of Rontgen rays in 1895 and including the basic studies first at the Cavendish Laboratory, Cambridge, under J.J. Thomson, and then at Oxford under J.S.E. Townsend. My own understanding of this period has come from various sources including original papers from the Proceedings of the Royal Society, the Cambridge Philosophical Society, the Philosophical Magazine, text books of the period, research and lectures given by visiting scientists including Einstein, Bohr and de Broglie, as well as from many remarks and discussions with Townsend during the decade after 1925. The concluding years of the last century saw an explosive advance in establishing the atomic nature of electricity in gases, properties of the motions of atomic particles and their transport coefficients, and, later, the important theory of ionization by collision. In this period, too, the fundamental principles of the spatio-temporal growth of ionization leading to the electrical breakdown of gases and of discharges in general, were established.

I have been given to understand that when J.J. Thomson was Cavendish Professor of Physics at Cambridge, the first two young graduates arriving to work with him in 1895 were Ernest Rutherford from New Zealand and John Townsend from Trinity College, Dublin. Others who were to work there in the same field included F.W. Aston, Paul Langevin of Paris, O.W. Richardson, C.T.R. Wilson, H.A. Wilson and J. Zeleny, later of Yale; while in Germany, Wilhelm Wien, Walter Kaufmann and E. Wiechert were prominent in the same field.

It is not easy for a modern scientist to appreciate the intellectual atmosphere of that period, so circumscribed by the inadequacy of the materials and techniques then available to meet the requirements of experimental approach. The development of the researches in this field of ionization, as presumably in other areas, falls into two approximately equal periods - namely, those prior to and since World War II; and it is about the former on which you have asked me to comment.

Collisional processes involving atoms, electrons, ions and photons involve a range of time intervals throughout a factor of about $10^8$, from

†This paper was presented as the Conference Dinner Talk

the long diffusion times of metastable molecules in gases down to the short inter-electrode photon transit times.   Today, as a result of advances in electronics originally developed from War-time radar operations and later solid-state research, we have available today circuits capable of producing high field delta-function pulses, oscilloscopes capable of recording intervals of less than nano-seconds, laser sources capable of emitting short light pulses of extremely high energy, as well as fast computers for complicated analyses, as an audience such as this is fully aware.  Before World War II, no such techniques were available, and a particular deficiency was the absence of time-resolution techniques, so that most research had to be undertaken in quasi steady-state conditions, although, later, time intervals controlled by megahertz oscillators were utilised.    In the early work with steady-state cases, electrode potentials were maintained by banks of secondary cells, but this apparent limitation proved later to be of crucial importance and an advantage.    However, many of those considerable diagnostic difficulties were overcome by brilliant design of experiment coupled with outstanding practical skill, and it seems that, if for this reason alone, much can be learned even today from study of these early achievements.

In the last decades of the 19th century views on the fundamental nature of electricity were brought to a head by experimental studies of the discharge of electricity through rarified gases, and particularly by the work of W. Hittorf in Germany and of William Crookes in England. They concluded that their "cathode rays" consisted of minute negatively charged particles, but it was a view so contested by strong supporters of Maxwell's electro-magnetic field theory that Schuster later commented that "the view that a current of electricity was only a flow of aether appealed generally to the scientific world and was held almost universally".    However, experiments on the deflection of cathode rays by electric or magnetic fields, especially those by E. Wiechert, Walter Kaufmann and J.J. Thomson, established the particle nature of the rays and produced reliable values for the ratio e/m of the particle charge to its mass; but their individual values were still unkown.

Now, Faraday's work on electrolysis with the concept of charged ions in salt solutions supported the idea of elementary ionic charges. Then W.Nernst's analysis in 1889 of electrolytic phenomena, using methods not strictly applicable for gases, led to an expression for the product Ne, of Avogadro's number N to the mono-valent ionic charge e, in terms of the ratio of the diffusion coefficient of the ions to their mobility; but neither N nor e were accurately known separately. An analogous concept for the properties of electricity in gases was not generally accepted; and, as was previously the case with cathode rays, strong supporters of Maxwell's theory of the electro-magnetic field formed the view that concentrations of charges like ions were really of electro-magnetic origin to which "gas statistics" did not apply, and the concept of a material particle was unnecessary.

Nevertheless, Thomson and Rutherford set about examining the consequences of the electrification (i.e. ionization) of gases by Rontgen rays, by the emanations from radio-active substances, or by the light from spark discharges.    They investigated how the resulting electric current depended upon the applied field intensity, or, when not swept away by a field, the electrification decayed spontaneously (due to recombination). These important processes of mobility and recombination to form neutral particles were more carefully analysed, and Rutherford's experiments in 1897 were probably the first specifically made to measure their coefficients.   These proved to be the forerunner of numerous experiments designed to measure ion mobilities with increasing accuracy, because of their significance in collision theory, right up to modern times.

In the absence of means of recording very short times, Rutherford employed long path distances using a pendulum to record corresponding time intervals; while Zeleny used a quasi steady-state technique in which ions were drawn longitudinally down a tube by a streaming ambient gas as they were also driven radially to the walls by a weak field.   Later, Langevin devised a method in which field-reversal was obtained by operating successive switches by fast falling weights.   Experimental data were normally analysed on the basis of Langevin's classical theoretical treatment; the wave-mechanical calculations were first carried out in 1934 by H.S.W. Massey and C.B.O. Mohr.

Now Townsend held the same view of the material nature of ions in gases as did Thomson and Rutherford; but he also considered that the excess charge on a gaseous molecule (which thus constituted an ion) in fact did little to interfere with the ordinary inter-molecular collisional process, other than to produce a uni-directional mechanical force which deviated each ionic free path, thus causing a general drift along the direction of the external electric field.   In effect, the cloud of charged ions could be regarded as another gas mixed with the ambient gas of normal molecules and taking part in all their usual collisional activity.   In other words, this "ion gas" would be subject to the Maxwell-Boltzmann statistics applicable to ideal gas molecular collisions, and so exhibit the macroscopic transport phenomena of diffusion, drift, and, when ions of opposite signs are present, of re-combination as well.   The adoption of this view was a step forward of fundamental importance, as we shall see.

In a Bakerian Lecture Rutherford had discussed the identity of gaseous ions, and he pointed out that their material existence could be established decisively if their individual electric charge and their coefficient of diffusion in gases were measured, and added that this would be difficult to do.   It was not long, however, before both these quantities were determined experimentally by Townsend, who took the view that the ions contained in a gas could be regarded as themselves forming a gas which obeyed Maxwell-Boltzmann statistics.

First, Townsend used the fact that gases liberated by electrolysis carried electric charges and possessed the property of forming a cloud in passing through aqueous vapours without any cooling or expansion; and also that the weight of the cloud was proportional to the charge in the gas. He then measured the rate of fall of the cloud, its weight and total charge, and, using Stokes' law, deduced the individual drop size and so their total number, finally giving the value of $5 \times 10^{-1}$ e.s.u. for the ionic charge.   This brilliant work in 1897 and 1898, contained, in Millikan's view, essential elements of some subsequent improved determinations by J.J. Thomson, H.A. Wilson and of Millikan himself.

In order to tackle the problem of diffusion experimentally, Townsend passed ionised gas through a metal tube which then acquired some of the ionic charge from the radial diffusion of ions to the tube walls.   By treating the process theoretically on the basis of kinetic theory of gaseous diffusion, the diffusion coefficient could be found in terms of the measured radial loss of charge of the ion stream.   Further, in a general analysis of ion motion in an electric field using Maxwell's equations of transport, Townsend deduced that the ratio of the diffusion coefficient to the ionic mobility was proportional to the product Ne where e is the ionic charge.  This was of the same form as the equation which Nernst had deduced for electrolytes, the form being known later, for no discernible reason, as the Einstein equation.   Then, using Rutherford's value for the mobility with his own definition of it for diffusion coefficient, Townsend in 1899 obtained a value of Ne for ions which was sufficiently close to the value accepted for monovalent electrolytic ions as to establish the equality of electric charge on the two types of ions. Consequently this was taken to be the elementary atom of electricity, which was also accepted as the charge on the negative particles – the electrons of cathode rays. Another important result followed immediately, since the electronic mass could be calculated from the measured bending of cathode rays in a magnetic field.

Important experimental diagnostic methods arose from the studies of cathode and the corresponding anode rays; these were the cathode-ray oscilloscope and the mass spectrometer of Aston.   Also, the observations on gases evolved in electrolysis showed that clouds could be formed by condensation in charge centres, and this led to the development by C.T.R. Wilson of the cloud chamber.

We come now to the next important step in the elucidation of ionization phenomena – that of the generation of electrons and ions by collisional processes in gases.   Thomson and Rutherford and others had studied the increase of ionization produced in a gas in an electric field between parallel plates when irradiated from external sources.   They had observed that these increases depended on gas pressure, electric field and even linear dimensions of the apparatus.   A current view was that these confusing phenomena were produced by the existence on electrodes of the so-called "double-layers" of electricity, a view which had been suggested in the 18th century in connection with electro-static machines; an applied electric field was thought to be able to "peal off" layers and so liberate new charges.   Indeed, it was also considered that the charges thought to exist in all molecules were held there partly by the bombardment of adjacent molecules.   Thus, if the bombardment rate is reduced, as at reduced pressure, then charges could become detached or "leak" off; the lower the pressure, the greater was the leak!

Concerning this problem, again treating gaseous ions as a small gas additive to the ambient gas, Townsend directed attention to the action of the electric field on the gaseous ions themselves.   He realized that in the free flight of an ion between collisions with gas molecules, the ion must gain energy from the field, while the average loss of energy in ideal gas collisions with molecules remained largely unaffected.  Energy gained in this way could eventually attain values sufficient to produce damage to a gas molecule in collision, causing the break off of an electron to produce a new free positive ion and with an additional electron. In the same way, these two new particles would attain more energy and themselves repeat this disruptive process, and general amplification continue. Townsend soon realised that the most effective ionizing agent must be the electron and not the ion so his assumptions were valid. Thus the vital electric field intensity E and the mean free path $\ell$, which in any given gas at constant temperature is inversely proportional to the gas pressure p, to give E/p as the essential energy parameter.

Consequently, in his own experiments on ionization growth, he maintained E/p constant while he varied the distance travelled by an electron cloud, thus obtaining the well-known curves of exponential growth.  From such curves, Townsend and collaborators, working at Oxford since 1900, disclosed additional processes of ionization resulting from the primary electron process, and he specified coefficients, corresponding to these collisional processes involving the positive ions and photons as well as the nature of the electrode surfaces.  In this way the important concept of an electron avalanche was evolved.

Further, his theoretical deduction of an expression for the primary coefficient $\alpha$ for ionization of neutral molecules in electron collisions involved specification of a critical atomic potential, different for different gases, and the magnitudes of these ionization potentials were obtained by comparing theoretical expressions with measured values of $\alpha$. This appears to be the first example of a critical potential being attributed to atoms and molecules, many years before the introduction of the Bohr atomic theory of which a critical energy level was an essential feature.

Another result obtained from this theory of the $\alpha$-coefficient was that the atomic cross-section operative in an ionizing collision could be considerably smaller than that effective in a gas-kinetic elastic collision; but the full significance of this concept was not generally appreciated at the time.  In Gottingen, J. Franck and G. Hertz, using a beam of electrons from a hot filament in a gas at low pressure, found that large energy losses attributed to ionization of atoms appeared to occur at or very near a critical ionizing potential.   Some controversy naturally followed until it was realised that mono-energetic electron beams, especially from heated cathodes, were not easy to obtain and the nature and magnitude of gas scattering were not then fully appreciated, so that the required accuracy was not so readily obtained as was first thought.

The discovery of ionization by collision proved the death-knell of any "double-layer" theory of electrification, because increase of ionization was obtainable merely by increasing inter-electrode distance when the electric field at electrode surface was kept constant.

The theory of ionization by collision and the statistics of collisional phenomena lie at the basis of our understanding of the electrical discharge through gases, and of course, of plasma physics. The theory of growth by primary and secondary ionizing processes at once led to an explanation of the measured values of the sharply defined static breakdown, or sparking, potentials of gases, and of Paschen's law and the general Similarity Principle in discharges. These subjects are today of great practical importance especially in the field of high-voltage power technology as well as in space science. The modern form of the energy parameter E/p is E/n, where n is the molecular gas concentration and its unit is called the Townsend. Extensive data on ionization coefficients for most gases are used today with numerical methods and computers to produce simulation of extremely rapid ionization growth, such as that which occurs with the sudden voltage collapse of a spark gap. These calculations cannot be done with formal analytical methods owing to the rapid spatio-temporal field variation due to space charge at high current densities.     The early work on low pressure discharges between co-axial cylinders elucidated the action of Geiger counters, and ionization theory also produced the spark chamber as another diagnostic tool in nuclear physics.

In all this work Townsend realised the need for accurate data on collisional processes involving electrons, ions, photons and electrode surfaces; in particular, he was concerned with mobilities, free paths (i.e. cross-sections), losses of energy in collisions, and mean energies of agitation in electric fields.     Consequently, he developed at Oxford a successful technique, which in refined and more sophisticated forms is in use today throughout the world and known as the "swarm" method.

In essence this procedure is a development of his original method of measuring the product Ne for ions based on the Nernst-Townsend equation. A stream of electrons (or ions) in equilibrium is driven through a gas by an electric field, and its lateral diffusion measured; with electrons a transverse magnetic field defects the beam to appropriate electrodes. The theory of the ionic or electronic motion was based on a solution of Maxwell's transport equation from which the required atomic data are derived as functions of the parameter E/p.     The method proved extraordinarily comprehensive and accurate; and particularly helpful data were obtained for the average fractional energy loss of electrons in atomic collisions, and their mean energies of agitation as a function of E/p, and many of his determinations are used today. In 1924, Townsend was invited to give an account of this work at the Centenary celebrations of the Franklin Institute in Philadelphia, and the paper was published by the Clarendon Press.

An important result, which followed from determination of the mean electron energy of a swarm as a function of E/p, was that in a helium glow discharge, for instance, where atomic ionization and excitation requiring energies exceeding about 20 electron volts clearly took place, the mean electron energy could be as low as about 4 electron volts. Unfortunately, this conclusion was the source of some controversy at the time. Nevertheless, the result drew Townsend's attention to the energy distribution function, which was then investigated by F.B. Pidduck and later by M.J. Druyvesteyn, Townsend and many others. Today, it is appreciated that this function is still of great importance in analyses of collisional processes and of plasma physics in general. The modern approach is through solution of the Boltzmann equation using the latest data on cross-sections, as, indeed, the comprehensive discussions at the NATO ASI meeting at Bourges-St. Maurice, France, in 1982 showed.

Another result obtained in 1913 by Townsend and Henry Tizard with this swarm method was that the atomic cross-section in collisions with slow electrons appeared to depend upon the electron velocity; a result quite inexplicable at the time. However, this research was closed down by the outbreak of World War I, but, fortunately, was resumed by Townsend and V.A. Bailey after the war. About the same, time C. Ramsauer and R. Kollath in Germany were investigating electron scattering by using collimated beams of electrons passing through a gas at pressures low enough adequately to reduce the number of collisions; electron energies lay over a wide range above about 30 e.v. Variation of atomic cross-section with electron velocity was also found, but of forms somewhat different from those obtained by swarm methods, which were not then considered to be accurate. However, differences were cleared up when it was realised that the two methods were in fact, complementary, one being more accurate for the higher energies and the other for slow electrons; in fact, the swarm method, when based on a full examination of the electron energy distribution function, is today the only one available for use with very low and even thermal electron energies. The phenomenon of the variation of atomic cross-section in electron collision is accordingly now known as the Townsend-Ramsauer Effect; and its explanation had to await the advent of wave mechanics. One may remark here that the controversies, which arose at the time over comparisons of data obtained by statistical swarm methods with those found by the apparently more direct beam methods, were both unnecessary and unfortunate. On reflection today, it seems that they were based upon a lack of appreciation of the limits of experimental accuracy of the two very different approaches. Criticism of the swarm method, allegedly based on the quantum theory, tended to encourage Townsend to feel somewhat sceptical about that theory as presented to him by some of the critics but the whole matter really illustrated Eddington's comment, "that a doctrine is not to be judged by the follies that have been committed in its name"!

A property of a Bohr atom which, however, had a profound effect on our understanding of ionization phenomena was that of the long-life metastable state. The long-life time was due to the general non-occurrence of spontaneous radiative transitions to return the atom to a lower or ground state; destruction requires the action of another body. Striking experiments of F.M. Penning in Holland showed how high-energy metastable atoms could ionize molecules of a different gas of lower ionization potential, especially when that value was near the metastable energy of the colliding atom. The importance of the effect, known as a collision of the second kind, was due to the fact that during a long life-time of some m.sec. a metastable atom had a high probability of colliding with an impurity molecule even at relative concentrations less than $10^{-5}$. The effect emphasised the necessity of using the highest possible gas purity in, for example, determination of ionization coefficients.

This point, then, is a suitable place briefly to describe the general experimental technique for ionization measurements used by Townsend and S.P. MacCallum at Oxford in the late 1920's and early 1930's.

Ionization currents ~ $10^{-13}$ A, which avoided space charge problems, were measured with a Dolezalek quadrant electrometer in conjunction with the Townsend electrostatic balance which maintained electrodes at constant potential. Internal electrodes were usually mounted on quartz, while external condensers connected to electrodes were mounted on amber or ebonite using split insulation and appropriate guard rings to avoid leakage. Where possible, electrodes themselves were made of pure nickel, but in some cases gold, silver and copper were employed; they were usually pre-heated but finally out-gassed by heating in vacuo by radiant heat from external gas burners or electric heating coils wound around the chamber. Copper and molybdenum gave trouble during such out-gassing owing to sputtering which could reduce insulation. External connection to internal electrodes was usually made by tungsten-hard glass seals or by copper stranded wire sealed by molten lead into long quartz side tubes sealed to a quartz ionization chamber. All electrical connections, condensers, etc. were completely screened by earthed metal covers.

Evacuation was carried out using Langmuir diffusion pumps, with mercury or, later, the then new, low vapour pressure oil, and backed by oil rotary pumps. The diffusion pumps were suitably isolated by liquid-air traps and charcoal traps.

Noble gases required most careful purification, especially helium owing to its high ionization and metastable potentials. This gas was sometimes actually prepared in the laboratory from thorianite, which was first dried and freed from many adsorbed gases by heating in a quartz vessel connected to traps containing phosphorus pentoxide and activated charcoal under evacuation. On raising the temperature to red heat the thorianite released occluded helium, which was continuously circulated over heated copper oxide (to remove hydrogen) stored over cooled charcoal.

A difficult problem was that of isolation of the ionization chamber from the final stop-cork of the gas-system. This was necessary because, before the introduction of the Alpert all-metal bakeable tap, the stop-cock was of vacuum grease-lubricated glass, and so could not be out-gassed. The solution adopted therefore lay in provision of an adequate gettering system. Initial diffusion of grease vapour was prevented first by liquid-air traps and then by a succeeding getter system, consisting of a long narrow quartz tube leading into a pair of quartz liquid-air traps, one of which contained pure calcium. When this was heated while its companion cooled, calcium vapour from the heated trap condensed into the cooler trap, so producing a getter action which adsorbed impurities; this action was also assisted by the cooling of the previously out-gassed long lead-in tube. This calcium transfer process was of course reversible.

Traces of gas impurity possibly evolved by electrodes under local bombardment of ions and electrons during measurements were removed, or certainly reduced, by using the "clean-up" effect of an 20 MHz high frequency electrodeless discharge through the ambient gas in an adjacent wide side tube. Accurate measurement of the sparking potential of the ambient gas proved to be a good indication of the degree of the purity attained. To judge from accounts in contemporary published papers, it seems safe to conclude that the gas purity, attained in the measurement of ionization coefficients by Townsend and MacCallum at that time, was at least as good as that used in similar work elsewhere. I make this comment because some contemporary criticism, even in text-books, of their ionization data was based on supposed grounds of inadequate gas purity!

Research work in ionization physics always seemed to me to provide excellent training for graduate students, and even for work afterwards in other fields, and this leads me to conclude this brief survey by referring to the achievement of Robert J. van de Graaff.

Born in Birmingham, Alabama, he graduated in engineering at the University and obtained a post with the Alabama Power Company. At home he had read some physics in his father's library and became fascinated with early accounts of atomic particles. To learn more of this subject he gave up his post in 1924 and left to spend a year at the Sorbonne in Paris, a famous research centre where the Curie's and others were actually working with atomic particles and radio-active examinations. This experience seemed to decide him henceforth to devote his energies to work in this field in general and on the motions of ions in particular. Consequently, on his return home he applied successfully for a Rhodes Scholarship to Oxford, and started as a graduate student to work there under Townsend in 1925, later obtaining the research degrees of B.Sc. and D. Phil. for work on the mobilities of positive ions in gases.

It seems that, although van de Graaff always had in mind his ultimate object of using ions as controlled atomic projectile, Townsend proposed investigation of more accurate measurements of ionic mobilities for the doctorate project, as this was still an experimental problem with important theoretical implications. The experimental project involved, not only techniques of current measurement, screening and manipulation of electro-static fields, but also studies of sources and of control of ions themselves. His experience in this work was such that van de Graaff in 1928 was able to publish his time-of-flight shutter method for measuring mobilities, just prior to a paper by A.M. Tyndall and L.H. Starr and C.F. Powell of Bristol describing a similar technique which became the basis of their distinguished research on precision mobility measurement for many years.

In 1929, van de Graaff left Oxford to take up a National Research Fellowship in K.T. Compton's Department at Princeton.  By this time, now after years of most valuable experimental work with ions, his concept was now so clearly formulated, that within only five months he had designed, constructed and successfully operated his 80 kV electro-static generator. As is well-known, this proved to be the forerunner of reliable machines which eventually found use throughout the world in hospitals, industry and nuclear research laboratories.  It is interesting to note that practically every essential aspect of this machine had been considered and used at one time or another during two centuries of experiment involving distinguished scientists, but it was van de Graaff who was able to put ideas together and get his design to operate successfully first time.

In conclusion, may I express the hope that this limited survey of the earlier period of ionization research will encourage anyone who has not already done so to read some of the earliest papers and text books written by those who, by outstanding experimental work, laid the basis of the subject in which we are all working today.

*Inst. Phys. Conf. Ser. No. 84: Section 9*
*Paper presented at RIS 86, Swansea, Wales, 7–12 Sept. 1986*

315

# Analysis of solid samples by laser ablation ICP-MS

D J Hall, J S Gordon, J E Cantle
VG Isotopes Limited, Ion Path, Road Three, Winsford, Cheshire, CW7 3BX

ABSTRACT

There are many analytical applications in the materials field where the element selectivity afforded by the RIMS technique is not required. Whilst a number of analytical techniques address this area, few, if any, have the versatility and simplicity of laser ablation into an inductively coupled plasma mass spectrometer (ICP-MS). The practical advantages over other techniques are that the laser can be used on insulating and conducting samples alike, whilst the high sensitivity coupled to the rapid scanning facility of the quadrupole/MCA combination allows multi-element analysis to be carried out in a few minutes on the transient signal from a single laser shot.

The striking feature of laser ablation is its simplicity. A Nd-YAG laser is focussed onto the surface of a sample within a cell. The ablated material is then carried on a stream of argon into the ICP. Using a single 0.5 J pulse in fixed Q mode, trace elements at 10 ppb levels have been detected although accurate quantitation generally requires standards of comparable composition to the sample of interest. The use of this system and its application to ceramics, metals and minerals are presented.

Through the analysis of a series of NBS steel standards (SRM's 661-665) using single fixed-Q laser pulses of 0.3 J (at 1064 nm), calibration curves were plotted for a range of elements at levels varying from 1-1000 ppm. Sensitivity factors derived from these curves were found to vary by almost two orders of magnitude, highlighting the necessity for standards in quantitative analyses.

The analysis of glasses is a problem area which has been addressed by laser ablation - ICP-MS. Provided that the sample has an absorption band in the region of the laser wave length, then direct ablation of the sample is possible, with minimal sample preparation. Often, however, the coupling between laser beam and glassy sample is poor; in these instances, a layer of colloidal graphite can be applied to the surface to help initiate the ablation step. Results have been presented for the analysis of boric oxide, and also for the analysis of diamond. (Although not a glass, the analysis of diamond does suffer from related problems.)

In conclusion, it was found that the laser ablation ICP configuration is ideally suited to qualitative studies or sample screening, with (a) typical analysis time of 60 seconds or less for complete mass range coverage, (b) memory between successive shots at 2 minute intervals less than 0.1% of peak count rate, and (c) ng/g detection limits for limited range scans. With the operating conditions investigated so far the degree of quantitation is strongly dependent upon the quality of standards. Repeatability is dependent upon sample type and operating conditions, with around 5% being typical for traces (greater than 5 ppm) in compacted minerals and ceramics, but metals have given poorer values.

The ability to analyse refractory, ceramic and metal samples directly makes laser-ablation ICP MS a particularly powerful technique.  Whilst still in a development stage, initial results have been encouraging, and further investigation is in progress.

*Inst. Phys. Conf. Ser. No. 84: Section 9*
*Paper presented at RIS 86, Swansea, Wales, 7–12 Sept. 1986*

317

# Multiphoton laser spectroscopy of the ion-pair and Rydberg states of HCl

M.A. Brown, R.J. Donovan, P.R.R. Langridge-Smith and K.P. Lawley
Department of Chemistry, University of Edinburgh,
West Mains Road, Edinburgh EH9 3JJ (U.K.)

## ABSTRACT

The Rydberg and ion-pair states of HCl have been studied using [2+1] multiphoton ionisation techniques. Previous high resolution vacuum ultra-violet absorption studies [1] and recent ab initio CI calculations [2], show that these high lying excited state potentials are complex, being dominated by considerable mixing between the "Valence" (Ion-pair $\sigma* \leftarrow \sigma$) and "Rydberg" states of the same symmetry. As a result the otherwise regular Rydberg pattern is heavily perturbed. Recently Zimmerer et al [3] have reported single photon excitation of HCl using tunable vacuum ultraviolet synchrotron radiation. Our approach has been to use laser multiphoton excitation, with tunable ultraviolet radiation in the region 230-240 nm, produced by optical non-linear frequency doubling and mixing techniques (Quantel YG581+TDL50).

Figure 1 shows a [2+1] multiphoton ionisation spectrum of HCl for two typical vibronic levels near 10.5 eV. One vibronic level being $v' = 11$ of the "Ion-pair" state and the other, the $v' = 0$ level of a "Rydberg"-like state that is strongly vibronically coupled to the "Ion-pair" state. Dispersed fluorescence spectra following excitation at the Q(3) feature of each vibronic band clearly show that the two states are strongly mixed: both spectra are characteristic of fluorescence from the outer ion-pair well to both the vibrational continuum (i.e. bound-free fluorescence) and high vibrational levels of the ground electronic state.

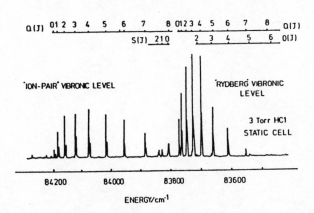

Figure 1

[2+1] Multiphoton ionisation spectrum of HCl (3 Torr HCl).

The appearance of XeCl exciplex ($B^2\Sigma^+ \rightarrow X^2\Sigma^+$) chemiluminescence is seen following excitation of HCl in the presence of Xe (10 Torr). The XeCl exciplex action spectrum (monitored at 308 nm) follows that of the HCl MPI spectrum, showing that electronically excited HCl is responsible for the reactive channel.

1.  A.E. Douglas and F.R. Greening, Can.J.Phys., <u>57</u> (1979), 1650.
2.  M. Betterdorff, S.D. Peyerimhoff and R.J. Buenker, Chem.Phys., <u>66</u> (1982), 261.
3.  G. Zimmerer in "Photophysics and Photochemistry above 6 eV"; Ed. F. Lahmani. Elsevier (Amsterdam), 1985, p. 357.

*Inst. Phys. Conf. Ser. No. 84: Section 9*
*Paper presented at RIS 86, Swansea, Wales, 7–12 Sept. 1986*

319

# Multiphoton ionization of $CF_3I$ and $CH_3Br$

R. D. Kay*, E. Borsella, and L. Larciprete
ENEA, TIB, Divisione Fisica Applicata
P.O. Box 65, 00044
Frascati (Roma), Italy

## Abstract

Multiphoton ionization of $CF_3I$ and $CF_3Br$ by excimer laser photons has been studied. These molecules belong to the 3Cv symmetry group and their electronic energy states are similar to those in $CH_3I$. The MPI fragmentation patterns we observed vary substantially for the different excimer photonic energies employed.

MPI spectra were obtained using a differentially pumped TOF mass spectrometer of 1amu resolution. A Lambda Physics EMG 103 excimer laser produced 15ns FWHM pulses which were focused by a 25cm lens to intensities from $10^7$ to $10^9 W/cm^2$. Ions were detected by a tandem channel-plate detector and the signal transferred to a fast oscilloscope or transient recorder. Data were transferred to a minicomputer and averaged over several tens of laser shots (50 to 100).

Using XeCl (4.02eV) photons on $CF_3I$, only the one mass peak of $I^+$ was found and the production goes as fluence raised to the 2.5 power. This indicates that $I^+$ is produced by a quasiresonant 1 + 2 MPI process, where the first photon is partially resonant with the A band.

With KrF (4.97eV) photons on $CF_3I$, $C^+$ and $CF^+$ are dominantly produced, with $CF_3^+$ and $I^+$ created to a lesser extent. KrF photons are resonant with the A band, and one would expect a large fraction of the $CF_3I$ to be dissociated into $CF_3$ and $I$. Production of $CF_3^+$ and $C^+$ vs fluence yield slopes around 3.5. A fourth order process with the first being the dissociation of $CF_3I$ (nearly saturated) followed by a $3^{rd}$ order MPI of $I$ and fragmentation of $CF_3$ can explain these results, as the appearance potentials of $CF_3^+$ and $I^+$ are known to be about 9.2eV and 10.45eV respectively. Appearance potentials are not known for $C^+$ or $CF^+$ production from $CF^3$. Because the energy to separate C (or $CF^+$) from other $CF_3$ compounds is some 30eV in the case of $C^+$ and some 20 eV in the case of $CF^+$ it is quite difficult to understand the strength of the $CF^+$ and $C^+$ mass lines on the basis of energetics. One suspects that a near resonance is involved which enhances the production of $CF^+$ and $C^+$.

Two ArF photons (6.41eV) can ionize $CF_3I$. A strong $I^+$ fragment is found along with weak peaks for $C^+$ and $CF^+$. Production of $I^+$ vs. fluence shows a slope of 1.7. The $CF_3-I$ dimer appears to be dissociated into $I^+$ and $CF_3^-$ in a non resonant two photon process.

*Department of Physics, The American University, Wash. D.C., 20016, USA

MPI was only seen for $CH_3Br$ using ArF excimer photons.    A typical spectrum revealed    primarily $C^+$,    followed by $H^+$, and also a small quantity of $CH_3^+$.    $CH_3Br$ is much more difficult to fragment    with XeCl    photons than is $CF_3I$.    Both    KrF and ArF single photons are resonant with the A band of $CH_3Br$.   One might expect ArF photons to be more effective at MPI since only two are required energetically, which seems to be the case. More work needs to be done on the production of fragments vs fluence  for this  molecule. A RMPI study   could shed   much light on how resonances with this dissociative channel may affect the fragmentation.

*Inst. Phys. Conf. Ser. No. 84: Section 9*
*Paper presented at RIS 86, Swansea, Wales, 7–12 Sept. 1986*

# Laser enhanced ionization spectroscopy in flames. Elimination of optical interferences. Excitation profiles

O.Axner, M.Lejon, I.Magnusson, H.Rubinsztein-Dunlop and
Sten Sjöström. Department of Physics, Chalmers University of
Technology, S-412 96 Göteborg, Sweden.

In Laser Enhanced Ionization Spectroscopy (LEI), thermal
ionization of an analyte atom in a flame is enhanced by a pulsed
dye laser tuned to an absorption transition. The enhanced
ionization rate which follows the excitation is detected
non-optically by applying an electric field across the flame and
measuring the current. Two step LEI, where the atoms are
selectively excited by the light from two dye lasers, gives an
increase in the sensitivity and selectivity of the method.

LEI spectroscopy has been used in our laboratory as an
ultra-sensitive method for trace element analysis utilizing both
one- and two-step laser excitations. The detection limits for some
30 elements have been determined. These detection limits, when a
flame is used as an atomizer, are for most elements far below those
which can be reached with other flame techniques.

A unique feature of two-step LEI compared to other
spectroscopic methods for trace element analysis is that problems
associated with spectral interference can be avoided. The signal
from the element under study can be increased by several orders of
magnitude through excitation by a second laser, while the signal
from the interfering element remains unchanged. Although it can be
of considerable magnitude in a one-step experiment, the one-step
signal with the unwanted contribution from an interfering element
can be neglected compared to the two-step signal or else subtracted
from it. This type of background correction is done by performing
three different measurements: the signal from two-step excitation
as well as signals from the two one-step excitations are measured
and then the two one-step signals are subtracted from the signal
for two-step excitation. The background correction method was
demonstrated in the measurement of Mg (10 ng/ml) in Na matrix
(10000 ng/ml) and Fe (0.5 ng/ml) in Ga matrix (50000 ng/ml).
Continuous background correction can be performed using a
mechanichal arrangement for blocking the respective laser output
(in turns). The subtraction can be performed electronically.

A theoretical model for description of the two-colour LEI
signal is now under progress. It is based on a strong-field
steady-state solution of the optical Bloch equations for a three
level system (Salomaa 1977)*.Incoherent two step excitation as we
as coherent two photon processes are included. Although the media
is collisionally dominated the coherent processes contributes
significantly to the LEI signal. The inclusion of the coherent pa
strongly alters the population of the levels compared to the
populations given by the ordinary rate equations.

* J.Phys.B.Atom.Molec.Phys., 10, 3005-3021 (1977)

*Inst. Phys. Conf. Ser. No. 84: Section 9*
*Paper presented at RIS 86, Swansea, Wales, 7–12 Sept. 1986*

# A comparison between laser enhanced ionization and laser induced photoionization in flames—absolute determination of state specific collisionally assisted ionization yields for excited atoms

O. Axner, T. Berglind and S. Sjöström. Department of
Physics, Chalmers University Of Technology, S-41296
Göteborg, Sweden.

Photoionization cross-sections are generally much smaller (orders of magnitude) than transition probabilities to bound states. In addition, the transition probability A-factors scale as $1/n^3$ among excited states. Therefore, excitation processes to bound excited states are much more efficient than photoionization processes.

In order to ionize excited atoms some kind of ionization process is needed. The field-ionization process is such an example. It is, however, not possible to field-ionize other states than those very close to the ionization limit. (n = 20 requires 2-3 kV/cm while the critical electrical field for field-ionization scales as $1/n^4$.) For such states, however, the transition probability A-factor, and thereby the excitation rate is still very small.

In a flame atoms ionize efficiently by collisions with thermally excited flame molecules. The probability that an excited atom ionize before quenching down to the ground state, i.e. the Ionization Yield, has been measured for Li and Na atoms in an air/acetylene-flame in this work. The determinations of the yields were done by measuring the number of electrons and ions produced when exciting the atoms to various excited states, as well as to the ionization limit.

By carefully *relating the ionization signal at the ionization limit,* when all excited atoms ionize (i.e. unity ionization yield), *to the signal from excitations to bound states,* and by relating the photoionization cross-section at the ionization limit to the transition probability of absorption from the ground state to bound excited states, *ionization yield values could be determined* for the p-series of Na and Li in an air-acetylene flame.

The ionization yields were found to be mainly increasing as the energy difference to the ionization limit was decreasing both for Li and Na. The ionization yields were found to increase slower than the increase in Boltzmann factor, $e^{-\Delta E/kT}$, when decreasing the energy difference to the ionization limit. For Li, the ionization yields were found to be larger than 50% for states closer than 2 kT (np > 6p) from the ionization limit  while the ionization yields for Na were found to be larger than 50% for states closer than 3 kT (np > 5p) from the ionization limit.

Since the ionization yields are found to be quite high in the flame due to both a high transition probability [A(3s-6p) = 100·A(3s-20p) in Na] and a high ionization yield (>50%), the atoms can be ionized very efficiently (three orders of magnitude more efficient when excited to 6p than photoionization).

*Inst. Phys. Conf. Ser. No. 84: Section 9*
*Paper presented at RIS 86, Swansea, Wales, 7–12 Sept. 1986*

# Laser enhanced ionization in graphite furnace

I. Magnusson, M. Lejon, H. Rubinsztein-Dunlop and S. Sjöström.
Department of Physics, Chalmers University of Technology,
S-41296 Göteborg, Sweden.

Two-colour Laser Enhanced Ionization (LEI) in graphite furnace was used for determination of traces of Co, Cr, Mn, Ni and Pb. Detection limits in the pg region were obtained for all elements. The high sensitivities for Mn and Pb indicate that the detection limits for these elements can be further lowered orders of magnitude if impurities in the blank solution were eliminated.

The samples of analyte in 50 μl water were dried, charred and atomized in a conventional Perkin Elmer graphite furnace ( HGA 72). The atoms were excited by resonant light from two dye lasers synchronously pumped by an excimer laser. The charged species created in collisional processes following a laser excitation, or by photo-ionization, were detected on an immersed tungsten wire with a bias voltage relative to the furnace tube at ground potential. The atomization was carried out at the maximum temperature of 2900 K. Electrical interference originating from the heating current of the furnace was totally avoided by firing the pump laser during the periods where the thyristor regulated heating current was zero.

Since the signal during an atomization only lasts for a few seconds it is difficult to check if optical interferences are present in the signal by scanning the laser wavelengths. Correction for spectral interferences can, however, be made by comparing the two-step excitation signal with the one-step signals with either of the two laser beams blocked. When the two-step signal is subtracted by both the one-step signals the remaining part of the signal is a measure of the analyte amount in the sample.

The obtained LEI signals were linear with the amount of analyte but the reproducibility was some times poor, possibly due to deformation of the electrode with time. The tungsten electrode has recently been replaced by a rigid rod of graphite. This type of electrode seems to solve the problems associated with reproducibilty and at the same time maintains the high sensitivity and linearity of the method. Investigations concerning the optimal conditions for LEI in graphite furnace are presently in progress. The LEI signal depends upon such parameters as electrode bias voltage, protection gas environment etc. The electrode can also be used as a platform for the sample.

LEI in graphite furnace has a potential to become an attractive alternative for trace element analysis when ultra high sensitivity is needed. The method could also be used for more fundamental studies of for example collision, excitation and ionization processes or for mobility measurements.

*Inst. Phys. Conf. Ser. No. 84: Section 9*
*Paper presented at RIS 86, Swansea, Wales, 7–12 Sept. 1986*                    327

# Quasi-RIMS of GdI for wavelength range 430–450 nm

J.M. Gagné, Ecole Polytechnique de Montréal, Dépt de Génie
Physique, Montréal, Québec H3C3A7, Canada.
T. Berthoud and A. Briand, CEA/IRDI/DERDCA/DCAEA/SEA
Centre d'Etudes Nucléaires de Fontenay-aux-Roses
B.P. N°6-92265 F.A.R. Cédex. France

## ABSTRACT

In this paper we present the results of our work on Quasi
Resonance Ionization Mass Spectroscopy (QRIMS) of Gadolinium using a
single Nd-YAG pumped dye laser and a magnetic sector mass spectrometer with a modified thermo-ionization source.

Resonance Ionization Mass Spectroscopy of Samarium and Gadolinium mixtures led us to choose the simplest resonance ionization
scheme : namely single laser excitation. We use a two step photoionization process in the wavelength range 430-450 nm.This process has
been found to be very efficient for Samarium.

With our experimental set-up and in our experimental con-
ditons we observed an unexpected, highly selective resonance ionization signal associated with Gadolinium. This signal is easily observed with an unfocused laser beam at energies of approximately 0.1 mJ
and for different wavelengths.

Following the observation of this phenomenon, we deemed it
important to investigate the possible mechanisms responsible for
this resonance ionization process. In the spectral range investigated the two photons total energy is far below the ionization potential (~4000 cm$^{-1}$). The ionization observed for Gadolinium must then
proceed via a scheme different from the usual two step photoionization process.

To explain this unexpected ionization we have taken into
account the physical properties of the mass spectrometer's thermoio-
nization chamber. Its heated filament can be considered, in first
approximation, as a black-body radiation source characterized by a
temperature of approximately 1800°C. The interaction between atoms
in highly excited levels and this radiation field appears to be an
efficient method for the ionization of the Gadolinium atoms. Moreover the static electric field in the ion-optic system can also contribute to the final step of the ionization process, but in our case
it does not seem to be of importance as far as ioniozation is considered.

Lastly, a scheme with three-step photoionization, with
photons of the same energy, has been evaluated, and in our conditions is far less efficient than the ionization by the black-body.

Analyses based on these results, have led to the identification of five quasi-resonant ionization lines for Gadolinium over
the 430-450 nm spectral range. The lower levels for the first step
are the first three levels of GdI $4f^{7}5d6s^{2}$ ($^{9}$D) ground term.

Five new high odd energy levels have been identified.
The line shifts observed have been interpreted, as the consequence
of the role of near-resonance intermediate excitation steps.

QRIM Spectroscopy appears to be a simple and powerful tool
either for spectroscopic or analytical studies : single laser running at low power represents the major advantage of this technique.

*Inst. Phys. Conf. Ser. No. 84: Section 9*
*Paper presented at RIS 86, Swansea, Wales, 7–12 Sept. 1986*

# Measurement of the diffusion coefficient of Li in argon using resonance ionization

John P. Judish and Rainer Wunderlich
Chemical Physics Section, Oak Ridge National Laboratory,
Oak Ridge, Tennessee 37831, USA

Resonance ionization of diffusing atoms has been used to measure the diffusion coefficient, D, of reactive atoms in a gas (Hurst 1978). Here we apply this method with some modifications to the system Li in Argon. The diffusing free Li atoms are produced in a line source at time t=0 by pulsed laser photodissociation of LiI in an isothermal cell. The temporal and spatial evolution of the initial distribution is monitored by resonance ionization of the free atoms at a later time, t, with a laser beam aligned concentric with the initial line source. The concentric alignment provides ionization signals large enough ($10^6$ – $10^7$ charges per pulse) to be measured without gas amplification. The ion signal is measured in dependence on the delay time t and for two different diameters of the ionization laser. The ratio of two such measurements is given by:

$$R_{12}(t) = \int_o^\infty r dr \exp(-r^2/4Dt) P_1(r) \Big/ \int_o^\infty r dr \exp(-r^2/4Dt) P_2(r)$$

and does not depend on the initial line density or the rate of chemical reactions which might be present. The $P_i(r)$ represent the radial ionization distributions in the two experiments. $P_i(r)$ has been calculated, according to the three-photon ionization process employed, using known atomic data and measured spatial intensity profiles of the ionizating lasers. D is obtained by a fit of the above equation to an experimental ratio.

At T = 563 K we obtain D = 12.30 N/s which is in very good (5%) agreement with a calculation based on a semiempirical method and also in good (20%) agreement with a calculation based on a modified Buckingham potential with parameters obtained from a measured Li-Ar interaction potential.

**Reference**

Hurst G S, Allman S L, Payne M G, and Whitaker T F 1978 Chem. Phys. Lett. **60** 150

---

*
Research sponsored by the Office of Health and Environmental Research, U.S. Department of Energy under contract DE-AC05-840R21400 with Martin Marietta Energy Systems, Inc. and supported in part by the Deutche Forschungsgemeinschaft-DFG through the Max-Planck Institute fur Kernphysik.

*Inst. Phys. Conf. Ser. No. 84: Section 9*
*Paper presented at RIS 86, Swansea, Wales, 7–12 Sept. 1986*

331

# Cross section determination of an autoionizing transition in atomic uranium through laser induced photoionization in a hollow cathode lamp

P.Benetti and A.Tomaselli
Departments of General Chemistry and Electronics
University of Pavia, Italy

P.Zampetti
ENEA, ISPAR Cre-Casaccia, Italy

In a hollow cathode lamp, an impedance drop occurs within few nanoseconds from a perturbing ionizing radiation [1]. Therefore - unlike the classical optogalvanic effect - we can assume, to a first approximation, that the associated electronic signal would not be affected by collisional interferences. In our experiment, two pulsed tunable dye lasers ( 5 ns FWHM ) irradiate a hollow cathode lamp filled with 238 U-Ar. The two lasers are synchronized with low jitter and variable ( typical 10 ns ) delay.

The first, low irradiance resonance pulse at 389.4 nm, prepares the atomic system in an intermediate excited level, while the second, with higher irradiance at 411.4 nm, brings atoms from that level to an autoionizing one. The signal appears only when both radiations shine the system and is measured by a boxcar detector.

Under the likely constraint that the tested autoionizing process is fast itself, kinetic considerations lead to manage the ionization as a leakage channel for the spectroscopic system and we can write:

$$\sigma = \frac{1}{t_0} \; \frac{\lim\limits_{I \to 0} \dfrac{dN_i}{dI}}{\lim\limits_{I \to \infty} N_i} \tag{1}$$

where $\sigma = \int \sigma(\nu)d\nu$ $\qquad$ $cm^2$

$t_0 = \lambda_2$ laser pulse duration, $\qquad$ s

$N_i =$ linear density of produced ions, $\quad cm^{-2}$

$I = \lambda_2$ laser photons, $cm^{-2}s^{-1}$ with bandwidth overlapping the absorption profile of the examined transition. Laser spectral mode effects are disregarded in view of the large homogeneous broadening [2].

The two limits in eqn 1 are experimentally obtained, as shown in fig.1, from the above mentioned fast optoelectronic signal, interposing neutral density filters only in the ionizing laser beam.

The data fitting the non linear (saturation) curve are found in agreement with the kinetic model and the obtained (first-approximation) cross-section value - for instance $\sigma = 1.6 \cdot 10^{-16}$ cm$^2$ in this case - falls in the expected range.

   This   method   is rather simple and doesn't require the knowledge   of the atomic levels population densities.

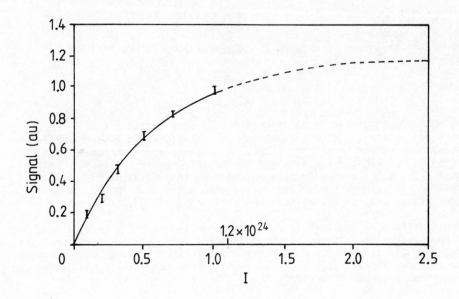

References.

1) M.Broglia, F.Catoni, P.Zampetti
   J. de Phys. ( Paris ) C7 44 479 (1983)

2) D.K.Killinger, C.C.Wang and M.Hanabusa
   Phys. Rev. A 13, 2145 (1976)

*Inst. Phys. Conf. Ser. No. 84: Section 9*
*Paper presented at RIS 86, Swansea, Wales, 7–12 Sept. 1986*

# Resonance ionization spectroscopy in hollow cathode discharges

M. Broglia and P. Zampetti
ENEA CRE-Cassaccia, COMB/ISPAR/FOTO
P.O. Box 2400 - 00100 ROMA A.D. - ITALY

Resonant laser excitation of an atomic species in a hollow cathode discharge results in a discharge impedance change that can be detected as a voltage drop, across a ballast resistor in the lamp feeding circuit (optagalvanic effect). Apart from some neon transitions involving excitation from metastable states, non ionizing laser excitation indirectly enhances the discharge net ionization rates. Short pulsed laser excitation produces signals in the us range, whose characteristic times have been ascribed to the mechanisms of perturbation and restoring of the discharge steady state.

In a previous work /1/, we first presented detection and discrimination of signals much faster than the conventional optogalvanic effect, and connected them with some effects of direct laser photoionization. The striking evidence of the signal dependence on the laser tuning and connection with the region near the cathode walls have been also pointed out.

In this work the resonant photoionization signal in the cathode dark space has been closely examined, with special regard to its time and amplitude behaviours: two consecutive time components can be detected, which are functions of the laser beam position in the dark space. A theoretical model of the charge production and collection in the high electric field region has been worked out. The excellent agreement between experimental and theoretical results supports our interpretation of the signal as a fast charge collection.

The very simple detection system and the signal independence from the complex discharge phenomena involving optogalvanic effects, make this technique extremely attractive to study photoionization and its spectroscopic and analytical applications (RIS).

/1/ M.Broglia, F.Catoni and P.Zampetti, Journ. de Phys. 44, C7-479 (1983)

*Inst. Phys. Conf. Ser. No. 84: Section 9*
*Paper presented at RIS 86, Swansea, Wales, 7–12 Sept. 1986*

335

# Laser bandwidth effects in resonance ionization spectroscopy of Nd

Gene A. Capelle

EG&G Energy Measurements, Inc. Santa Barbara, CA  93117, and

J. P. Young, D. L. Donohue, and D. H. Smith

Oak Ridge National Laboratory, Oak Ridge, Tennessee  37830

If one reviews the published literature on resonance ionization spectroscopy (RIS), it becomes apparent that there is not perfect agreement between workers on the various wavelengths or efficiencies at which ions are generated. This variance in the reported data for a given ion (Nd) prompted this further study of the experimental changes that may account for the differing results.

The RIS signal from Nd was recorded as a dye laser was scanned from <5800 Å to >6000 Å. Measurements were made using four different laser line widths, ranging from 4-6 Å to <0.1 Å FWHM. A two-color experiment was also carried out, using the dye laser and a portion of the XeF (pump laser) pulse. Since the ionization potential of Nd is 5.46 eV, to reach ionization starting with ground state Nd requires the absorption of three red photons. For the two-color experiments, the energy of a XeF photon (3.53eV) plus a photon of <6000 Å (>2.07 eV) is more than enough to ionize Nd, so only two photons are required for this process.

It was found that the number of RIS-active wavelengths for Nd, constituting the optical processes possible, was dependent on the bandwith of the laser. The RIS active wavelengths were tabulated and, where possible, an identification of the intitial state and the first transition involved was made. Some of the wavelengths appeared to be shifted from the energy of the initial transition, and the shift appeared to be a function of laser band-width.

The relative intensity of several peaks reported in a former work are quite different than what was observed in this work. This difference probably stems from the different lasers used and may well be dependent on the laser pulse width (1 μ s and 15 ns in the former and present works, respectively) and/or mode structure associated with the individual lasers. The RIS data obtained when the XeF laser beam was combined with the dye laser beam was also tabulated. In this case, RIS ions were seen for all possible transitions from the first two even initial states of neodymium. In comparing the data, there are some energy shifts for the same initial transition between the two sets of data. These can be explained by Stark shifts that are a necessary part of the process involved in the single-color experiment. These shifts are unnecessary, and because of the reduced dye laser power were not observed, in the two-color experiments.

It may seem surprising that the number of RIS-active wavelengths increases as the bandpass narrows. If the ultimate pulse power were reduced as the bandwith was decreased, one would expect fewer active wavelengths at

---
*This work was performed under the auspices of the U.S. Department of Energy under Contract No. DE-AC08-83NV10282 and DE-AC05-840R21400.

narrow bandwidths for several reasons. The power would be reduced so that i
would be less likely to promote Stark shifts. Secondly, closer transitio
matchups would be required to complete the RIS process. As it happens, th
power at a given bandpass does not decrease with bandwidth in this experi
mental setup; it remains essentially the same. One sees, then, the grea
potential of Stark shifting to promote electron transitions, since many mor
transitions, leading to ions, are seen as the bandwidth decreases.

It has been shown that the number of RIS processes available fo
neodymium in a three-photon ionization process over the wavelength range o
5800 to 6000 Å is a function of bandwidth of the laser. The narrower th
laser bandwidth, given constant output pulse power, the more transitions ar
observed. When comparing RIS data between several experimental arrangements
this fact should be kept in mind.

*Inst. Phys. Conf. Ser. No. 84: Section 9*
*Paper presented at RIS 86, Swansea, Wales, 7–12 Sept. 1986*

# Ion counting technique for studying multiphoton ionization processes

M.Armenante, R.Bruzzese, N.Spinelli, A.Sasso, and S.Solimeno

Dipartimento F.N.S.M.F.A., Università di Napoli

Pad.20 Mostra d'Oltremare, 80125 Napoli (Italy).

ABSTRACT

Laser induced multiphoton ionisation (MPI) and fragmentation of molecules have been studied by analysing energy and mass spectra of charged particles. The usual detection devices share the common feature of requiring the production of a rather large number of ions, thus putting a lower limit to the laser intensity and gas pressure.

In this communication we describe an apparatus designed for detecting and analysing the production of single ionic fragments. It consists of a time-of-flight (TOF) spectrometer with a three-gap ion source: in the first region ions formed and pushed off by a static field through a narrow hole into a second, field-free region. In the third region ions are accelerated by a static voltage into the field-free drift tube.

Ions are detected, and their times measured, at the rate of a single ion per laser shot; all the events in which more than one ion reaches the detector within the chosen time window are rejected in order to prevent any discrimination against heavy masses. Moreover, the detector response throughout the entire mass spectrum of parent and fragment ions is mass independent and proportional to the relative abundances. All difficulties connected with nonlinear detector response and/or deconvolution of the output current signal are avoided. The collection efficincy of the apparatus was evaluated by a Montecarlo program and collection efficincies ranging from 75% to 15% were found for ion kinetic energies ranging from 0.1 to 1.0 eV.

This apparatus has been used for studying four-photon (in the visible) ionisation and fragmentation of benzene molecules down to a pressure of $10^{-6}$ mbar, by using laser pulses with an intensity of up to 1 GW/cm$^2$.

We shall discuss the influence of laser beam characteristics on this specific MPI process. In particular, we shall present results on the influence of the transverse spatial distribution of the laser beam on the estimate of the two-photon ionisation cross section from the $^1B_{2u}$ electronic state of benzene. We shall also report on the influence of laser power intensity fluctuations on the same MPI process.

It is presently under way an implementation of the electronics of the above apparatus so to detect up to eight ionic fragments per laser shot. This will allow us to study the correlation function for the appearence of different ionic fragments of benzene in different experimental conditions.

*Inst. Phys. Conf. Ser. No. 84: Section 9*
*Paper presented at RIS 86, Swansea, Wales, 7–12 Sept. 1986*

339

# Resonant multiphoton ionization cross-sections.
# Application to CO

G. Sultan and G. Baravian

Laboratoire de Physique des Gaz et des Plasmas ( Associé au CNRS )
Université de Paris-Sud 91405 Orsay - France

This communication contains two parts. In the first part, results are presented for the calculation of a simplified model assuming constant in time and space a light beam interacting with a population of neutral particles in their ground state. This calculation leads to an expression giving the number of created ions as a function of the light beam. This rough approach is useful for the understanding of the RMPI (Resonant Multiphoton Ionization) in the real case of a focused laser beam where the intensity is a function of time and space. In the second part the results of an experiment for RMPI of CO by a Nd-Yag laser [1] are presented and the determination of the cross-section of RMPI by using a more realistic model taking into account the spatio-temporal distribution of laser intensity is made.

1-Simplified model.

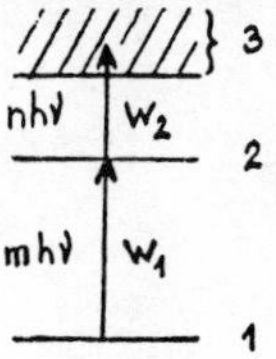

Fig.1

In the RMPI schematized in fig.1 the level 1 is the ground state, the level 2 the intermediate resonant level, and the level 3 the ionized state. The calculation shows that at the time t the population of the ionized particles n(t) may be written

$$n_3(t) = n_1(0)[1+(a/b)sh(bt) \exp(at) - ch(bt) \exp(at)]$$

where $n_1(0)$ is the population of the ground state at t=0 , $a = -(w_1+w_2+\beta)/2$ and $2b = \sqrt{[ (w_1+w_2+\beta)^2 -4w_1 w_2 ]}$ with $w_1 = \sigma_m \phi^{m-1}$ and $w_2 = \sigma_n \phi^n$ where $\sigma_m$ and $\sigma_n$ are respectively the cross-sections for the transition $1 \to 2$ and $2 \to 3$. $\beta$ represents all the terms of depopulation of 2 other than $2 \to 3$, $\phi$ is the laser intensity. If $\beta$ is weak compared to $w_1$ and $w_2$ we may write

$$n_3(t) = [n_1(0)/(w_1-w_2)][w_1 (1-\exp(-w_2 t))-w_2 (1-\exp(-w_1 t))]$$

-for weak values of $\phi$ (let $\tau$ be the laser pulse duration) we have $w_1\tau \ll 1$, $w_2\tau \ll 1$ and $n_3(t) = n(0) w_1 w_2 \tau^2 /2 \propto \phi^{m+n}$ the slope of the curve ln $n_3$ vs ln $\phi$ is $m+n$.
-for intermediate values of $\phi$, if $m > n$ , only the transition $2 \to 3$ is saturated ($w_1\tau \ll 1$ and $w_2\tau \# 1$, which leads to $n_3(t) = n_1(0)w_1\tau /e \propto \phi^m$ the slope is m.
-for high values of $\phi$, $w_2\tau \# w_1\tau \# 1$, $n_3(t)$ does not depend on $\phi$ the slope is 0.

2-Application to CO
In fig. 2 are represented the experimental values of $n(CO^+)$ vs $\phi$ (extracted from a previous work [1] and the calculated curve, taking

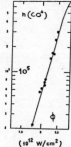

into account the spatio-temporal distribution of the laser intensity, by introducing elementary volumes bounded by isophote surfaces [2] . This is obtained for m = 7 and n = 5. The best fit between experimental values and calculated curves is obtained for

$$\delta_5 = 3.2 \times 10^{-55} \; s \; x(cm^{-1}/W)^5 \quad and$$

$$\delta_7 = 3 \times 10^{-84} \; s \; x(cm^{-1}/W)^7 \; .$$

Fig.2

$(10^{12} \; w/cm^2)$

References

[1] G.Sultan and G.Baravian  Chemical Physics 103, 417   (1986)
[2] G.Baravian and G.Sultan  Physica 128C, 343   (1985)

*Inst. Phys. Conf. Ser. No. 84: Section 9*
*Paper presented at RIS 86, Swansea, Wales, 7–12 Sept. 1986*

341

# Avoidance of contamination in trace element analysis by RIS

R. Zilliacus, E-L. Lakomaa, I. Auterinen and J. Likonen

Technical Research Centre of Finland, Reactor Laboratory
Otakaari 3 A, SF-02150 Espoo

The ultra trace element analysis calls for strict control in the sample handling environment/1,2/. Numerous erroneous results have been reported, when elements common in the environment such as Al, Mn, Cr and Pb have been determined in biological samples. This causes controversy in the evaluation of results and in the comparison between different studies.

The resonance ionization spectrometric (RIS) method is one of the most promising analytical techniques for the determination of very low element concentrations in samples of different kinds. The development of this technique for practical analytical purposes demands the simultaneous development of easy and accurate sample handling techniques so that contamination of the samples can be excluded already at the early stage of the development work. This is essential for the quick acceptance of this technique for routine analytical laboratories calling for sensitive analytical methods.

The main sources of the contamination are dust from air, the vessels and the instruments used for sample handling. The dust in ordinary laboratory air contains for example Al 0.1-10 ng/l. Stainless steel instruments have Fe, Cr and Ni in high amounts and the use of these in sample handling or near the sample handling areas can bring about severe contamination when the above mentioned elements are analyzed unless precautions are taken.

In order to develop a suitable sample handling environment in connection with RIS-analysis a dust poor room of class 100 in the working areas has been built. Class 100 of clean environment is defined so that there are 4 particles or less in one liter of air. In normal air the amount of particles can be more than $10^6/l$. The air of the room is changed 180 times an hour and the filtration is effected by means of absolute filters. The walls of the room were painted with non-metallic paint and the floor was covered with a vinyl surface. The fume bench and the laminar bench were built without any metallic parts. Contamination of the vessels can be avoided by careful cleaning. All the vessels used for the preparation of the samples and standards are washed with 2N nitric acid followed with a 6g/l-solution of EDTA and rinsed several times with deionized water. All the vessels and instruments used are made of plastic.

The samples are atomized in our system with electrothermal heating in a graphite furnace in vacuum/3/. The atomizer is isolated with an air-tight plastic cover to prevent contaminated laboratory air from entering the chamber. The graphite crucibles are cleaned before use by heating up to the atomization temperature. The crucibles are transported between the laboratory and the RIS -device in an air-tight box which can be attached straight to the device. The crucibles are moved from the transportation box into the atomizer by using manipulator and pneumatic piston.

In working with Al analysis by RIS in the ng region random contamination was noticed when the sample was inserted into the crucible in ordinary laboratory. Gallium which is not widely spread contaminant in air could be analyzed even in femtogram region without working in a clean room/4/.

REFERENCES:
1. Versiek J, Cornelis R 1980 Anal. Chim. Acta <u>116</u> 217
2. Tschöpel P, Tölg G 1982 J. Trace and Microprobe Tech. <u>1</u> 1.
3. Bekov G I et al, Proceedings of this conference.
4. Likonen J et al, Proceedings of this conference.

*Inst. Phys. Conf. Ser. No. 84: Section 9*
*Paper presented at RIS 86, Swansea, Wales, 7–12 Sept. 1986*

343

# Resonance ionization spectrometric analysis of gallium by use of electrothermal graphite atomizer

J. Likonen[a], I. Auterinen[a], G. Bekov[b], V. Radaev[b], E-L. Lakomaa[a], J. Ojanperä[a] and R. Zilliacus[a]

a     Technical Research Centre of Finland, Reactor Laboratory, Otakaari 3 A, SF-02150 Espoo, Finland

b    Institute of Spectroscopy, Academy of Sciences of the USSR, Troitzk, SU-142092 Moscow Region, USSR

Laser resonance ionization has been used in the analysis of low gallium concentrations. The measurements have been made at the Technical Research Centre of Finland {a} and at the Institute of Spectroscopy {b}. The crustal abundance of Ga is fairly low thus allowing easy analysis without contamination under normal laboratory conditions. The excitation scheme used for gallium is 4 p $^2P_{3/2} \rightarrow$ 5 s $^2S_{1/2} \rightarrow$ 17 p $^2P_{1/2,3/2}$, the wavelengths being 403,3 nm and 430,0 nm, respectively. The saturation of the ion signal as a function of energy fluence was investigated. The saturation energy fluences are 70 $\mu J/cm^2$ for the first and 15 $mJ/cm^2$ for the second step. The excited atoms are ionized after laser excitation by an electric field pulse of 7,5 kV/cm, which is well above the threshold value 6,3 kV/cm.

When studying the thermal atomization of gallium compounds it was found that the integral signal was independent of the crucible heating conditions, provided that the final temperature was high enough (> 2000 $^0C$ in our case). Ga standard solutions prepared of different compounds {$GaCl_3$, $Ga(NO_3)_3$, $Ga_2(SO_4)_3$, $NH_4Ga(OH)_4$} containing the same amount of Ga gave equal analytical signals in all cases. The standard deviation was less than 15 % when analyzing 27 samples of 50 $\mu l$ of a $GaCl_3$ solution containing 50 pg Ga.

The calibration curve was found to be linear in the concentration range studied ($10^{-4}$ to $10^{-8}$ %). The smallest amount of Ga used in analytical experiments with the setup {a} described in [1] was $9 \cdot 10^{-14}$ g. According to the signal obtained in this experiment the detection limit is on $10^{-16}$ g ($10^6$ atoms) level. The detection limit when using the setup {b} described in [1] was two orders of magnitude higher because it used a less powerful excimer laser for pumping.

Ga in high purity Ge has also been analyzed. The sample size was 10 – 120 mg. At temperatures over 1550 $^0C$ a Ga signal was observed and the optimum temperature for registering the analytical signal was found to be 1650 $^0C$. The Ga concentration in Ge samples was $8 \cdot 10^{15}$ atoms/$cm^3$ which was determined by electrophysical methods. The analytical signals from the Ge samples and diluted $GaCl_3$ samples containing the same amount of Ga turned out to be equal. This means that Ga atomizes completely both in the Ge and in $GaCl_3$ samples under our experimental conditions. The concentration determination carried out with the spectrometer {b} by the reference beam technique [2] yielded a value of $6 \cdot 10^{15}$ atoms/$cm^3$ which is equal to the result of the electrophysical method. The comparison measurements between the two spectrometers {a} and {b} yielded comparable results.

REFERENCES:

1   Bekov G I, Kudryavtsev Yu A, Auterinen I and Likonen J, Proceedings of this conference.

2   Akilov R, Bekov G I, 1982, Sov. Tech. Phys. Lett. 8, 225.

*Inst. Phys. Conf. Ser. No. 84: Section 9*
*Paper presented at RIS 86, Swansea, Wales, 7–12 Sept. 1986*

345

# The potential of resonant ionization mass spectroscopy for detecting environmentally important radioactive nuclides

M.H.C. Smyth, S.L.T. Drysdale, R. Jennings, K.W.D. Ledingham,
D.T. Stewart, M. Towrie
Department of Physics and Astronomy, University of Glasgow,
Glasgow G12 8QQ, Scotland

C.M. Houston
Department of Chemistry, University of Glasgow,
Glasgow G12 8QQ, Scotland

M.S. Baxter and R.D. Scott
Scottish Universities Research and Reactor Centre,
East Kilbride, Glasgow G75 0QU, Scotland

## ABSTRACT

The monitoring of effluents from the nuclear power industry is of great importance. Of particular interest are the low energy β–emitting nuclides since they may not be of negligible radiological significance. They are difficult to detect by conventional radiometric methods at specific activities levels of $1pCi(3.7 \times 10^{-2}Bq)/g$ of sample. This corresponds to a concentration of about parts in $10^9$ for a radionuclide with half life of $\sim 10^7y$ in e.g. sediments. For shorter half lives the concentrations are even smaller.

RIMS has a sensitivity level for traces in solids of ppb. This sensitivity is justified for a RIMS system using ion ablation and a time of flight mass spectrometer. By increasing the ion ablation current to 1mA and the laser rep rate to several hundred/sec perhaps parts in $10^{12}$ sensitivity is possible.

Even at this sensitivity many of the important radionuclides cannot be detected in environmental samples without recourse to chemically concentrating. However new and quick techniques of acid dissolution of rocks and sediments in a microwave oven make such concentrating possible.

---

*Three of us (CMH, SLTD and MT) wish to thank SERC for support, while MHCS acknowledges support from DENI.*

*Inst. Phys. Conf. Ser. No. 84: Section 9*
*Paper presented at RIS 86, Swansea, Wales, 7–12 Sept. 1986*

# Multiphoton transitions in caesium vapour

C.M. Houston
Department of Chemistry, University of Glasgow,
Glasgow G12 8QQ, Scotland

S.L.T. Drysdale, K.W.D. Ledingham, C. Raine, K.M. Smith,
M.H.C. Smyth, D.T. Stewart and M. Towrie
Department of Physics and Astronomy, University of Glasgow,
Glasgow G12 8QQ, Scotland

## ABSTRACT

Multiphoton ionisation has been used for the detection of gas-phase atoms above a room temperature sample of metallic caesium. An excimer-dye laser system, with harmonic generation capability, was used in conjunction with a proportional counter filled to 30 Torr with a high purity, (70:30) $Ar/CH_4$ gas mixture.

Two photon ionisation spectra (i.e. one photon excitation, one photon ionisation) of caesium were obtained in the regions 455-460nm, 387-390nm and 360-362nm. The 6S-7P, 6S-8P and 6S-9P transitions, which occur in the above spectral ranges, were well resolved. The effect of laser fluence on the widths and relative intensities of the two 6S-7P transitions was investigated.

Using frequency doubled light, two photon ionisation spectra were also recorded in the region 315-332nm. Many sharp transitions, 6S-Rydberg levels, were observed. There was an overall trend for these transitions to decrease in intensity as the caesium single photon absorption edge at 318nm was approached. However, when the intermediate level of the photoionisation process was very close to the continuum, considerable collisional enhancement of the ionisation yield was apparent.

Three photon ionisation spectra (ie. two photon resonant (via a virtual state), one photon ionisation) of caesium in the 630-644nm region were obtained. In these, 6S-higher S level transitions and 6S-D transitions could be seen, but not 6S-P which are forbidden via two photon excitation.

Four of us (CMH, SLTD, CR and MT) wish to thank SERC for support, while MHCS acknowledges support from DENI.

*Inst. Phys. Conf. Ser. No. 84: Section 9*
*Paper presented at RIS 86, Swansea, Wales, 7–12 Sept. 1986*

349

# Two different types of hollow-cathode discharges used for high resolution laser spectroscopy on copper

H. Bergström, H. Lundberg, W.X. Peng*, A. Persson, S. Svanberg,
C.-G. Wahlström,
Department of Physics, Lund Institute of Technology, P.O. Box 118,
S-221 00 Lund, Sweden.
*Department of Physics, Jilin University, Changchun, P.R. China.

The spectra of two different Cu I transitions, 578.2 and 570.0 nm, have been recorded with two different types of high-resolution laser spectroscopy.

Employing a hollow cathode of the design developed by Lawler et al, /1/, we have used saturation spectroscopy for studying the hyperfine structure of the $3d^{10}4p$ $^2P$ fine structure levels of $^{63}Cu$ and $^{65}Cu$. With Ar at a pressure of ~0.5 mbar as the carrier gas and a discharge current of 150-200 mA, copper atoms were efficiently produced in the metastable $3d^9 4s^2$ $^3D_{3/2}$ fine structure level ( ~12000 cm$^{-1}$ above the ground state) by sputtering. Using a single-mode dye laser operating with Rhodamine 110 dye the Cu atoms were subsequently excited to the $3d^{10}4p$ $^2P_{3/2,1/2}$ levels.

High modulation frequencies are advantageous for obtaining a good signal-to-noise ratio; see e.g. /2/. For the $^2P_{1/2}$ state a preliminary evaluation gives the magnetic-dipole interaction constants a($^{63}Cu$)-500(8) MHz, a($^{65}Cu$-539(10) MHz and for the 4p $^2P_{3/2}$ state values in agreement with previous, more accurate level crossing results /3/ were obtained.

For the 4p $^2P_{1/2}$ (and 4p $^2P_{3/2}$) state we also used a large bore low pressure hollow cathode /4,5/ as an atomic source for Doppler reduced spectroscopy. Ar at a pressure of 0.5 mbar was used as the buffer gas. A hole in the cathode bottom plate served as a nozzle for forming the atomic beam which effused into a high-vacuum (10$^{-4}$ mbar) scattering chamber where the beam was collimated before intersecting the laser beam. Typical modulation frequencies were limited to 1-10 kHz, since the noise is created by plasma emission rather than by direct laser fluctuations.

The experiments are proceeding and we are aiming at measuring the 5p levels as well. A very interesting way to do this would be to use a second laser for ionizing the atom resonantly from the 5p levels to an autoionization level. It is expected that the optogalvanic signal will be extremely strong in this case. Theoretical calculations of the hfs for the levels discussed are in progress using many-body perturbation theory.

REFERENCES:

1. Lawler, J.E., Siegel, A., Couillaud, B. and Hänsch, T.W., J. Appl. Phys., 52, (1981), 4375.
2. Kröll, S. and Persson, A., Opt. Comm., 54, (1985), 277.
3. Ney, J., Z. Physik, 196, (1966), 53.
4. Hannaford, P. and Lowe, R.M., J. Phys. B., 14, (1983), L43.
5. Bergstrom, H., Lundberg, H., Persson, A., Schade, W. and Zhao, Y.Y., Phys. Scr. 33, (1986), 513.

*Inst. Phys. Conf. Ser. No. 84: Section 9*
*Paper presented at RIS 86, Swansea, Wales, 7–12 Sept. 1986*                                        351

# Optogalvanic observation of autoionization spectrum of Kr

Akihide Wada and Chiaki Hirose
Research Laboratory of Resources Utilization, Tokyo Institute of
Technology, 4259 Nagatsuta, Midori-ku, Yokohama, 227, Japan

## ABSTRACT

Autoionization is a well known phenomenon investigated by multi-photon
and VUV absorption spectroscopy and results from the interaction of discrete state with continuous states. Recently, the observation by optogalvanic spectroscopy (OGS) of one-photon transitions to the autoionizing
levels from the excited states of Xe in rf-discharge by using a visible
laser light has been reported (Grandin and Husson 1978).

In the optogalvanic spectrum of Kr in hollow-cathode discharge, when
the laser beam was passed through the cathode dark region near the cathode
surface, several one-photon transitions to the 7d' or 9s' autoionizing
levels from the 5p' excited levels were observed in the region from 16700
to 17300 cm$^{-1}$. Mainly used discharge tube was a Kr filled hollow cathode
lamp with the pressure of 6 torr (Hamamatsu Photonix Inc.). The cathode
cylinder had 3 mm i.d. and was 17 mm in length. The discharge current
ranged from 2 to 15 mA.

The intensities of the optogalvanic double-resonance (OGDR) signals
of these lines showed strong dependence on the choice of the pumped 5p'
sublevel. Their assignment has been given on the basis of the OGDR and
the available term values (Kaufman and Hamphreys 1972: Yoshino and Tanaka
1979), and the results are listed in Table 1. The profile of observed
spectrum is symmetric and shows good agreement with the overlapped Lorentz
profile in contrast to the ordinary autoionization line which is asymmetric
due to the interference between the autoionizing transition and the direct
photoionization (Fano 1961). This implies that the direct photoionizations

Table 1. Assignment of Autoionization
Lines of Kr Observed in Visible Region.

| Position of Spectral Lines (cm$^{-1}$) | Assignment |
|---|---|
| 16783 | 7d'[5/2]2 - 5p'[3/2]2 |
| 16888 | 7d'[3/2]2 - 5p'[3/2]2 |
| 16933 | 7d'[5/2]3 - 5p'[3/2]2 |
| 17074 | 7d'[3/2]1 - 5p'[3/2]2 |
| 17134 | 7d'[5/2]2 - 5p'[3/2]1 |
| 17190 | 9s'[1/2]1 - 5p'[3/2]1 |
| 17204 | 9s'[1/2]0 - 5p'[1/2]1 |
| 17216 | 9s'[1/2]1 - 5p'[1/2]1 |
| 17237 | 7d'[3/2]2 - 5p'[3/2]1 |

from the 5p' sublevels to the $P_{3/2}$ continuous levels are negligible.
When the pumped level is changed in the OGDR, we find a significant discrepanc
between the observed spectrum and the overlapped Lorentz profile if the
linewidths of the component Lorentzians are kept invariant by changing
the pumped level.  This disagreement can be explained by introducing
the correction by higher-order perturbation theory.  This means that the
autoionizing levels which are close to each other are mutually coupled by
their configuration interaction with the continuous levels.  By using
a computer fitting considering the higher-order interaction, the term
values of 9s' and 7d' sublevels are determined.
We also estimate the number of ions produced by the autoionization to
suggest that the autoionization may be adopted to the temporally and
spatially controlled ionization of gases and be applied to the stable
manipulation of a pre-ionization of a excimer laser by using a
conventional dye laser.

References
Fano U 1961 Phys. Rev. 124 1866
Grandin J-P and Husson X 1981 J. Phys. B: At. Mol. Phys. 14 433
Kaufman V and Hamphreys C J 1972 J. Opt. Soc. Am. 59 2614
Yoshino K and Tanaka Y 1979 J. Opt. Soc. Am. 69 159

*Inst. Phys. Conf. Ser. No. 84: Section 9*
*Paper presented at RIS 86, Swansea, Wales, 7–12 Sept. 1986*

# The first ionization potentials of several second-row transition elements determined by resonance ionization spectroscopy of laser produced metal beams

Peter A. Hackett, David M. Rayner, Steven A. Mitchell and
Orson L. Bourne
Laser Chemistry Group, National Research Council Canada
100 Sussex Drive, Ottawa, Ontario, K1A OR6

In general ionization potentials have been determined from spectroscopic term data or mass spectrometrically via appearance potential measurements. Both techniques have been difficult to apply to transition metals. Spectroscopic methods have been inaccurate as Rydberg levels are difficult to identify in the midst of a dense manifold of valence excitations and interpretation of appearance potential data has been impeded by the presence of many low lying electronic states of the neutral and the ion. These problems have been particularly severe for the second row transition metals and beyond. Indeed, published values for the first ionization potential of zirconium differ by some 2000 $cm^{-1}$. We have a particular interest in this problem as resonance ionization of zirconium vapour offers a potential route to zirconium-91 depleted zirconium; a material of relevance for thermal-neutron nuclear power plants.

We have determined accurate values of the first ionization potentials of zirconium, niobium and molybdenum using a versatile atomic beam source employing a pulsed supersonic helium expansion through a laser produced metal vapour. This produces a cold atomic beam which is interrogated by various pulsed dye lasers in the ionization region of a time-of-flight mass spectrometer. Following the laser pulses, high lying Rydberg states are ionized by a pulsed field which also serves to extract the ions. Ionization potentials are determined from these Rydberg spectra in a conventional way.

IONIZATION POTENTIALS/CM$^{-1}$

| | Laser, field ionization[a] | Vacuum UV absorption[b] | Long extra-polation[c] | Inter-polation[d] | Appearance potential[e] |
|---|---|---|---|---|---|
| Y | | 50144(1) | 52650 | 51447 | 50813 |
| | | | +2486 | +1283 | +649 |
| Zr | 53506.0(3) | | 56077 | 55145 | 53152 |
| | | | +2571 | +1639 | -354 |
| Nb | 54513.4(5) | | 54600 | 55511 | 53152 |
| | | | +87 | +998 | -1361 |
| Mo | 57203.9(3) | | 57260 | | 58230 |
| | | | +56 | | +1029 |

a) This work, b) W.R.S. Garton, E.M. Reeves, F.S. Tomkins and
B. Ercoli, Proc. R. Soc. London A 333 17 (1973), c) C.E. Moore
Atomic Energy Levels NBS 35 11 (1952), 111, d) M.A. Catalan and F.R.
Ricc An. Real Soc. Esp. Fis. y. Quim. A 48 328 (1952), e) E.G. Rauh
and R.J. Ackermann. J. Chem. Phys. 70 1004 (1979).

*Inst. Phys. Conf. Ser. No. 84: Section 9*
*Paper presented at RIS 86, Swansea, Wales, 7–12 Sept. 1986*

355

# Photoelectron spectroscopy applied to a study of laser induced photoionization of lanthanides and actinides

J. P. Young, D. L. Donohue, and D. H. Smith
Analytical Chemistry Division
Oak Ridge National Laboratory
Oak Ridge, TN 37831

Photoelectron spectroscopy has been utilized to gain a better understanding of the optical processes involved in resonance ionization mass spectrometry (RIMS) applied to lanthanides and actinides. To this end, the energy distribution of the photoelectron which results from the laser induced ionization has been determined. The resultant photoelectron spectra (PES) yield information concerning the initial state of the atoms which are ionized and the possible transitions involved in the RIMS process. We have investigated a number of lanthanides, including Pr, Nd, Sm, Eu, Ho, and Tm, and U at several wavelengths known to generate ions by RIMS. The power dependence of some of the PES was also measured.

From an evaluation of the resultant electron energy spectra, characteristics of the optical processes involved in RIMS have become apparent, and spectroscopic constants for elements have been confirmed or identified. For example, two possible wavelengths for carrying out RIMS of samarium, 580.16 and 593.71 nm yield $PES$ that match a calculated PES which is based on an initial state of $^7F_4$ (2273 cm$^{-1}$) and $^7F_3$ (1490 cm$^{-1}$), respectively. Evidence has been amassed, therefore, that confirms an earlier assumption for the optical process involved in these RIS 1+1+1 routes. A number of neodymium RIMS transitions have been found to correlate with the experimental PES in the same fashion. One neodymium RIMS wavelength, 587.1 nm, does not. This particular process probably involves an initial transition, from $^5I_4$ ground state, followed by two possible routes in the second bound-bound transition; both of these lie within the band pass of our laser, 3 cm$^{-1}$. We have been unable to obtain PES spectra for either Eu or Ho that would confirm an earlier "hybrid resonance" process for ionizing these elements. Many of the ionization routes for these two elements can be better explained as resulting from a 2+1 process.

An interesting result is obtained in the characterization of the PES from ionizing uranium at 591.54 nm. The optical route for this RIS process is known to be a 1+1+1 process originating from the ground state. There are two different values in the literature for the first ionization potential (I.P.) of uranium, 49928 and 48799 cm$^{-1}$. The experimental PES we obtain for uranium matches the PES calculated for the former value and does not match the PES calculated for the latter value.

*Research sponsored by the U. S. Department of Energy, Office of Basic Energy Sciences, under Contract DE-AC05-84OR21400 with Martin Marietta Energy Systems, Inc. J. P. Young received partial support from the Office of Health and Environmental Research.

*Inst. Phys. Conf. Ser. No. 84: Section 10*
*Paper presented at RIS 86, Swansea, Wales, 7–12 Sept. 1986*

357

# Summary of the Third International Symposium on Resonance Ionization Spectroscopy (RIS)

Keith Boyer

Physics Department, University of Illinois at Chicago

The Third International Symposium on Resonance Ionization Spectroscopy (RIS) differed from previous symposia in several significant ways demonstrating that this is a rapidly maturing discipline with great future importance. Papers presented centered on accomplishments rather than plans. Many developments reported at the symposium lead to this conclusion. A number of devices which were in the planning or development stage in previous RIS symposia are now in operation on a routine basis with very impressive results such as the noble gas time of flight spectrometer of Atom Sciences which achieved the quantitative determination of 1000 atoms of $^{81}$Kr with high reliability in the presence of $10^{15}$ atoms of other krypton isotopes, and their siris device which detects trace elements in semi-conductor substrates at the 2 parts per billion level with 10% accuracy. Even more exciting was the discovery that RIS techniques could be applied to both light and very large complex molecules by the use of nozzle cooling techniques. Although this possibility had not been considered in previous conferences, achievements reported here open up an enormous new field to RIS with applications in biology, medicine, chemical analysis and in fundamental studies of chemical processes.

After opening remarks, a number of excellent scientific papers explored the physics of multiphoton ionization processes. Of particular interest was the use of strong external electric fields to distort atoms in the Rydberg state and thus to change their chemistry, a technique referred to as "Atomic Engineering." Theoretical predictions and experiments on the way spectra are modified by the AC stark shift and measurements of thermal contributions to the ionization process provided important information for RIS.

Noble gas RIS experiments were used to date ground water from deep acquifers and ice from the polar caps by determining isotopic ratios of krypton. A detection sensitivity for isotopic ratios of $10^{12}$ now achieved is expected to be increased many times which will be important for studies of deep ocean circulation patterns and many other applications.

Devices and their performance were described for measuring impurities in solids and on surfaces at the 2 parts per billion level with parts per trillion projected using a variety of techniques for extracting neutral atoms from solids and from surfaces. Techniques included ion beam sputtering and electron beam and photon induced desorption of atoms from surfaces. Theoretical and experimental studies combined to show that both momentum transfer and bond breaking were the underlying mechanisms responsible for atom desorption and that neutral atoms far exceeded the numbers of ions released. Thermal evaporation was found to be quite effective for bulk sampling of many materials including the special case of laser ablation which was treated for the first time in this symposium. Laser ablation is probably the most flexible process with the widest range of applications of any method so far used to provide sources of either atoms or molecules for RIS.

Reports were also given on other subsystems employed in RIS including mass spectrometry, laser systems and detection devices. Time of flight mass spectrometers have proven to be very simple and effective devices when combined with pulsed lasers. Further improvements in selectivity and sensitivity resulted when they were also synchronized with pulsed sources of atoms, and pulsed detectors. While magnetic and electrostatic mass spectrometers combined to play an important role in the applications reported, there were also new techniques demonstrated for isotopic selectivity including hyperfine splitting and double resonances. The "Reflecton" device was described as were examples of its use. It increases the resolution of time of flight spectroscopy, extending the effective flight path by reflecting the ions several times. The various solid state and surface analytical devices presented demonstrated the ability of these systems to detect and to provide quantitative analysis of impurity levels at the 2 parts per billion level in semiconductor substrates, and in samples of biological material for both research and medical and diagnostic use. Further developments of these techniques promise an increased sensitivity by several orders of magnitude according to predictions made by experimenters.

The ability of RIS to work with very small samples in which picogram quantities of trace elements could be identified and measured is very important for medical diagnosis, for biological research and for environmental monitoring. Reports presented included the measurement of picogram quantities of copper in one tenth milliliter of blood, of nanogram quantities of uranium in urine and the analysis of airborne particles. The small size of samples needed has a secondary advantage when chemical steps are required to prepare the sample of RIS as is necessary for example to measure the amount of uranium in urine, since this greatly reduces the purity requirements for reagents.

The applications of diode lasers in the near infrared were described. This device holds great promise for future applications of RIS. In molecular RIS, the diode laser development could be crucial as the identification and isotopic selection step could be accomplished with the diode laser supplying the infrared or red excitation step. However, diode laser arrays produced by photolithography techniques could also provide low cost, high average power outputs for pumping yag or similar solid state laser devices which could then be harmonically converted into visible or ultraviolet wavelengths. The developments in solid state laser technology under way today makes it probable that efficient, long-lived, low cost, and reliable tunable laser systems will be available for RIS applications in the next few years. An example is titanium doped sapphire which can be flash lamp pumped or doubled yag pumped to give an output tunable from 620 to 1000 nanometers. It is the complexity, difficulty in use and expense of laser systems which limits RIS applications today. The projected developments should remove these limitations.

The most exciting developments of this symposium were the extensions of RIS technology to studies of light and heavy molecules, developments not previously expected. Experiments on relatively small molecules such as xylene and acetamide were accomplished by nozzle cooling in the gaseous phase to simplify the absorption spectra by eliminating rotational structure and reducing vibrational complexity. Isomeric components of xylene could be measured at $10^{-3}$ concentrations relative to the other isomers. The energetics of the amide unit bonds, a building block of proteins and peptides, was investigated by measuring the ionic fragmentation patterns, yields and power laws produced by different ultraviolet wavelengths.

Large, complex molecules such as chlorophyl, meosporphyrin and tripeptide, as reported by Boesl of the Munich group, were laser evaporated into a low temperature stream of nozzle cooled argon. Their absorption spectra were measured by using time of flight in a Reflecton Detector. A selectivity of $10^4$ was obtained for mass analysis. The ability to identify sharp lines in such complex molecules by nozzle cooling opens up a wide segment of the field of chemical compounds to RIS techniques, a development of great significance.

Isotope separation, a special application of RIS techniques, has the potential for large scale commercial application but it places heavy demands on laser technology. Atomic vapor systems for large scale separation of uranium and mercury were reported on and arguments presented to show these could be cost-effective processes. Large scale separation of isotopes of metallic elements used in the nuclear energy industry and of light elements such as sulphur used for agricultural tracers are two examples of processes which would be industrially important if made cost effective. In addition to the atomic vapor method, molecular RIS processes using nozzle cooling have been under investigation for a number of years. In retrospect this technology should have extended RIS techniques to molecules as well as to atoms, very early in RIS history.

Each of the RIS symposia have ended with a session on application of RIS in the physics of elementary particles, and a number of interesting papers on this topic were presented this time. The search for free quarks and for very heavy stable particles left over from the "Big Bang" continues and although the sensitivity of the experiments has been increased in the past two years, the results are still negative. A solar neutrino experiment using the $^{81}B_r(\nu,e^-)^{81}Kr$ reaction to measure a different set of thermonuclear reactions from the Chlorine–Argon experiment of Davis and an experiment to measure the mass of the electron neutrino by the double Beta decay of $^{128}T_e$ by RIS were also discussed.

In summary, RIS has moved from the planning and construction phase to the laboratory research and demonstration stage with commercial and medical analytical and diagnostic services expected to follow soon. The technology has been extended to involve the world of molecules. While its importance as a research and analytical tool can hardly be overestimated, it is also true that present RIS devices are complex, expensive and require expert personnel just as is true of the scanning electron microscope. New developments, particularly in the laser field may result in specialized equipment on the scale of the gas chromatograph or the photospectrometer in the future, but it will take a few years of further development to achieve this.

*Inst. Phys. Conf. Ser. No. 84*
*Paper presented at RIS 86, Swansea, Wales, 7–12 Sept. 1986*

361

# Double beta decay of 128Te and RIS

T. Kirsten
Max-Planck-Institut für Kernphysik, P.O. Box 103980
D-6900 Heidelberg, W. Germany

## 1. Introduction

A non-vanishing neutrino restmass would enhance the total rates of double beta decay (DBD), a very weak type of natural radioactivity. DBD was first observed for the transition $^{130}Te \to {}^{130}Xe$ (Kirsten et al 1968) by detecting $^{130}Xe$-excess ($^{130}Xe*$) in geologically old tellurium minerals. Extending such studies to the case $^{128}Te \to {}^{128}Xe$, an upper limit of

$$< m_{\nu_e} > \leq 5.6 \text{ eV} \quad (2\sigma) \tag{1}$$

has been inferred from high sensitivity Xe-mass spectrometry (Kirsten et al 1983a,1983b). Improving this limit would require to identify $O(10^5)^{128}Xe*$-atoms causing an isotopic shift of only a few permil to within $\sim \pm 20\%$. With the presently employed techniques, this is not possible. However, laser-induced RIMS (Schneider 1986) has the potential to

(i)   increase sensitivity by providing a degree of ionization much higher as in conventional electron impact ion sources and
(ii)  to discriminate mass 128 background ions from $^{128}Xe$-ions due to the chemical selectivity of the resonance ionization process (Schneider 1986).

In this paper, we first outline the theoretical background and briefly describe the detection method. Then, we quantitatively compare

(i)   the best data already available
(ii)  hopefully realistic goals one has to aim for in RIMS detection to make the effort awarding
(iii) principal theoretical limitations.

In the treatment we have made certain simplifications in favour of practical transparency.

## 2. Theoretical background

One of the most important input parameters in grand unified theories is the electron neutrino restmass. So far, all attempts to unambiguously identify a non-vanishing value of $m_{\nu_e}$ have failed (Kündig 1986). Among the best limits obtained so far are those deduced from DBD measurements (Kirsten et al 1983a, Bellotti et al 1984, Caldwell 1986, see also Doi et al 1985).

DBD with neutrino emission ("$2\nu$-DBD") is a rare type of spontaneous radioactive decay

$$N_Z^A \longrightarrow N_{Z+2}^A + 2e^- + 2\bar{\nu}_e \tag{2}$$

It constitutes a second order weak interaction process and conserves leptonic charge. In practice, hope for observation exists only if the single beta decay is energetically forbidden (e.g. $^{128}Te$, $^{130}Te$, $^{76}Ge$,...). Decay

rates are extremely small ($\lambda < 10^{-20} \text{yrs}^{-1}$) due to the small available phase space.
Neutrinoless DBD ("o$\nu$-DBD")

$$N_Z^A \rightarrow N_{Z+2}^A + 2e^- \qquad (3)$$

enjoys a phase space $\sim 10^6$ times larger than $2\nu$-DBD, but it violates lepton number conservation by two units. It is forbidden if lepton conservation is rigorous, yet it could occur with a relatively high rate even for small violations of lepton conservation (amplitude $\varepsilon$). Neutrinoless DBD requires the neutrino to be a two component (Majorana) particle ($\nu \equiv \tilde{\nu}$). Only then can the virtual neutrino emitted by a neutron bound in the nucleus be reabsorbed by a second bound neutron to trigger DBD without neutrinos actually leaving the nucleus. However, even if the neutrino is a Majorana particle, o$\nu$-DBD could still not occur in case of complete polarization. The initial neutrino is emitted right handed, yet for reabsorption a left handed neutrino is required. There are two possibilities to overcome this helicity mismatch:

(i)   Partial **acceptance** of the wrong helicity by an explicit right handed current admixture (amplitude $\eta$) to the generally left handed weak current.
(ii)  Partial **correction** of the wrong helicity through helicity flips of the virtual neutrino. Such flips become possible as a consequence of a non-zero Majorana mass of the neutrino.

Both effects can drive DBD. For the decay rate we have the proportionality

$$\lambda_{o\nu} \sim \varepsilon^2 = \left[ c_1 m_\nu^2 + c_2 m_\nu \, \eta + c_3 \eta^2 \right] \qquad (4)$$

($c_i$=constants for a given nucleus).
In particular, if $\eta > 0$, it follows $m_\nu > 0$ (Kayser et al 1986) and therefore, from $\lambda_{o\nu} > 0$ it follows $m_\nu > 0$. Otherwise, the absolute rate depends on the decay energy and on the Gamov-Teller matrix element of the particular case (see Doi et al 1985).
In summary, experiments measuring or limiting $\lambda_{o\nu}$ can sensitively test

(i)   lepton conservation
(ii)  the nature of the neutrino (Majorana vs. Dirac-$\nu$)
(iii) right handed (V+A) admixtures to the weak current
(iv)  the Majorana neutrino restmass.

## 3. Detection method

Inspired by the importance of the neutrino properties in grand unification, many improved DBD experiments have been performed during the last few years. The direct counting experiments (for summary of results see Caldwell 1985, Doi et al 1985, Haxton and Stephenson 1984, Kirsten 1984a) have the potential to directly distinguish between $2\nu$-DBD and o$\nu$-DBD but till now there is still no unambiguous evidence for DBD of any kind from these experiments. Here we restrict ourselves to the "geochemical" detection technique. This method has turned out unambiguous evidence for $2\nu$-DBD of $130$Te and $82$Se (Kirsten et al 1968, for summaries see Kirsten 1984a, 1984b, Doi et al 1985, Kirsten et al 1986) but o$\nu$ remains undetected.
Nevertheless, meaningful upper limits on o$\nu$-DBD rates and hence on $m_\nu$ have been deduced.
The method benefits from the accumulation of DBD product nuclei during geological time ($O(10^9 \text{yrs})$) in ores or minerals which are rich in prospective DBD parent nuclei. Here, we restrict our discussion to tellurium ores since the best limits on o$\nu$ DBD result from the decay

128Te → 128Xe, and the decay 130Te → 130Xe is the classical example for which 2 ν-DBD was first detected. Both tellurium isotopes occur in approximately equal abundances in natural tellurium (31.79% and 34.48%, respectively). During geological time the DBD-products 128Xe and 130Xe are added to the usually small but not vanishing quantities of dissolved Xe of normal isotopic composition (e.g. 1.92% 128Xe, 4.08% 130Xe, 26.89% 132Xe,..) and cause an isotopic shift in the mass spectrum. Then, from a mass spectrometric Xe-analysis the number of DBD daughter atoms $N_d$ can be calculated. With the gas retention time t determined by radiometric dating techniques (Kirsten 1984b) and the tellurium parent isotope concentration $N_p$, $N_d$ can be converted into a DBD constant $\lambda$ according to

$$\lambda_{DBD} = \frac{N_d}{N_p} \cdot \frac{1}{t} \tag{5}$$

(mean lives $\tau = \lambda^{-1}$ are always $\gg$ t).
The measured decay constant $\lambda_\Sigma$ constitutes the sum of both DBD modes:

$$\lambda_\Sigma = \lambda_{2\nu} + \lambda_{0\nu} , \tag{6}$$

the two parts cannot be distinguished experimentally. However, writing

$$\lambda_\Sigma = \lambda_{2\nu}(1 + \frac{\lambda_{0\nu}}{\lambda_{2\nu}} ) \tag{7}$$

we obtain from theory

$$\lambda_\Sigma = A(1 + \Lambda_0) \tag{8}$$

with

$$A = F_{2\nu} |M_{2\nu}^{GT}|^2 \tag{9}$$

and

$$\Lambda_0 = c_1 m_\nu^2 f^2 \xi^2 \tag{10}$$

A equals $\lambda_{2\nu}$,theor. for a given parent nucleus as calculated from the relevant nuclear matrix element $|M_{2\nu}^{GT}|$ and the "kinematical factor" $F_{2\nu}$ which mainly depends on the decay energy (see Doi et al 1985). f and $\xi$ are ratios $F_{0\nu}/F_{2\nu}$ and $|M_{0\nu}^{GT}| / |M_{2\nu}^{GT}|$, respectively, $c_1 m_\nu^2$ is defined in eq. 4. (To obtain conservative limits on $m_\nu$ we have set $\eta = 0$.)
We distinguish:
(i)    $\Lambda_0 \ll 1$, $\lambda_\Sigma$ dominated by $\lambda_{2\nu}$.
(in particular, for $m_\nu = 0$   $\Lambda_0 = 0$ and $\lambda_\Sigma = \lambda_{2\nu}$ ). If the decay energy is large (e.g., 2.53 MeV for 130Te), $\lambda_{2\nu}$ dominates even if $m_\nu$ were as large as say, 50 eV. Therefore, this nuclide is useless to limit $m_\nu$ but it can test the theory for $A = \lambda_{2\nu}$ .
(ii)    $\Lambda_0 \gg 1$. This requires large $m_\nu^2 f_\nu^2$.
f2 increases strongly with decreasing decay energy, hence, everything else being equal, smaller neutrino masses can be tested for cases with lower decay energy (128Te! E=869 keV only). However, lower decay energy implies small A, hence fewer daughter atoms available for detection.
(iii)    $\Lambda_0 = 0(1)$. Both modes contribute to $\lambda_\Sigma$ .
For any given geochemical DBD system, this transition from case (ii) to case (i) limits for principal reasons the sensitivity to test $m_\nu$ with the geochemical method.

In deducing $\Lambda_0$ and hence, $m_\nu$, not only the experimental ability to distinguish $(1+\Lambda_0)$ from 1 but also the uncertainty in the theoretical prediction for $A=\lambda_{2\nu}$ must be considered. If this uncertainty, mainly in $|M|$, is a factor $10^{\pm a}$, to firmly establish $m_\nu > 0$ we must have $\lambda_\Sigma, \exp > 10^a \cdot A$nominal. This is a severe restriction since for $^{128}$Te and $^{130}$Te, indeed a $\sim$ 1-2. This became clear from comparing $A$theory to $\lambda_{2\nu}$,exp for $^{130}$Te (knowing already that at least $m_\nu < 50$ eV, it follows for $^{130}$Te that $\Lambda_0 \ll 1$, hence one can safely take $\lambda_\Sigma \approx \lambda_{2\nu}$ ,exp in this case). Under these circumstances, it is rather remarkable and illustrates the advantage of small decay energy that for $^{128}$Te, nevertheless a limit of $m_\nu < 5.6$ eV has been deduced from the present best measurement (see below).

The problem of large uncertainties in the theoretically predicted rates could be solved if one is allowed to follow Pontecorvo (1968) in his assumption of similar nuclear matrix elements for the similar nuclei $^{128}$Te and $^{130}$Te. In this case, the matrix elements cancel and $m_\nu$ can be calculated from the ratio

$$\rho = \frac{\lambda_\Sigma^{128}}{\lambda_\Sigma^{130}} = \frac{F_{2\nu}^{(128)}(1+\Lambda_0^{(128)})}{F_{2\nu}^{(130)}(1+\Lambda_0^{(130)})} \tag{11}$$

If the presumption is correct, then the prediction is accurate to $\sim$ 20% (rather than a factor of 10) and any $\Lambda_0^{128}$ larger than $\sim$ 0.2 would become distinguishable as an increase of $\rho$. In good approximation, $\Lambda_0^{130}$ can be neglected and we obtain

$$\rho_\Sigma = \rho_{2\nu} \ (1 + \Lambda_0^{(128)}) \tag{12}$$

## 4. Data and further potential

For a consideration of the present experimental situation and for illustration of the potential of further improvements we take the best existing data (Kirsten et al 1983a, 1983b). They were obtained from conventional Xe-mass spectrometric analysis of a $1.31 \times 10^9$ yr old native tellurium mineral from the Goodhope mine in Colorado. The available sample consisted of 4.5 g pure tellurium.

## 4.1 Absolute evaluation from $^{128}$Xe

Following the theoretical treatment of Doi et al (1985) eq. 8 becomes, with $m_\nu$ in eV,

$$\lambda_\Sigma^{(128)} = 5.3 \ 10^{-26} \text{yr}^{-1}(1 + 0.02 \ m_\nu^2) \tag{13}$$

The authors have scaled the absolute value for $A=\lambda_{2\nu}^{128}$,theor. to the matrix element as deduced from experimental data on $\lambda_{2\nu}^{130}$, a conservative procedure enforced by the obvious contradiction between observed $^{130}$Te-$2\nu$ decay rates and purely theoretical nuclear matrix elements. If $N(^{128}$Xe) is the number of excess $^{128}$Xe atoms accumulated in one gram of Goodhope ore, eq. 13 yields

$$N(^{128}\text{Xe}) = 1.06 \cdot 10^5 \ (1 + 0.02 \ m_\nu^2) \tag{14}$$

The experimental result is $N(^{128}\text{Xe})=(0.54^{+0.59}_{-0.54}) \times 10^5$, that is, there is no positive identification of $^{128}$Te-DBD of any kind. Taking the $2\sigma$ confidence limit, we obtain

$$1.06 \times 10^5(1 + 0.02 \ m_\nu^2) \leq 1.72 \times 10^5 \tag{14a}$$

and consequently, $m_\nu \leq 5.6$ eV ($2\sigma$). If we were to improve this result, we would have to be able to measure distinct values for $N(^{128}Xe)$ between 1.06 and $1.72 \times 10^5$ with reasonable accuracy. For example, the difference between $m_\nu = 0$ and $m_\nu = 3$ eV would correspond to $1.06 \times 10^5$ atoms/g vs. $1.25 \times 10^5$ atoms/g of $^{128}Xe^*$ to be measured. Sample weights cannot be increased at will since tellurium minerals are generally rare. With the special requirements for our purpose: rich in Te, very old, pure in actinides (to prevent side reactions) and, most importantly, low in dissolved Xe-concentrations, only gram-sized samples are realistically obtainable. In any case, the mixing ratio of DBD produced radiogenic $^{128}Xe^*$ and non-radiogenic $^{128}Xe$ contained in the sample is a limiting factor since one must have a detectable isotopic shift caused by $^{128}Xe^*$ relative to the atmospheric isotope pattern of dissolved and adsorbed Xe. This shift is conveniently expressed as permil deviation $\delta^{128}_{132}$ with the definition

$$\delta^A_{(132)} = 1000 \left\{ \left[ \left(\frac{^A Xe}{^{132}Xe}\right)_{sample} \Big/ \left(\frac{^A Xe}{^{132}Xe}\right)_{atm} \right] - 1 \right\} \quad (15)$$

whereby $^{132}Xe$ is the normalizing isotope for atmospheric Xe.

4.2 Evaluation from $^{128}Xe^*$-$^{130}Xe^*$ ratio
    Eq. 12 reads numerically (Doi et al 1985)

$$\rho_\Sigma = 1.972 \times 10^{-4} (1 + 0.021 \, m^2_\nu) \quad (16)$$

In measurable quantities,

$$\rho_\Sigma = 1.085 \frac{^{128}Xe^*}{^{130}Xe^*} = 1.085 \cdot 0.47 \cdot \frac{\delta^{128}_{(132)}}{\delta^{130}_{(132)}} = 0.51 \frac{\delta^{128}_{(132)}}{\delta^{130}_{(132)}} \quad (17)$$

($1.085 = {}^{130}Te/{}^{128}Te$ isotope abundance ratio, $0.47 = ({}^{128}Xe/{}^{130}Xe)_{atm.}$). We note again that only ratios rather than absolute concentrations enter the determination of $m_\nu$. The excess of $^{130}Xe$ in old tellurium ores is so large that its determination is rather accurate. The critical quantity is $\delta^{128}_{132}$ , therefore we write

$$\delta^{128}_{(132)} = 3.87 \cdot 10^{-4} (1 + 0.021 \, m^2_\nu) \, \delta^{130}_{(132)} \quad (18)$$

The experimental results for Goodhope are

$$\delta^{128}_{(132)} = 5.5 \pm 6, \qquad \delta^{130}_{(132)} = (2.72 \pm .01) \times 10^4.$$

Taking the $2\sigma$ -confidence limit, $\delta^{128}_{(132)} \leq 17.5$, we obtain

$$10.5 (1 + 0.021 \, m^2) \leq 17.5 \quad (19)$$

leading to $m_\nu < 5.6$ eV.
To substantially improve this limit, it would be necessary to achieve an accuracy in $\delta^{128}_{(132)}$ of order $\pm 1$ at an absolute level of $\delta^{128}_{(132)} \approx 10$ corresponding to $\sim 10^5$ atoms of $^{128}Xe^*$. For illustration, the difference between $m_\nu = 0$ amd $m_\nu = 3$ eV corresponds (in our example) to $\delta^{128}_{(132)} = 10.5$ and $\delta^{128}_{(132)} = 12.5$. If $m_\nu < 2$ eV, this would become undetectable. With other words, at this level the $2\nu$ -DBD overwhelms $0\nu$ -decay even for $^{128}Te$ with its low decay energy. However, in this case of $\rho_\Sigma \approx \rho_{2\nu}$ , the accurate measurement of $\rho$ would be particularly rewarding since it would test the appropriateness

of the Pontecorvo-assumption $| M_{2\nu}^{128}| = |M_{2\nu}^{130}|$. This would make an achieved upper limit for $m_\nu$ intrinsically independent of any theoretical uncertainties. To establish the definite occurrence of $2\nu$ -DBD of $^{128}$Te, that is, to measure $\delta_{(32)}^{128}\nu$ 10 rather than $\leq$17 is certainly a realistic goal with improved detection technique. Once this is achieved, it becomes extremely important to rule out the possibility of observing mass 128-excess due to hydrocarbon or HI-background. The chemically selective RIMS technique is ideal for this purpose. We note in this context that there have been repeated claims of positive $^{128}$Te-DBD detection at a level which would call for neutrino masses of $\sim$ 10 eV (Hennecke et al 1975, Hennecke 1978), but these false results were caused by hydrocarbon background as later admitted (Richardson et al 1986).

## 5. Concluding remarks

The determination of the electron neutrino restmass via DBD of Te below the upper limit of $\sim$ 5.6 eV which has already been established is limited by
 (i)   the absolute number of radiogenic $^{128}$Xe atoms
 (ii)  the mixing ratio of radiogenic and non-radiogenic $^{128}$Xe
 (iii) the precision of isotopic ratio determinations at the permil level
 (iv)  possible background on mass 128
 (v)   the validity of the Pontecorvo-assumption.
All these aspects (except ii) can benefit from RIMS. However, below $\sim$1 eV the method fails for principal reasons because of the then dominant $2\nu$ -DBD even for $^{128}$Te.

## References

Bellotti E, Cremonesi O, Fiorini E, Liguori G, Pullia A, Sverzellati P and
    Zanotti L 1984 Phys Lett 146B 450
Caldwell D 1985 6th Workshop Grand Unification, Minneapolis, April 1985
Caldwell D 1986 UCSB-HEP-86-8 and Proc. 12th Internat.Conf. on Neutrino
    Physics, Sendai, Japan, June 1986 (to appear)
Doi M, Kotani T, Takasugi E 1985 Progr.Theor.Phys. Suppl.83 1
Haxton W and Stephenson G 1984 Progr. Particle and Nucl.Phys. 12 40
Hennecke E, Manuel O, Sabu D 1975 Phys.Rev. C11 1378
Hennecke E 1978 Phys.Rev. C17 1168
Kayser B, Petcov S, Rosen P 1986 private communication
Kirsten T 1984a Proc. AMCO-7, Darmstadt, ed. O.K. Klepper, p 63
Kirsten T 1984b Fifth Workshop on Grand Unification, ed. K.Kang, World
    Sci.Publ. p 268
Kirsten T, Schaeffer O, Norton E, and Stoenner R 1968 Phys.Rev.Lett. 20
    1300
Kirsten T, Richter H and Jessberger E 1983a Phys.Rev.Lett. 50 474
Kirsten T, Richter H and Jessberger E 1983b Z.Phys.C16 189
Kirsten T, Heusser E, Kaether D, Oehm J, Pernicka E and Richter H 1986
    Proc. 12th Intern.Conf. on Neutrino Physics, Sendai, Japan, June 1986 (to
    appear)
Kündig W 1986 Phys.Bl. 42 380
Pontecorvo B 1968 Phys.Lett. 26B 630
Richardson J, Manuel O., Sinha B and Thorpe R 1986 Nucl.Phys. A453 26
Schneider K 1986 this volume

# Author Index